6th International Conference on Hydrodynamics, Energy and Electric Power System (HEEPS 2022)

Journal of Physics: Conference Series Volume 2271

Xiamen, China
25 – 27 March 2022

ISBN: 978-1-7138-5524-8
ISSN: 1742-6588

Printed from e-media with permission by:

Curran Associates, Inc.
57 Morehouse Lane
Red Hook, NY 12571

Some format issues inherent in the e-media version may also appear in this print version.

Institute of Physics
Dirac House, Temple Back
Bristol BS1 6BE UK

Phone: 44 1 17 929 7481
Fax: 44 1 17 920 0979

techtracking@iop.org

Additional copies of this publication are available from:

Curran Associates, Inc.
57 Morehouse Lane
Red Hook, NY 12571 USA
Phone: 845-758-0400
Fax: 845-758-2633
Email: curran@proceedings.com
Web: www.proceedings.com

TABLE OF CONTENTS

Journal of Physics: Conference Series **2271** (2022) 011001 doi:10.1088/1742-6596/2271/1/011001

Preface

Due to recent pandemic, 2022 6th International Conference on Hydrodynamics, Energy and Electric Power System (HEEPS 2022) was held virtually online during March 25-27, 2022. The decision to hold the virtual conference was made in compliance with many restrictions and regulations that were imposed by countries around the globe. Such restrictions were made to minimize the risk of people contracting or spreading the COVID-19 through physical contact. There were 100 individuals who attended this on-line conference, represented many countries including China, India, Mexico, Australia and Italy.

The conference is an international conference for the presentation of technological advances and research results in the fields of Hydrodynamics, Energy and Electric Power System. The conference brings together leading researchers, engineers and scientists in the domain of interest from around the world. We warmly welcome previous and prospected authors submit your new research papers to HEEPS 2022, and share the valuable experiences with the scientists and scholars around the world.

During the conference, the conference model was divided into three sessions, including oral presentations, keynote speeches, and online Q&A discussion. In the first part, some scholars, whose submissions were selected as the excellent papers, were given about 5-10 minutes to perform their oral presentations one by one. Then in the second part, keynote speakers were each allocated 30-45 minutes to hold their speeches. In the second part, we invited four professors as our keynote speakers. Prof. Abdul Hasib Chowdhury, Bangladesh University of Engineering and Technology, Bangladeshi, performed a speech: *Improving frequency resilience of power systems through dynamic-adaptive load shedding.* And then we had Prof. Chuanchang Li, Changsha University of Science & Technology, China. He delivered a speech: *Emerging composite phase change materials for thermal and cold energy storage.* Prof. Zong Zhi, Dalian University of Technology, China. He shared a speech: *Configuration Hydrodynamics of Trimaran ships.* Lastly, we were glad to invite Assoc Prof. Pengcheng Lin, Guangdong University of Technology, China. He made a wonderful speech: *Phase change gel: from organic to inorganic, from foundation to application.* Their insightful speeches had triggered heated discussion in the third session of the conference. Every participant praised this conference for disseminating useful and insightful knowledge.

The proceedings are a compilation of the accepted papers and represent an interesting outcome of the conference. Topics include but are not limited to the following areas: Hydrodynamics, Water Resources and Hydropower, Energy, Electric Power and other related topics. All the papers have been through rigorous review and process to meet the requirements of international publication standard.

We would like to acknowledge all of those who supported HEEPS 2022. The help and contribution of each individual and institution was instrumental in the success of the conference. In particular, we would like to thank the organizing committee for its valuable inputs in shaping the conference program and reviewing the submitted papers.

We sincerely hope that the HEEPS 2022 turned out to be a forum for excellent discussions that enabled new ideas to come about, promoting collaborative research. We are sure that the proceedings will serve as an important research source of references and knowledge, which will lead to not only scientific and engineering findings but also new products and technologies.

The Committee of HEEPS 2022

Journal of Physics: Conference Series **2271** (2022) 011001 doi:10.1088/1742-6596/2271/1/011001

Committee member

Conference Chair
Chong Kok Keong, Universiti Tunku Abdul Rahman, Malaysia

Technical Program Committee Chair
Prof. Sudip Basack, MAKA University of Technology, Kolkata, India

Organizing Committees
Prof. Fangfang Li, China Agricultural University, China
Prof. Zhenxiang Xing, Northeast Agricultural University, China
Prof. Genquan Qin, Strong Technology Co.,LTD., China
Prof. Yunxia wang, College of Civil Engineering, Hefei University of Technology, China
Prof. Junying Rao, College of Civil Engineering, Guizhou University, China

Technical Program Committees
Prof. Sudip Basack, MAKA University of Technology, Kolkata, India
Prof. Ing. Stefan Stanko, PhD., Slovak University of Technology, Bratislava Slovakia
Prof. Gagik H. Torosyan, National polytechnic university of Armenia, Republic of Armenia
Prof. Edén Bojórquez, Autonomous University of Sinaloa, Mexico
Prof. Qing Quan Liang, Victoria University, Australia
Prof. Bahram Rezayibana, Islamic Azad University, Ardabil branch, Ardabil, Iran
Prof. Giuseppe Failla, University of Reggio Calabria, Italy
Prof. Kuan Yew Wong, Universiti Teknologi Malaysia, Malaysia
Prof. Fangfang Jian, Henan University of Science and Technology, China
Prof. Chi-Wen Lin, National Yunlin University of Science and Technology, Taiwan, China
Prof. Fangfang Jian,Henan University of Science and Technology, China
Prof. Mariana Dumitrescu, University Dunarea de Jos Galati, Romania
Prof. Khalil El-Hami, University of Hassan 1st, Faculty of Khouribga, Morocco
Prof. Osiris Canciglieri Junior, Pontifical Catholic University of Paraná, Brazil
Assoc. Prof. Jiang Zhu, Tokyo Institute of Technology, Japan
Engineer Junwen Chen, China Petroleum Engineering & Construction Corporation, Southwest Branch, China
Engineer Cheng Jin, New United Group Company Limited, China

Journal of Physics: Conference Series **2271** (2022) 011002 doi:10.1088/1742-6596/2271/1/011002

Peer Review Statement

All papers published in this volume have been reviewed through processes administered by the Editors. Reviews were conducted by expert referees to the professional and scientific standards expected of a proceedings journal published by IOP Publishing.

- **Type of peer review:** Double Anonymous
- **Conference submission management system:** Morressier
- **Number of submissions received:** 83
- **Number of submissions sent for review:** 75
- **Number of submissions accepted:** 33
- **Acceptance Rate (Submissions Accepted / Submissions Received × 100):** 39.8
- **Average number of reviews per paper:** 2
- **Total number of reviewers involved:** 20
- **Contact person for queries:**
  Name: Xuexia Ye
  Email: xx.ye@keoaeic.org
  Affiliation: AEIC Academic Exchange Information Centre

Research on hydrodynamic characteristics of circular aquaculture tank based on PIV technology

Fang Zhu[1a], Zekun Zhang[2b], Jiajun Hu[3c], Dejun Feng[3d*]

[1] School of Naval Architecture and Maritime, Zhejiang Ocean University, Zhoushan 316022, China

[2] School of Fisheries, Zhejiang Ocean University, Zhoushan 316022, China

[3] National Engineering Research Center for Marine Aquaculture, Zhejiang Ocean University, Zhoushan 316022, China

[a] 1425479414@qq.com , [b] 1105978182@qq.com, [c] 247809880@qq.com,

[d] fengdj@zjou.edu.cn

Abstract—The diameter-depth ratio of the water inlet pipe in the aquaculture tank (the ratio of the diameter of the aquaculture tank to the water depth, D/H) and the setting angle of the water inlet pipe (the acute angle formed by the water outlet direction and the tangent to the tank wall, α) are important factors that affect the hydrodynamic characteristics of the tank and directly affect the circulation. The comprehensive performance of the aquaculture system, in order to study the influence of the diameter-depth ratio of the aquaculture tank and the setting angle of the inlet pipe on the flow field of the aquaculture tank, this paper uses the particle image velocimetry (PIV) to measure the effect of different diameter-depth ratios and the setting angle of the inlet pipe on the circular shape. Hydrodynamic characteristics of circulating water aquaculture tanks. The test selects the single-pipe water intake mode commonly used in actual production for the circular aquaculture tank as the research object, and selects three groups of commonly used D/H conditions (3:1, 4:1, 5:1), Set up 8 groups of inlet pipe setting angles for each group of D/H conditions. In the experiment, the PIV technology was used to measure the flow field of the water layer 1 cm from the bottom of the tank under different working conditions, and the hydrodynamic characteristics of the culture tank were quantitatively analyzed through 2 aspects: the average flow velocity v_{avg} and the velocity uniformity coefficient DU_{50}.The comprehensive test results show that the D/H and the setting angle of the inlet pipe obviously affect the flow field and hydrodynamic characteristics of the aquaculture tank. With the increase of the D/H, the v_{avg} in the aquaculture tank gradually decreases, and the DU_{50} gradually decreases. The changing trends of the hydrodynamic characteristics of the aquaculture tank with the setting angle of the inlet pipe under different D/H conditions are similar. When the water inlet pipe is set at an angle of 50°~60°, the sewage collection and hydrodynamic characteristics in the aquaculture tank are optimal. The research results in this paper can provide a reference for the setting method of the inlet pipe of the circular aquaculture tank for industrial circulating water.

1. Introduction

Recirculating Aquaculture System (RAS) is to use engineering facilities and water treatment equipment to treat the sewage discharged during the breeding process to realize recycling. By building standardized breeding management technology, artificially controlling breeding conditions provides suitable growth for breeding organisms. Environmental conditions, to achieve high-yield, high-efficiency, high-quality and "zero" pollution systems [1]. Factory recirculating aquaculture has the advantages of saving resources and less pollution, which is in line with my country's national energy development policy of energy conservation and emission reduction. It is one of the solutions for the aquaculture industry to achieve carbon peaking and carbon neutrality [2]. Factory recirculating aquaculture has high density and large amount of feed, and it is very easy to produce residual bait and manure during the breeding process. If not treated in time, it will seriously affect the growth of breeding objects [3]. In actual aquaculture production, the flow field in the tank is generally adjusted by setting the water inlet and outlet methods to achieve the purpose of efficient sewage collection, but the setting method of the water inlet pipe is often only based on the subjective experience of farmers and lacks theoretical guidance.

Tvinnereim and Skybakmoen[4] compared the influence of four water inlet methods on the flow state of the pool, namely, single-pipe single-port water inlet, horizontal jet tube water inlet, side water inlet pipe water inlet, and horizontal and vertical combined pipe water inlet. The water inlet mode of the side inlet pipe drives the water body, and the calculation formula of the average flow velocity in the aquaculture tank is given. Oca and Masaló [5] used Particle Tracking Velocimetry (PTV) to measure the flow field distribution in a long rectangular aquaculture tank with four different inflow and outflow methods. The results show that there are problems such as poor water mixing ability and sewage discharge ability with one-side water inlet and one-side water outlet. Zhu Fang [6] studied the influence of the water inlet pipe setting method on the hydrodynamic characteristics of the aquaculture tank through the PIV technology. The research results show that the setting method of the water inlet pipe obviously affects the hydrodynamic characteristics of the aquaculture tank. Taking into account the average flow velocity v_{avg} and velocity uniformity coefficient U, the setting distance of the water inlet pipe is d=1/4 r, and the setting angle is 20°~30°, the hydrodynamic coefficient of the aquaculture tank is optimal.

2. Materials and methods

2.1. Experimental setup

The flow field measurement experiment in the culture tank was carried out in the ocean measurement laboratory of Qingdao GL Technology. The test device is shown in Fig 1. Experimental setup, which is mainly composed of four parts: model aquaculture tank, water circulation system, flow measurement system and laser particle image velocimetry (PIV) system. The model pool is round and made of transparent acrylic sheets. The inner wall is 50 cm high and the inner diameter is 98 cm. There is a drainage outlet with a diameter of 5 cm at the center of the bottom. The bottom of the pool has no slope. The water circulation system consists of a model aquaculture tank, a conical waste collection filter bucket, a water pump, a water pipe, etc. The direction of the water flow is shown by the arrow in the figure. There is a dial at the connection between the water inlet pipe and the bracket, and the water inlet pipe can be rotated to adjust the water inlet angle. . The flow measurement system consists of 2 flow meters (Keyence FD-Q20C) and valves installed in the water inlet pipeline. The flowmeter can detect the real-time input flow (L/min) of the water inlet pipe to the aquaculture tank, and the valve can adjust the water inlet. The PIV system consists of a laser transmitter, a high-definition camera (30fps), tracer particles, etc. The tracer particles are sprinkled into the flow field during measurement, and the particle velocity represents the velocity of the fluid at the correstanking position in the flow field. The light (sheet-shaped beam) shines into a test plane in the water field, and the particle positions of the two exposures are recorded by the imaging method, and the displacement of the particle group is obtained by the image analysis technology, which can be obtained by the displacement and the time interval of exposure. The velocity vector at each point in the flow field[7].

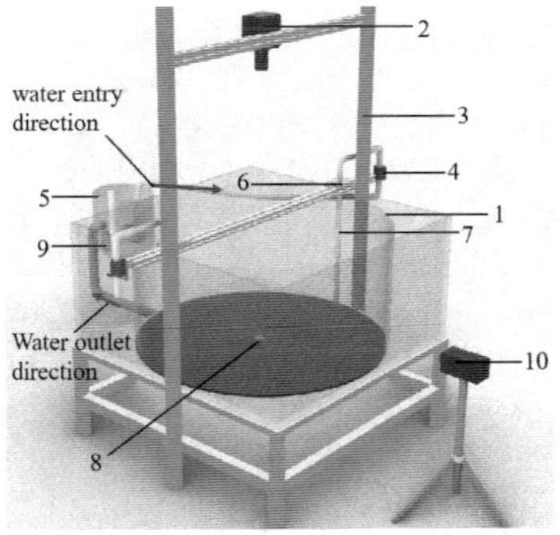

(a). Schematic of the experimental setup

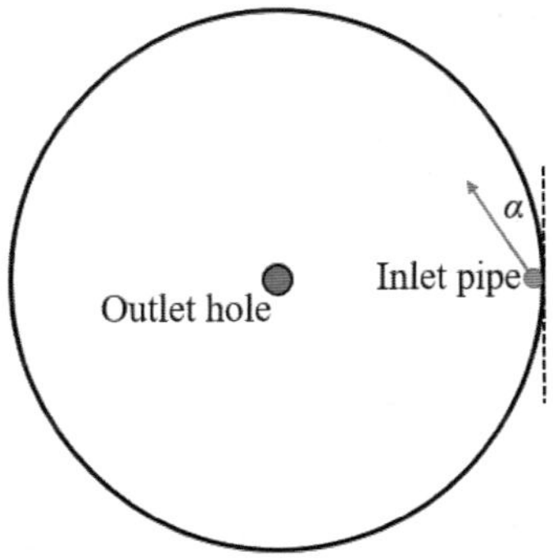

(b). Top view of aquaculture tank

1. Aquaculture tank; 2. Camera; 3. Aluminum bracket; 4. Flowmeter; 5. Cone-shape bucket; 6. Angle adjusting device; 7. Inlet pipe; 8. Outlet hole; 9. Water pump; 10. Laser

Fig 1. Experimental setup

2.2. Experimental design

Three sets of diameter-to-depth ratio conditions (D/H=3:1, 4:1, 5:1) were set up in the experiment, the water depths were 33, 25 and 20 cm, the hydraulic retention time was 30 min, and the influent flow of the inlet pipe was 8.6, 6.6 and 5.2 L/min. For each set of diameter-depth ratio working conditions, 8 sets of inlet pipe setting angles (α=0°, 10°, 20°, 30°, 40°, 50°, 60°, 70°) are set, a total of 24 sets of experimental conditions. Before the start of the test, install and debug the test equipment, then turn on the water pump, adjust the influent flow required for the test according to the flow meter, adjust the laser height, wait for about 30 minutes, and then turn on the laser after the water in the tank is stable, and operate the PIV system to start collecting images. Perform the next set of experiments.

2.3. Data analysis

The experimental data processing method is as follows: First, use MATLAB to calculate the flow velocity data at each point measured by the PIV software, quantitatively analyze and compare the hydrodynamic characteristics of the culture tank under different working conditions. conducted a specific analysis of the flow field with three hydrodynamic characteristic quantities, and then used the MATLAB software to draw the flow rate data into a flow field diagram, and qualitatively analyzed the influence of the setting method of the water inlet pipe on the distribution characteristics of the flow field in the aquaculture tank.

v_{avg} is the average flow velocity in the culture tank[9], that is The average value of the flow velocity at each measurement point in the aquaculture tank, m/s; DU_{50} [10]is the uniformity coefficient of the flow velocity distribution in the aquaculture tank, and the closer to 100, the more uniform the flow velocity distribution is. v_{50} is the average value of the lower speed of the top 50% of the speed of each monitoring point, m/s,; v is the average weighted speed, m/s; vi is the speed of the monitoring point, m/s; ri is the distance between the monitoring point and the tank center distance, m;

$$DU_{50} = \frac{v_{50}}{v} \times 100 \text{ , } \quad v = \frac{\sum_{i=1}^{n} v_i r_i}{\sum_{i=1}^{n} r_i} \tag{1}$$

3. Results

3.1. Effect of jetting velocity on the sewage convergence

The PIV system uses the movement of the tracer particles to measure the flow field in the culture tank. According to the movement information of each tracer particle, the hydrodynamic characteristics of the culture tank can be quantitatively analyzed . According to the PIV measurement, the v_{avg} under each test condition was obtained, and the results are shown in Fig 2(a). The flow rate determines whether the dirt in the aquaculture tank can be started and discharged smoothly, and a large flow rate has a positive impact on the collection and discharge of the dirt in the aquaculture tank. It can be seen from the figure that the D/H of the aquaculture tank obviously affects the v_{avg} in the aquaculture tank. The v_{avg} in the aquaculture tank gradually decreases with the increase of the D/H. When the D/H is 3:1, the v_{avg} in the aquaculture tank is the largest. In order to control the hydraulic retention time of aquaculture tanks to be 30 min, the flow rate of the inlet pipe gradually decreased with the increase of the D/H, resulting in a gradual decrease of the outlet flow rate of the inlet pipe. This is also the main reason that the v_{avg} in the aquaculture tank gradually decreases with the increase of the diameter-to-depth ratio. Under different D/H, the v_{avg} in the aquaculture tank has a similar trend with the increase of the setting angle of the inlet pipe. The v_{avg} in the aquaculture tank first increases and then decreases with the increase of the setting angle of the water inlet pipe. When the setting angle of the water inlet pipe is 50°, the maximum flow velocity is obtained, which is beneficial to the starting and collection and discharge of dirt. Based on the above experimental results, according to the principle that the larger the flow rate is, the more favorable it is for the discharge of dirt in the aquaculture tank. When the D/H is 3:1 and the angle of the inlet pipe is set to 50°, the v_{avg} in the aquaculture tank is the largest and the sewage collection capacity is the strongest.

According to formula (1), the uniformity coefficient DU_{50} of velocity distribution under each test condition is calculated as shown in Fig 2(b). The velocity distribution uniformity coefficient indicates the velocity gradient difference in the aquaculture tank. The higher the velocity distribution uniformity coefficient, the better the uniformity of the flow field in the aquaculture tank, which is beneficial to the growth of the breeding objects. The uniformity of the flow field varies with the setting angle of the inlet pipe under different D/H conditions. The uniformity coefficient of velocity distribution in aquaculture tank decreases gradually with the increase of D/H. Under different D/H conditions, the variation trend of the velocity distribution uniformity coefficient with the increase of the setting angle of the inlet pipe is roughly the same. The uniformity coefficient of velocity distribution increases gradually with the increase of the setting angle of the water inlet pipe, and the optimal uniformity coefficient of velocity distribution is obtained when the setting angle of the water inlet pipe is 60°. Based on the above experimental results, according to the principle that the larger the uniformity coefficient of velocity distribution is, the more conducive to the growth of cultured objects, the optimal working conditions are that the D/H is 3:1, and the setting angle of the water inlet pipe is 50°~60°.

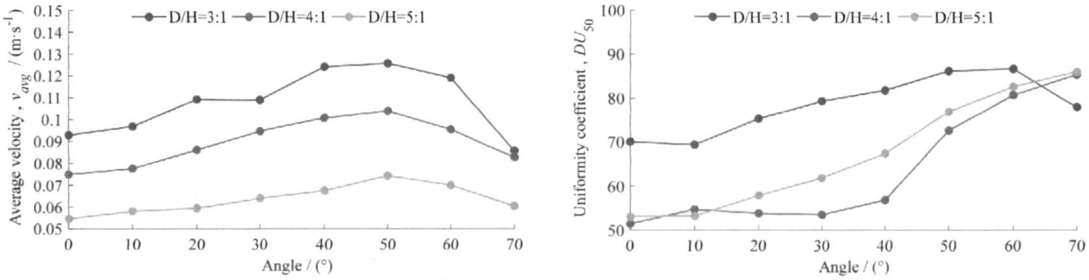

(a). The v_{avg} in aquaculture tank with different D/H and angles

(b). The DU_{50} in aquaculture tank with different D/H and angles

Fig 2 Hydrodynamic characteristics in aquaculture tank with different D/H and angles

4. Conclusion

In this study, the PIV flow velocity measurement technology was applied to the measurement of the flow field in the aquaculture tank, which expanded the research method of the hydrodynamic characteristics of the aquaculture tank and effectively improved the precision and accuracy of the research results. The main research conclusions are as follows: The v_{avg} and DU_{50} in the aquaculture tank gradually decreases with the increase of the D/H. The hydrodynamic characteristics of the aquaculture tank are optimal when the D/H is 3:1. The flow field and hydrodynamic characteristics of the culture tank under different D/H conditions The change trend of the setting angle of the water inlet pipe is similar. The DU_{50} in the aquaculture tank first increases and then decreases with the increase of the setting angle of the water inlet pipe. The setting angle of the water pipe increases gradually, and the optimal DU_{50} is obtained when the setting angle of the water inlet pipe is 60°. The above experimental conclusions show that the ratio of D/H of the aquaculture tank is 3:1, and the angle of the water inlet pipe is 50°~60°.

Acknowledgments
This work was financially supported by the National Innovation and Entrepreneurship Training Program for College Students (202010340011).

References
[1] Yang Hongsheng. Development ideas and implementation approaches of blue granary scientific and technological innovation in China[J]. Journal of Fisheries of China, 2019, 43(1): 97-104.
[2] Zhang J H, Liu J H, Zhang Y Y, et al. Strategic Approach for Mariculture to Practice"Ocean Negative Carbon Emission" [J], 2021, 36(03): 252-258.
[3] Tong X Y, Jiang S Y, Ge B M, et al. Development Status and Countermeasures of Tank Aquaculture Industry in China[J]. Journal of Agricultural Science and Technology, 2020, 22(08):5-13.
[4] Akeem Babatunde Dauda, Abdullateef Ajadi, Adenike Susan Tola-Fabunmi, et al. Waste production in aquaculture: Sources, components and managements in different culture systems[J]. Aquaculture and Fisheries, 2019, 4(3):81-88.
[5] Oca J, Masaló I, Reig L. Comparative analysis of flow patterns in aquaculture rectangular tanks with different water inlet characteristics[J]. Aquacultural Engineering, 2004, 31(3-4): 221-236.
[6] Zhu F, Gui F K, Hu J J, et al. Research on flow field of circular recirculating aquaculture tank based on PIV [J]. Transactions of the Chinese Society of Agricultural Engineering, 2021, 37(23):296-300. (in Chinese).
[7] Cavazzini G. The Particle Image Velocimetry-Characteristics, Limits and Possible Applications[M]. Rijeka: InTech, 2012.
[8] Oca J, Masaló I. Design criteria for rotating flow cells in rectangular aquaculture tanks[J]. Aquacultural Engineering, 2007, 36(1): 36-44.
[9] Oca J, Masalo I. Flow pattern in aquaculture circular tanks: Influence of flow rate, water depth, and water inlet & outlet features [J]. Aquacultural Engineering, 2013:65–72.

Experimental Investigation on Vortex Induced Vibration of Marine Risers with Different Wall Thicknesses

Ziqian Zhong[1], Peng Li[1*], Liping Feng[1], Zhen Liu[1], Chuanzhen Wang[1]

[1]College of Architecture and Civil Engineering, Shandong University of Science and Technology, Qingdao, China

*Corresponding author's e-mail: lipeng@sdust.edu.cn

Abstract-In this paper, the vortex-induced vibration experiment of the risers with different wall thicknesses in wave-flow combined water flume was carried out. The dynamic strain sensing technology was adopted. The sensitivity and micro evolution pattern of riser VIV response relative to wall thickness were examined by varying the wall thickness of various riser models, comparing the dimensionless root mean square of strain, vibration frequencies and trajectories by varying different wall thicknesses of the risers. The results indicate that the wall thickness does not have a large influence on the Strouhal numbers of the vibration of the risers, which are always around 0.18; the larger the wall thickness is, the earlier the lock-in region is jumped out, as wall thickness increases, the outflow velocity corresponding to the occurrence of an "8" or oblique "8" shape vibration trajectory continues to delay, and the velocity range for such trajectories gradually narrows.

1. Introduction

The forced excited in-line (IL) vibration of the riser induced by environmental loads such as waves and ocean currents, and the actions of ocean currents can produce vortexes alternately on the sides of the riser. Shedding of these vortexes will produce periodically variable forces that lead to VIV in the riser along the lift perpendicular to the cross-flow (CF) direction. When the frequency of the vortex shedding is close to the natural frequency of the riser structure, the vortex discharge process will be controlled by the vibration of the structure, leading to a "lock-in" [1,2]. Occurrence of this lock-in will cause violent riser vibration and a steep amplitude rise, greatly exacerbating the fatigue damage to the riser [3,4]. Recently, many studies have been reported on VIV. Xu et al. characterized the multimodal VIV of flexible cylinders subjected to uniform flow and examined the effects of rotational angle on the multimodal response of flexible cylinders model experiment [5]. Han et al. studied the multi-modal response of VIV excitation of flexible inclined cylinder under different inclination angles. For the parameters of the riser, the influence f boundary conditions, slenderness ratio and damping on the vortex induced vibration of the riser is analyzed [6]. Gao et al. experimentally studied the effect of surface roughness on vortex induced vibration of flexible cylinders. By comparing the resistance, lift, tension, vortex shedding frequency, vibration frequency and displacement response of cylinders with different roughness, it is found that compared with smooth cylinders, rough cylinders have smaller displacement response, a narrower lock-in area and higher vortex shedding frequency [7].

Wall thickness plays a fatal role in the service safety, cost, structural design, and technical feasibility of deep-sea risers. It is increasingly uneconomical to simply increase the wall thickness to increase the strength in terms of local buckling, extended buckling or overall buckling, and it is necessary to combine other engineering designs, such as a buckle arrestor, controllable lateral buckling, and so on. Existing studies have mostly relied on theoretical calculation to determine the minimum wall thickness of the riser, usually the riser is used as a pressure vessel to limit the annular stress value of the riser under the action of internal and external pressure, and limit the equivalent stress value of the riser under the action of combined load, but the design results are relatively conservative. With the advancement of deep-sea development, accumulation of pipeline operation experience and continuous improvement of quality control technology, the industry has put forward higher requirements for design accuracy, and the influence of wall thickness on the micro change of the dynamic response of the riser is an indispensable factor.

The outline of this paper is arranged as follows: Section 2 describes the experimental design of the model, mainly including experimental equipment, riser model and test technology. Section 3 is analysis and discussion, mainly involving the vibration frequency of different wall thicknesses riser models, time-frequency analysis, strain and vibration trajectory micro-evolution law. Section 4 summarizes the conclusions.

2. Experimental details

The experiment was carried out in the combined wave-current water flume of the Engineering Hydrodynamics Laboratory, Ocean University of China. The flume has a maximum water depth of 1.0 m, a maximum velocity of 1.0 m/s and a maximum wave height of 0.3 m. It is sided with toughened glass and bottomed with steel plate which can be used for fixing the riser support. The entire flume system provides the external environment required for our experiment. Fig. 1 shows the overall arrangement of the experiment.

The test apparatus was consisted of a support structure, a top tension application system, riser models, and dynamic strain sensor and acquisition system. The support structure comprises an assembly of aluminum profiles connected by right-angled corner fittings, high-strength T-bolts and flange nuts. The test details and local design are shown in Fig. 1.

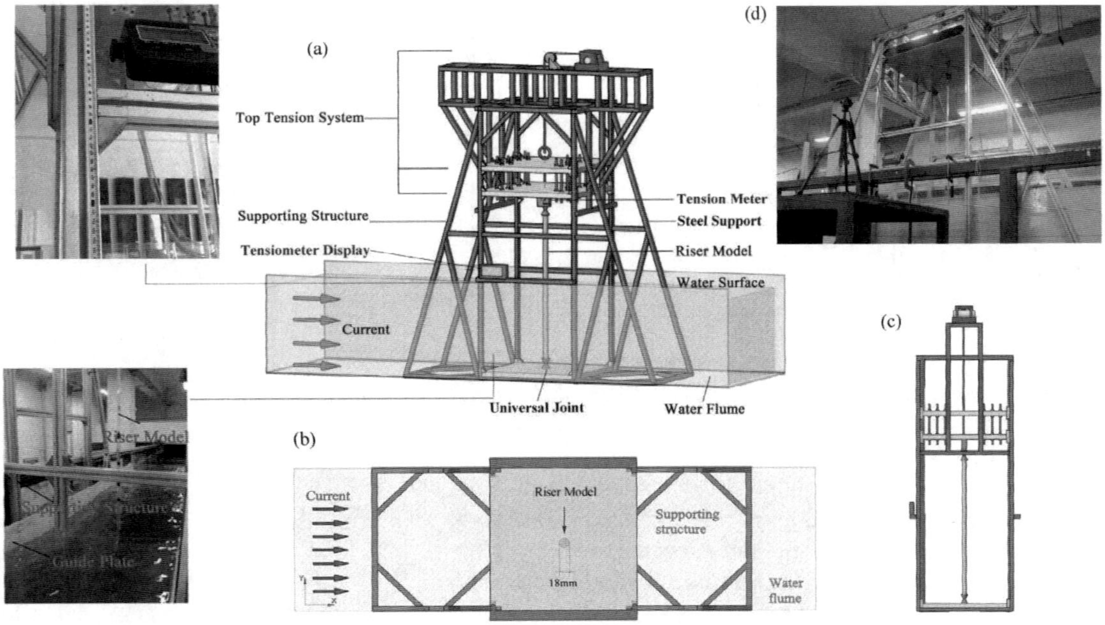

Fig.1 (a) Schematic diagram of the experiment device; (b) Top view of the device; (c) Side view of the device; (d) Actual diagram of the experiment.

The test riser models were constructed from organic glass (PMMA). Three wall thickness risers of 1.0mm, 1.5mm and 2.0mm are selected for the test. Each riser model was kept in a standing posture, with 0.8 m (40%) of its lo wer part placed in a uniform flow, with outflow velocity U = 0.1 m/s-0.7 m/s, 11 stages were loaded. The Doppler velocimeter was used to measure the outflow velocity. And dynamic resistance strain gauge was used to collect riser strain. Two measuring points were deployed along the length of the riser, strain gauges are arranged at intervals of 90°, and the dynamic strain in the CF and IL directions were respectively measured. The detailed parameters of riser are shown in Table 1. The layout of riser model and measuring points is shown in Figure 2.

Table 1 Riser model parameters

Riser Model	**1**	**2**	**3**	**Unit**
Material (M)	PMMA	PMMA	PMMA	–
Length in Water (L_s)	0.80	0.80	0.80	m
Outside Diameter (D)	18.00	18.00	18.00	mm
Aspect Ratio ($\lambda = L/D$)	111.11	111.11	111.11	–
Wall Thickness (δ)	1.00	1.50	2.00	mm
Inside Diameter (d)	16.00	15.00	14.00	mm
Flexural Rigidity (EI)	4.63	6.52	7.30	Nm^2
Mass per Unit Length (m)	0.07	0.09	0.11	Kg/m
Fundamental Nature Frequency (f_1)	3.41	3.39	3.34	Hz
Mass Ratio (m^*) (empty riser)	0.26	0.36	0.43	–

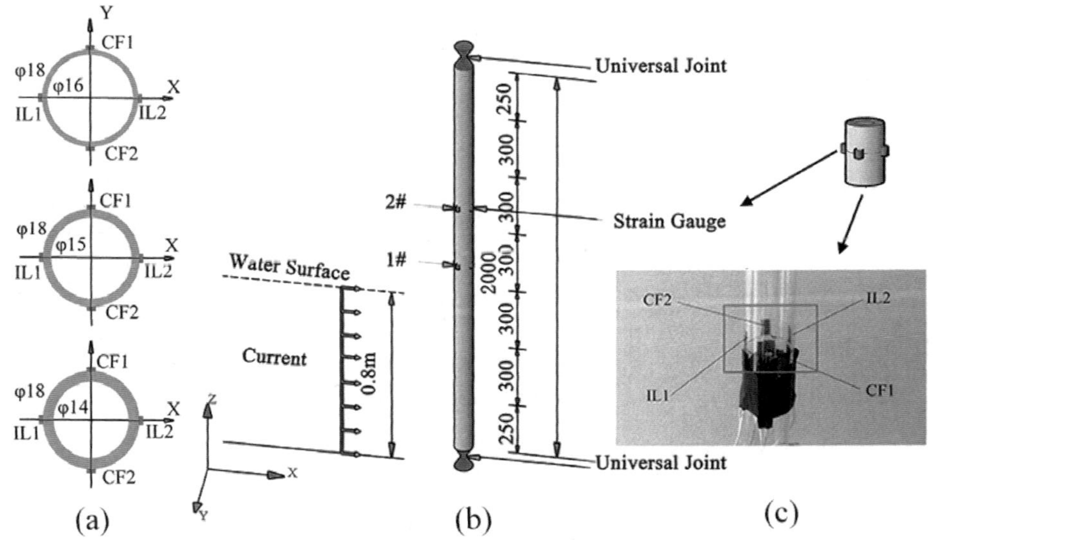

Fig.2 (a) Sectional view of riser; (b) Riser model; (c) Strain sensor measuring point layout

3 Results and discussions

3.1 Analysis of vibration frequency under different wall thickness

Strouhal (St) number is an important parameter reflecting the flow field in the wake region of riser. Fig. 3 shows the dimensionless dominant frequencies of the risers in the CF and IL directions as a function of reduced velocity among three different wall thicknesses. Here, f_1 is the first-order natural vibration frequency of the riser and f_d is the dominant vibration frequency of the riser, i.e., the frequency corresponding to the peak power spectral density. Ur is the reduced velocity, which is the dimensionless parameter of the velocity, and the calculation formula is $Ur=U/(f_nD)$, where U is the outflow velocity, f_n is the natural frequency of the riser, and D is the outside diameter of the riser. The dimensionless dominant frequencies of the risers in the CF and IL directions with wall thickness 1.0

mm, 1.5 mm and 2.0 mm were linearly fitted using least squares method. The slopes of the fitted lines in the CF direction are the *St* numbers, which are 0.19, 0.18, and 0.19; the slopes *k* values of the fitted lines in the IL direction are 0.37, 0.34, and 0.35. $k \approx 2St$, suggesting that the IL dominant frequencies are roughly twice the CF values, which follow the law of Strouhal number [8,9]. The *St* numbers of the three wall thickness risers are always around 0.18, which proves that the change of the wall thickness of the riser has no significant impact on the *St* number. It can be seen that the dimensionless dominant frequency of the three risers in the CF direction basically conforms to the linear fitting law, but it is out of the fitted straight line at high outflow velocities in the IL direction. At $Ur \geq 9.0$, the dominant frequency ratio in the two directions of the wall thickness 1.0mm and 1.5mm risers f_{dx}/f_{dy} is 1, while the wall thickness of 2.0mm riser shows a dominant frequency transition in the IL direction at high velocities.

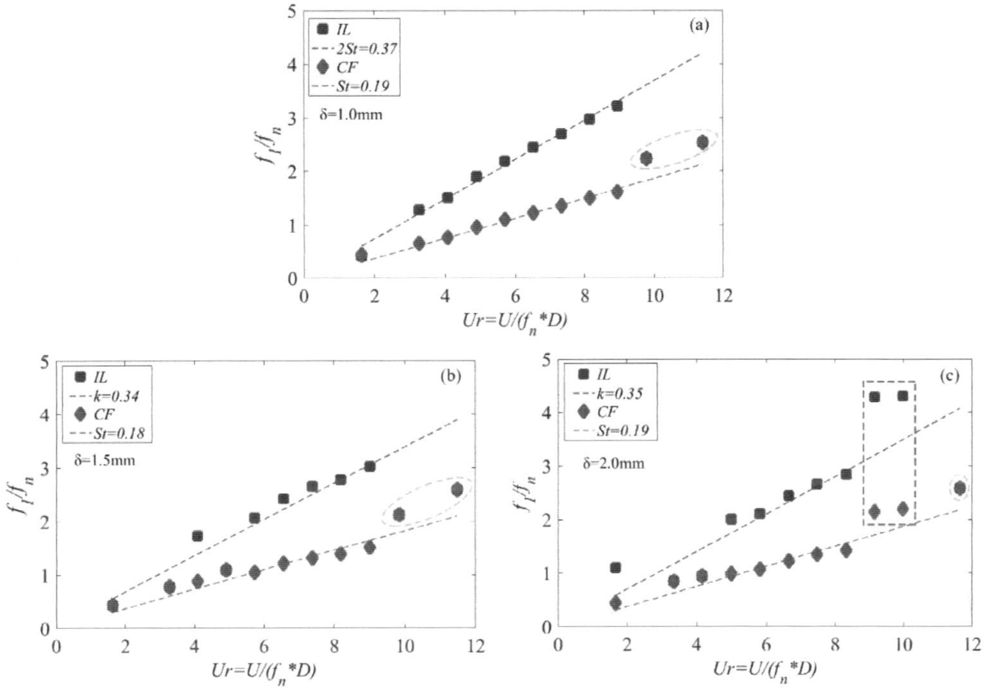

Fig.3 Wall thickness δ =1.0mm(a), 1.5mm(b), 2.0mm(c) risers CF and IL direction dimensionless leading frequency with reduced velocity

3.2 RMS strain analysis of three wall thickness risers at different flow velocities

In order to further explore the vibration response patterns for the riser of three wall thicknesses, Fig.4 compares the RMS strain variations as a function of velocity among different wall thicknesses. It can be clearly seen from the figure that the nonlinear characteristics of the RMS strain value of the three wall thickness risers, as the flow velocity increases, the vortex shedding frequency gradually increases, and the vibration of the riser gradually increases. The RMS strain of the risers with wall thicknesses of 1.5mm and 2.0mm in both directions gradually increases until it exits the lock-in region after the outflow velocity of 0.55m/s and 0.5m/s respectively. At this time, the RMS strain value of the riser decreases and reaches the lowest when the outflow velocity is 0.6m/s. Before the riser with a wall thickness of 1.0mm exits the lock-in region, there are same vibration response changes in the CF and IL directions., and the RMS strain value gradually increases. When *U*=0.55m/s, the RMS strain value of riser in the IL direction decreases, while the RMS strain value in the CF direction continues to increase. According to the RMS strain value in the CF direction of three wall thickness risers, the larger the wall thickness is, the greater the RMS strain value in the CF direction is, and the earlier it jumps out of the lock-in region.

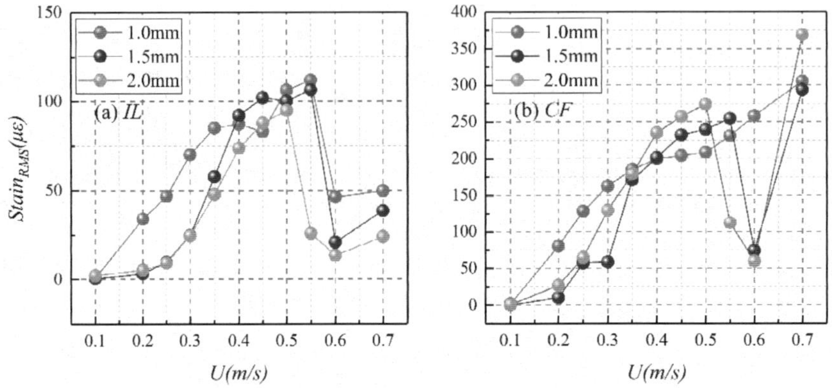

Fig.4 RMS strains in the IL direction(a) and in the CF direction(b) of the riser versus outflow velocity for different wall thicknesses

3.3 CF and IL coupled vibration trajectories as a function of outflow velocity for different wall thicknesses

Fig.5 compares the coupled vibration response diagrams of the risers in the CF and IL directions as a function of outflow velocity among different wall thicknesses. In each of the sub-diagrams, the x-axis represents the strain in the IL direction and the y-axis represents the strain in the CF direction of the riser. From these diagrams, although the risers display the same trajectory variations, their coupled vibration evolutions are not the same. As outflow velocity increasing, the trajectories of the risers with wall thickness 1.0 mm and 1.5 mm change from a crescent to an oblique "8" to an "8" to a rod or shuttle shape, whereas that of the riser with wall thickness 2.0 mm changes from a crescent to an "8" to a rod shape. As Kang, Z. et al. studied on the trajectories of the riser, the risers with three wall thickness have a crescent or "8" shape since the three wall thickness risers have a frequency ratio of 2 in the lock-in region in the IL and CF directions [10]. After jumping out of the lock-in region, the riser with wall thickness 1.5mm still maintains the track of "8" shape, but the trajectory is the superposition of "8" shape and rod shape when the outflow velocity of the riser with wall thickness 1.0mm is 0.55m/s, which is caused by multiple vibration frequencies in the IL direction. Similarly, at low outflow velocities, the risers with wall thickness 1.5mm and 2.0mm are in the mode conversion region, and multi order vibration frequencies are involved in the vibration response of the riser, so that the trajectories of the risers are rod or spindle. The light-colored regions in the figure are the track of "8" or oblique "8" shape of the riser, which is distributed in a ladder shape along the longitudinal direction. That is, as wall thickness increases, the outflow velocity corresponding to the occurrence of an "8" or oblique "8" shape vibration trajectory continues to delay, and the velocity range for such trajectories gradually narrows.

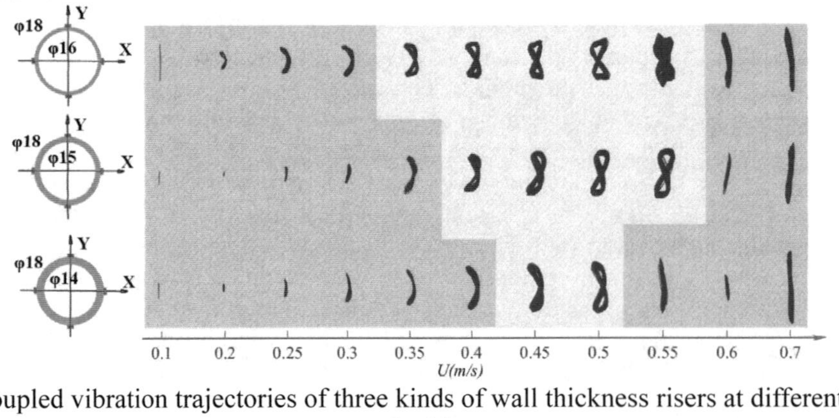

Fig.5 Coupled vibration trajectories of three kinds of wall thickness risers at different outflow velocities

4. Conclusions

In this paper, the VIV experiment of riser models with different wall thicknesses was carried out to study the sensitivity characteristics and evolution laws of VIV of riser with wall thickness from multiple perspectives, such as dominant frequency, dimensionless root mean square of strain, strain time history curve, and vibration trajectory evolution process. The main conclusions are as follows:

(1) Through the analysis of Strouhal numbers and dominant frequency of risers with different wall thicknesses, although the wall thickness of the three risers is different, the Strouhal numbers of the risers are always around 0.18, with no significant impact.

(2) When the risers are in lock-in region, the larger the wall thickness of the riser, the greater the mean square value of strain in the CF direction, and the earlier it jumps out of the lock-in region.

(3) The change of wall thickness has a great influence on the trajectory of the riser. As wall thickness increases, the outflow velocity corresponding to the occurrence of an “8” or oblique “8” shape vibration trajectory continues to delay, and the outflow velocity range for such trajectories gradually narrow.

Acknowledgement

This work was supported by the Key Research and Development Program of Shandong Province (2019GHY112061), Colleges and Universities of Shandong Province Science and Technology Plan Projects (J16LH04), Research and innovation team of ocean oil and gas development engineering structure, College of Architecture and Civil Engineering, Shandong University of Science and Technology (2019TJKYTD01).

References

[1] Chaplin, J.R., Bearman, P.W., Cheng, Y. (2005) Blind predictions of laboratory measurements of vortex-induced vibrations of a tension riser. J. Journal of Fluids and Structures, 21: 25-40.
[2] Thorsen, M.J., Sævik, S., Larsen, C.M. (2017) Non-linear time domain analysis of cross-flow vortex-induced vibrations. J. Marine Structures, 51: 134-151.
[3] Aswathy, M.S., Sarkar, S. (2019) Effect of stochastic parametric noise on vortex induced vibrations. J. International Journal of Mechanical Sciences, 153-154: 103-118.
[4] Kurushina, V., Pavlovskaia, E., Postnikov, A., Wiercigroch, M. (2018) Calibration and comparison of VIV wake oscillator models for low mass ratio structures. J. International Journal of Mechanical Sciences, 142-143: 547-560.
[5] Xu, W., Ma, Y., Ji, C., Sun, C. (2018) Laboratory measurements of vortex-induced vibrations of a yawed flexible inder at different yaw angles. J. Ocean Engineering, 154: 27-42.
[6] Han, Q., Ma, Y., Xu, W. (2018) Dynamic characteristics of an inclined flexible cylinder undergoing vortex-induced vibrations. J. Journal of Sound and Vibration, 394: 306-320.
[7] Gao, Y., Fu, S., Wang, J., Song, L., Chen, Y. (2015) Experimental study of the effects of surface roughness on the vortex-induced vibration response of a flexible cylinder. J. Ocean Engineering, 103: 40-54.
[8] Song, J.N., Lin, L., Teng, B. (2011) Laboratory tests of vortex-induced vibrations of a long flexible riser pipe subjected to uniform flow. J. Ocean Engineering, 38(11-12): 1308-1322.
[9] Huera-Huarte, F.J., Bangash, Z.A., LM González. (2014) Towing tank experiments on the vortex-induced vibrations of low mass ratio long flexible cylinders. J. Journal of Fluids and Structures, 48: 81-92.
[10] Kang, Z., Ni, W., Sun, L. (2016) An experimental investigation of two-degrees-of-freedom VIV trajectories of a cylinder at different scales and natural frequency ratios. J. Ocean Engineering, 126: 187-202.

Inundation analysis of Xiuzhen River based on one-dimensional hydrodynamic model

Zhenzhi Hao[1,a], Guoqing Sang[1,b*], Taili Li[2,c]

[1]School of Water Conservancy and Environment, University of Jinan, Jinan, Shandong, China

[2]Penglai Water Services Centre,Penglai,Shandong,China

[a]email: 1595127469@qq.com, [c]email: 2508825673@qq.com,

*Corresponding author: [b]email: sangguoqing@163.com

Abstract: In order to deal with the mountain flood disaster for small and medium-sized rivers and reduce the loss of life and property caused by mountain flood to the people, it is necessary to calculate the flood inundation risk of the region along the river. In this paper, taking Xiuzhen River in Lanshan District of Rizhao City as an example, an one-dimensional hydrodynamic model was established and D8 algorithm was coupled to simulate flood inundation in the target area, and the submerged depth, flood amount and submerged area under different frequency of flood were analyzed. The results show that the method can calculate flood inundation risk and scope quickly and accurately. It provides flood and drought disaster risk information and scientific decision-making basis for the sustainable development of social economy.

1. Introduction

Due to the influence of human activities, global warming and extreme weather have increased the frequency of natural disasters. While the flood is one of the ten natural disasters facing the world today, heavy rain and flash floods are likely to cause flooding, fast speed, water flood volume is big, the power of its vast rage tend to the affected area, to human life and property bring great losses[1], the world's current economic loss caused by flooding accounted for more than 30% of the total losses of all kinds of natural disasters[2], China has also invested a lot of energy in flood control, improve the flood control system construction of major rivers, and constantly improve the flood control capacity. However, severe floods caused by extreme weather may still occur. If existing protection capacity is exceeded, the existing flood control system is still unable to protect. Moreover, due to the climate, natural geography and other factors in Shandong Province, floods and droughts will happen constantly and have the characteristics of sudden, abnormal and uncertain.

To determine the inundation range corresponding to rainfall at different frequencies in advance enables people to analyze and judge the occurrence and detailed state of flood evolution in the visual system, which can provide decision support for flood control and disaster reduction work more effectively. Since 1981, the combination of aerospace and aerial remote sensing technology and geographic information system (GIS) technology can quickly obtain the general situation of flood inundated area and prove the feasibility of this method[3]. In 2014, Zhang Shanghong[4] proposed a new RCDC method to simulate flood submerging, and defined DEM source data with triangular network to improve the efficiency of flood submerging model in complex terrain areas. In 2016, Bruno Merz[5] proposed the coupled model chain RFM (regional flood model), which can deduce the information of discharge, submergence and damage mode. In 2021, Liu Yimeng[6] used CADDIES 2D model to simulate and verify the rainstorm and flood in Beijing urban area, and compared the simulation accuracy of FloodArea model and CADDIES 2D model in the case of input data with different resolutions.

To sum up, most of the previous studies on inundation simulation focused on urban areas with large population, and less on market towns and villages distributed on both sides of rivers in hilly areas. This paper is looking for quick and accurate simulation of the whole river basin under different rainfall intensity method of flood situation, therefore with embroidery needle river as an example, and take the flood simulation of the whole basin as the calculation unit, calculate the inundation range of floods with different frequencies, so that personnel and property transfer in advance, with non-engineering measures to reduce flood damage, is of great significance to evaluate the flood control and disaster mitigation and flood etc[7].

2. Simulation methods and model construction

2.1 One-dimensional hydrodynamic model

In order to simulate flood evolution in river channels, the 6-point Abbott-Ionescu discrete Saint Venant equations are used to simulate the river surface lines. The upper boundary condition is selected as the flood discharge time series of junan County in the upper reaches of xiuzhen River. The downstream boundary is the estuary, the influence of tidal fluctuation is considered, the relation between Q and H of water level and flow is chosen as the lower boundary condition, and the catch-up method is used to solve the problem. Saint-venant's equations are mainly used to describe the motion variation law of one-dimensional water flow, which mainly includes two parts: the flow continuity equation (Eq 1) and the momentum equation (Eq 2)

$$\frac{\partial Q}{\partial x} + B\frac{\partial Z}{\partial t} = q_1 \tag{1}$$

$$\frac{\partial Q}{\partial t} + \frac{\partial}{\partial x}\left(\alpha\frac{Q^2}{A}\right) + gA\frac{\partial Z}{\partial x} + g\frac{Q|Q|}{AC^2R} = 0 \tag{2}$$

Where, Q is the flow rate (m^3/s); A is the water cross section area (m2); B is the width of river surface (m); α is momentum correction coefficient; C is Chezy coefficient; R is hydraulic radius.

2.2 D8 algorithm model

In order to simulate the evolution of flood after overflow, D8 is used to simulate the submergence of flood after overflow. Namely in the eight neighborhood of the unit grid determine slope,the steepest slope is the direction of water flow, with 1, 2, 4, 8, 16, 32, 64 and 128 the eight values on behalf of the east, southeast, south, north, northwest, southwest and west and northeast eight directions for effective in the direction of flow, The flow direction of water is determined by judging the elevation or effective area of the grid around it, and then water flows to that grid at the micro scale, which is the flow direction of water[8].

3. Model applications

3.1 Overview of the simulation area

The Xiuzhen River is located in southern Shandong and northern Jiangsu on the coast of the Yellow Sea, with a drainage area of 412.0 km^2 and a main stream of 46 km. The main stream in the study area is 23.5 km long, and the height difference in the study area is 28.08 meters. The Xiuzhen River belongs to the coastal water system of the Huaihe River Basin and is an independent river flowing into the sea. The ravines are developed in the basin, and the creeks are vertical and horizontal, and most of them are perpendicular to the ground contour. The northwest is close to the slopes of hills and ridges, the source of the ditch and creek is short, the slope is steep and the current is rapid, and the downstream terrain is relatively flat. The northwest is close to the slopes of hills and ridges, the source of the ditch and creek is short, the slope is steep and the current is rapid, and the downstream terrain is relatively flat, and quickly converge to the river channel, with rapid water flow and steep water rise. It is easy to form the flood disaster of steep rise and steep fall, small peak height, short duration but great harm.

3.2 Model construction

3.2.1 Division of calculation units

Due to the Xiuzhen River is located in a typical general hilly area with a steep slope, the interval runoff during heavy precipitation has a greater impact on the flood-carrying capacity of the river. In order to make more accurate calculations of the flooding capacity of the downstream river, it is necessary to calculate the inter-zonal flow production at different design frequencies, so ARCSWAT was used to divide the whole basin into five sub-basins and five inter-zonal floods that feed into the main river channel as point sources and inter-zonals respectively.

3.2.2 Hydrological calculations

According to the analysis of the rainstorm and flood characteristics and runoff law in the Xiuzhen River Basin, the single peak of the flood in the Xiuzhen River Basin is about 24 hours, so the design rain period is determined to be 24 hours in this calculation. The Pearson P-III curve was fitted to the measured rainfall data of Dashan Station and Beikuo Station with a series length in the selected watershed, where Cs=3.5Cv, and the instantaneous unit line method was used for P=5%, 2%, 1% , 0.5% Calculation of yield and sink flow and flood peaks at four different frequencies.

3.2.3 Parameter setting

In this simulation, we designed measured sections at intervals of 100~200m, and mapped by extending 100~ 300m from the top of the dike to the left and right banks. According to the actual situation of Xiuzhen River, the roughness coefficient of main channel is 0.03-0.05, and the roughness coefficient of beach channel is 1.5~2 times of main channel. Now we can calculate the average slope by river bottom elevation and river length. The longitudinal section is shown in Figure 6, and the calculation of main river is shown in Table 1.

Table 1 Segmental specific drop in the main channel of the Embroidery Needle River

Serial number	Name	Main river section mileage/m	than the drop
1	Interval flooding 1	0~471	0.00179
2	Interval flooding 2	471~4405	0.00078
3	Interval flooding 3	4405~9057	0.00180
4	Interval flooding 4	9057~18362	0.00063
5	Interval flooding 5	18362~22428	0.00122

4. Flood inundation simulation

4.1 River simulation results

Based on the constructed one-dimensional hydrodynamic model, the flood evolution of the flood information obtained by the hydrological calculation can be obtained, and the water surface lines along the route at different frequencies can be obtained. Through the comparative analysis with the embankment tops on the left and right banks of the river, the positions of the embankment points of different frequencies along the Xiuzhen River are determined. The mileage of the overflow points is shown in Table 2. Figure 1 and Figure 2 show the comparison of the water surface line along the Xiuzhen River and the top of the left and right banks.

Table 2 Mileage of the overflow point locations

Kishibetsu	Reproduction period	Location mileage/m
Left Bank	20 years	10047-10533, 12669-13105
	50 years	9885-10533, 12488-13105, 14805, 20666
	100 years	9885-10533, 12488-13105, 14805-14949, 20666
	200 years	9885-10533, 12488-13105, 14805-14949, 15928-16499, 19388-19778, 20666-20755
Right Bank	20 years	9885-10533, 12488-13105, 15136, 16695-16858, 18866, 19503-19611
	50 years	9741-10649, 12360-13273, 15136, 16695-16858, 17482, 18866, 19503-19996, 20971
	100 years	9741-10649, 12360-13273, 15136, 16040-16858, 17361-17482, 18866-19005, 19503-19996, 20971
	200 years	9741-10649, 12360-13273, 15136, 15928-17482, 18866-19996, 20971, 22195

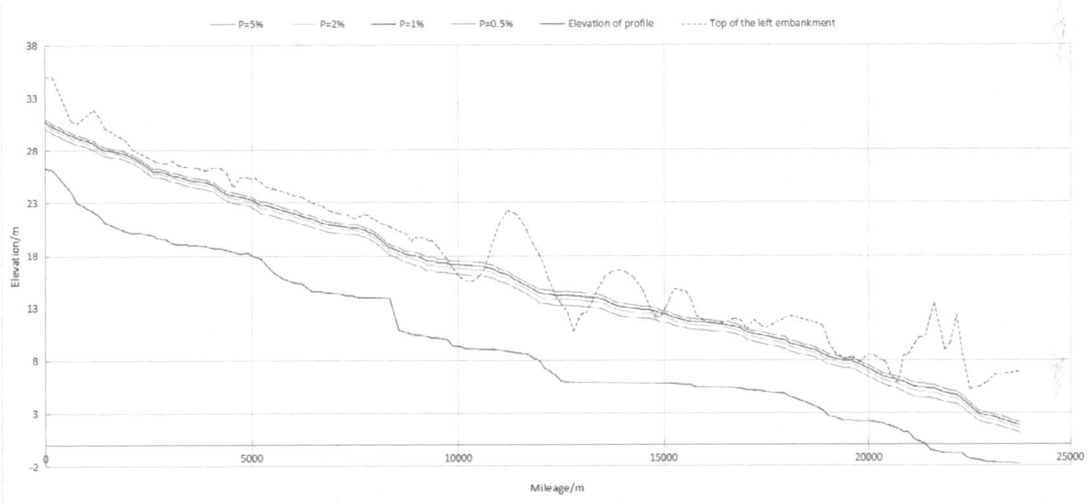

Fig. 1 Comparison of the top of the left embankment of the Xiuzhen River with the water surface line along its course

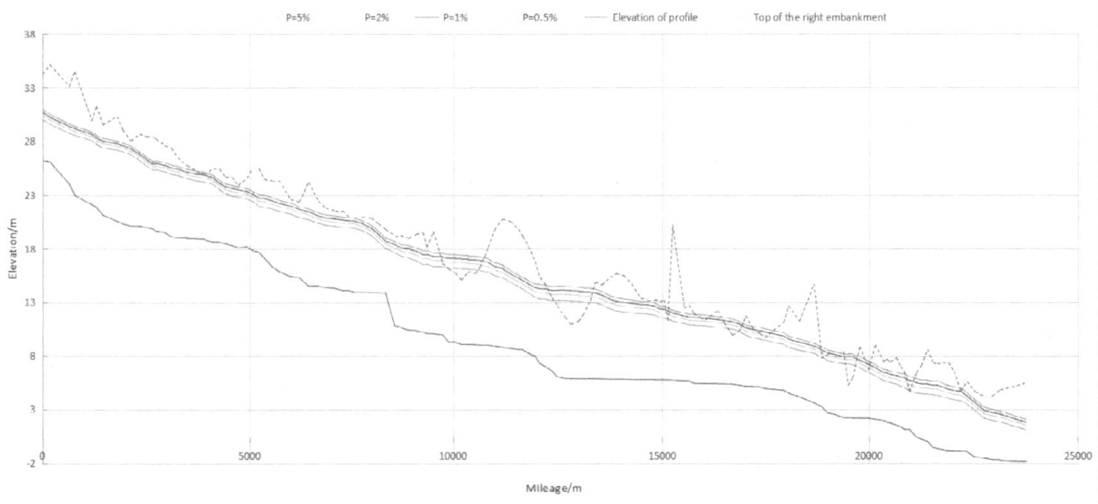

Fig. 2 Comparison of the top of the right embankment of the Xiuzhen River with the water surface line along its course

4.2 Inundation simulation results

Flood inundation is a short-term complex dynamic process. The nature of the flood itself and the landform of the submerged area are important factors affecting the inundation of the flood[9]. The main mechanism is the elevation difference between the flood source and the submerged area. Since passive inundation is not affected by topographical factors, as long as the area in the study area is lower than the given flood inundation level, it is regarded as an inundated area. Although this inundation analysis model is very simple, the submerged area and water depth can be obtained quickly and intuitively, but there is a big gap with the actual flood inundation range, so this simulation adopts active flood inundation.

Based on the mileage and water level of the overflow point obtained from the one-dimensional hydrodynamic model above, combined with DEM data in the study area, the overflow point was taken as the starting point of D8 algorithm to retrieve the eight-way grid of the neighborhood and determine whether it was flooded by judging the grid elevation. If it is submerged, the grid will be stored and retrieved to the surrounding grid as a new starting point. Repeat this process to traverse all grids. Finally, we adopted D8 algorithm to obtain the submerge range of Xiuzhen River with four different frequencies of 5%, 2%, 1% and 0.5%. The submerge range of each frequency is shown in Figure 3.

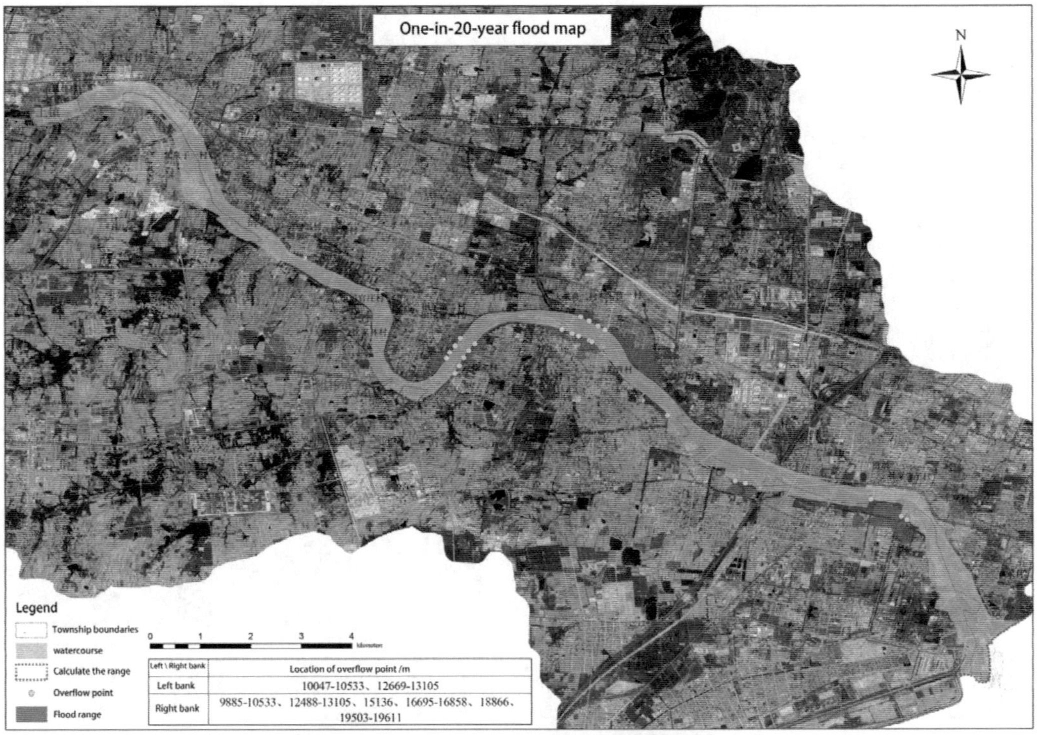

Left \ Right bank	Location of overflow point /m
Left bank	10047-10533、12669-13105
Right bank	9885-10533、12488-13105、15136、16695-16858、18866、19503-19611

(a) one-in-20-year flooding

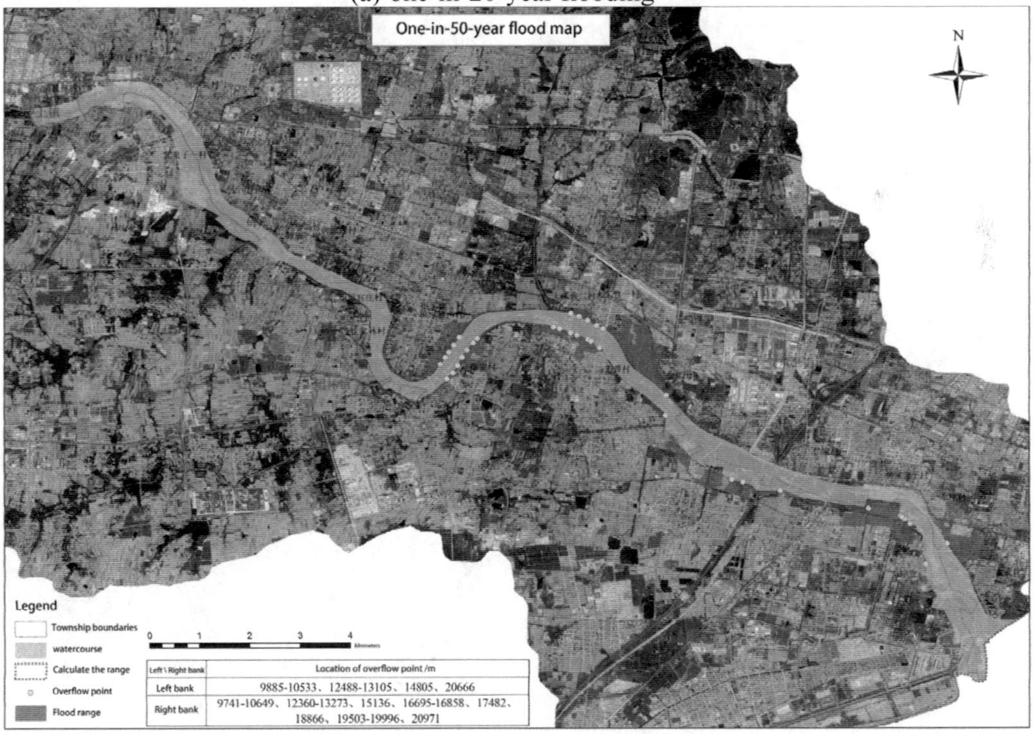

Left \ Right bank	Location of overflow point /m
Left bank	9885-10533、12488-13105、14805、20666
Right bank	9741-10649、12360-13273、15136、16695-16858、17482、18866、19503-19996、20971

(b) one-in-50-year flooding

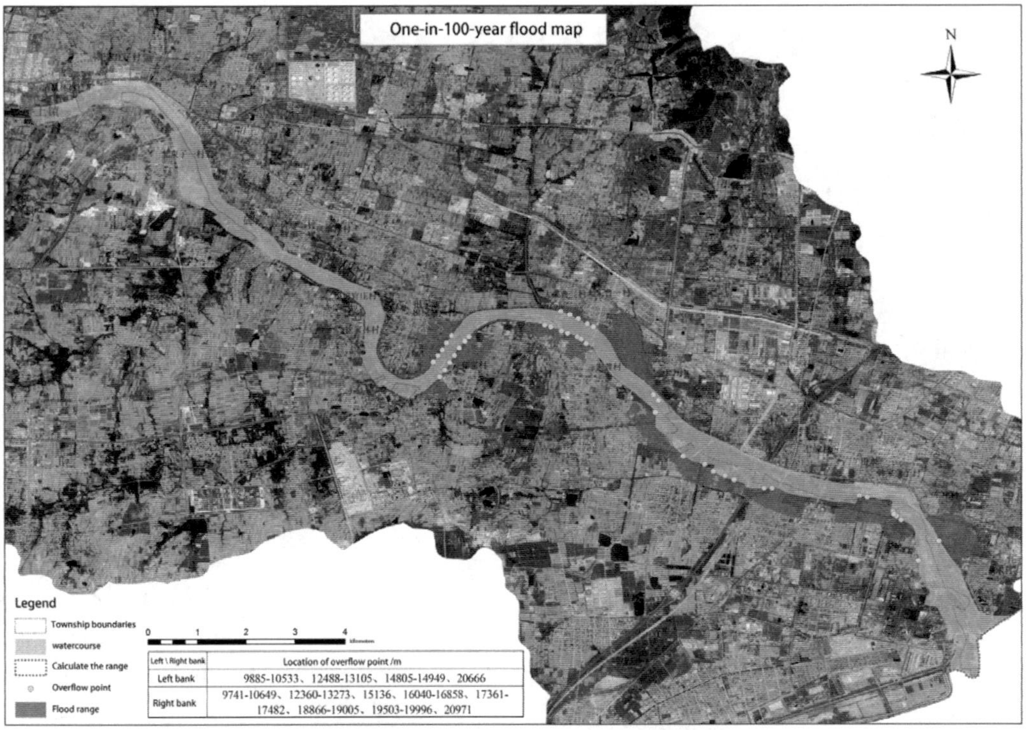

Left \ Right bank	Location of overflow point /m
Left bank	9885-10533、12488-13105、14805-14949、20666
Right bank	9741-10649、12360-13273、15136、16040-16858、17361-17482、18866-19005、19503-19996、20971

(c) one-in-100-year flooding

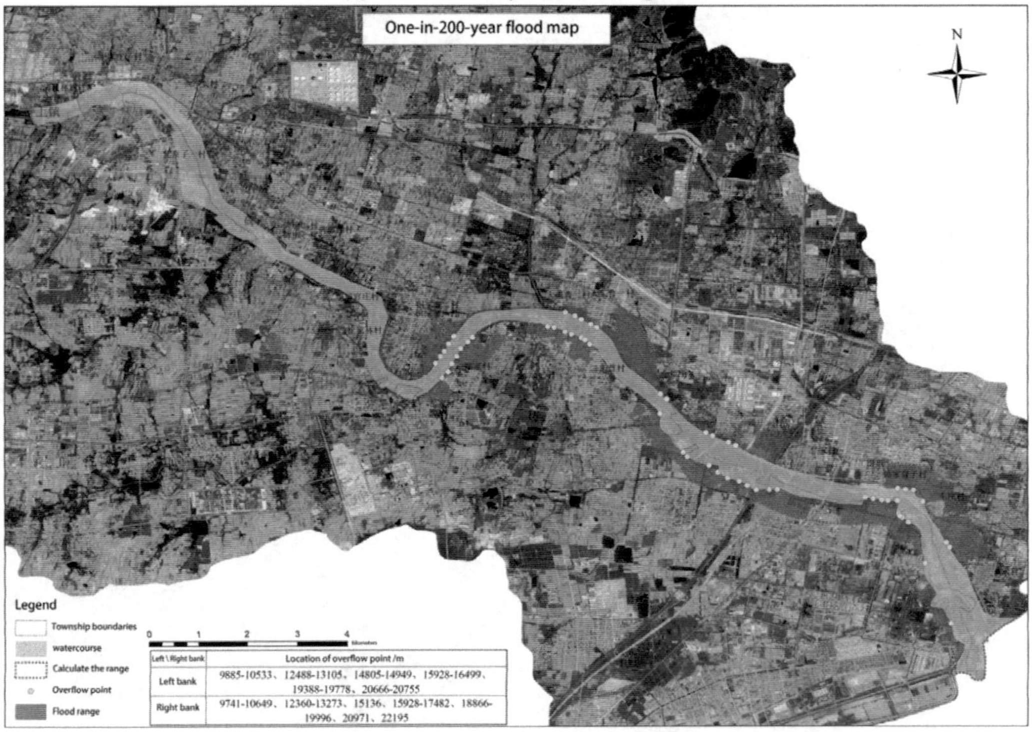

Left \ Right bank	Location of overflow point /m
Left bank	9885-10533、12488-13105、14805-14949、15928-16499、19388-19778、20666-20755
Right bank	9741-10649、12360-13273、15136、15928-17482、18866-19996、20971、22195

(d) one-in-200-year flooding

Fig 3 Schematic diagram of the extent of inundation at different rainfall frequencies

5. Analysis of results

(1) By comparing the gradient of the main reach of the Xiuzhen River and the location of the embankment break point, most of the river embankment break occurs in the range of 9,057 m to 18,362 m in the main river mileage. This section is mainly due to its proximity to the downstream and large catchment area. The drop dropped from 0.0018 to 0.00063 compared to the previous reach, which led to the slowdown of the flow rate, so the landing of the river and finally caused the risk of levees overflowing.

(2) Due to the large undulating area in the hilly area, the affected area in the downstream is much higher than that in the upstream without considering the special conditions such as blockage, dyke breach and siltation.

(3) With the increase of flood magnitude, the number of overflow levees also increases, but the inundation area of the same overflow levees does not change significantly with the increase of flood magnitude. In addition, the affected area is distributed along the river course. With the increase of flood magnitude, the submerged area extends poorly to the direction of the vertical river course, and mostly extends along the river course.

(4) The reason that the simulated results are larger than the actual inundation range may be that we mainly considered the influence of water level on the inundation range in this study, but water volume and other factors are not considered.

6. Conclusion

In this study, taking Xiuzhen River in Lanshan District, Rizhao as an example, through the combination of one-dimensional hydraulic model and D8 algorithm, the flood inundation simulation of Xiuzhen River Basin was carried out. In the case of considering the connectivity, the flood inundation range under different frequencies can be determined, and the result has good accuracy, practicability and scientificity. The research results show that the simulation method combining the one-dimensional hydraulic model and the D8 algorithm makes the flood simulation more accurate, scientific and fast; at the same time, combined with the disastrous weather forecast, the flood inundation simulation can be quickly carried out to determine the flood inundation range. Reduce the loss of life and property caused by flash floods to the people, and effectively provide decision support for flood control and disaster reduction.

However, the results of the submerged range are greatly affected by the DEM accuracy. If a digital elevation file with a lower resolution is used, due to the large grid, when using the D8 algorithm for submerged simulation, discontinuous submerged areas are likely to be generated, which will affect the final flood inundation. If a higher-precision DEM is adopted and the impact of villages and hydraulic structures on flood discharge and the impact of water volume on the submerged area are considered, better inundation results will be obtained.

Acknowledgments

This research work was financially supported by the Natural Science Foundation of Shandong Province (ZR2019BEE064).

References

[1] Liu, H. (2020) Simulation of flood inundation in the eastern block of Honghu floodplain based on DEM[D]. Huazhong University of Science and Technology
[2] Cheng W. (2004) Modeling method for flood evolution in river basins and inverse analysis of river network roughness[D]. Zhejiang University.
[3] Lin H, Ji T. (1986) Estimation of reservoir inundation area by space remote sensing.
[4] Zhang S, Wang T, Zhao B. (2014) Calculation and visualization of flood inundation based on a topographic triangle network[J]. Journal of Hydrology, 509:
[5] Bruno M, Heiko A, et al. (2016) Large-scale flood risk assessment using a coupled model chain[J]. E3S Web of Conferences, 7:

[6] Liu Y, Yang Saiyi, Wang Y, Liu X, Zhang X, Zhu Y. (2021) Simulation and validation analysis of storm flooding in Beijing urban area based on CADDIES-2D model[J]. Hydropower Energy Science,1,39(11):107-110+92.
[7] Sun A, Xu L, Shi Y, Shi C. (2009) Simulation of flood inundation extent based on GIS[J]. Journal of North China Institute of Water Resources and Hydropower,30(02):9-11.DOI:10.19760/j.ncwu.zk.2009.02.004.
[8] Cui Q,Wu M,Kong Xg,Zhang L. (2011) An application of a DEM-based digital river network system extraction algorithm[J]. Computer Technology and Development,21(06):204-207.
[9] Zhang X,Zhang W,Liu Y,Qiu S. (2006)Research on storm surge flood inundation calculation model[J]. Journal of System Simulation,(S2):20-23.

Experimental research on the characteristics of sewage convergence in rectangular RAS tank

Zekun Zhang[1a], Xunran Pan[1b], Bo Liu[1c], Xiaoyu Qu[1d*]

[1] School of Fisheries, Zhejiang Ocean University, Zhoushan 316022, China

[a] 1105978182@qq.com , [b] 1361328347@qq.com , [c] 372963740@qq.com, [d*] quxiaoyu@zjou.edu.cn

Abstract—The performance of sewage convergence of the aquaculture tank is directly related to the water quality in the tank, and it is the key to the success of aquaculture. This paper takes the rectangular aquaculture tank as the research object, and systematically explores the effect of hydraulic driving factors (jetting velocity, jetting position), tank bottom slope, and sewage weight on the sewage convergence performance of aquaculture tank by conducting model experiments to provide a scientific basis for the design of aquaculture tanks and optimization of jetting-driven device. The research results found that the higher the jetting velocity, the better the sewage convergence; the closer the jetting position is to the drain hole, the better the sewage convergence; the smaller sewage weight, the better the sewage convergence; when the small slope is 1:35, the sewage convergence is better. It is recommended to adjust the slope of the bottom of the tank to some degree, control the breeding density and feeding amount when constructing the tank. The jetting-driven device should be installed as close to the drain hole for large-size sewage. To obtain a better performance of sewage convergence, the flow velocity should be increased as much as possible under the condition of ensuring the healthy growth of the breeding objects.

1. Introduction

There are various types of tank shapes in aquaculture tanks, generally circular, raceway, rectangular, etc. and they have their advantages and disadvantages [1-2]. Circular tanks have the most uniform body exchange, with no dead ends and slow flow zones. For example, the bottom of the tank is designed as a conical slope bottom with a large taper, and a sewage outlet is set in the center of the tank, which can use water power and the gravity of the sediment to discharge the sewage, and the sewage discharge effect is ideal. Its disadvantage is that the circular tank is constructed continuously, the utilization rate of the ground is small, the tank walls cannot use each other, and the unit cost is high[3-5]. The driving method used in the raceway culture tank is mainly to set vertical water inlet pipes on the walls and four corners, and adjust the angle of the jet to drive the water flow in the tank to circulate in a specific direction. This driving method is easy to form a divergent water flow at the jet port. The water flow along the overall water flow direction is faster, while the bottom water flow at a certain angle or opposite to the overall water flow is slower or even a non-swirl zone occurs, so it cannot be drain from the tank. Thus, dead zone of sewage of various sizes are formed [6]. For rectangular tanks, there are two ideal flow regimes in rectangular tanks: plug flow and mixed flow. Plug flow, that is, the water body does not mix and diffuse in the direction from the inlet to the outlet, and the highest concentration of pollutants in the tank water occurs at the outlet; mixed flow means that the tank water has a mixing effect in the direction of the flow, and the water quality is uniform throughout the tank. However, in actual aquaculture practice it often deviates from these two ideal flow patterns. Due to the different settings of water inlet and outlet, the water flow pattern of the rectangular tank is very different. However, although there are problems such as poor water mixing capacity and sewage discharge capacity in rectangular tanks with one-side water inlet and one-side water outlet, they are still widely used in farms mainly based on running water aquaculture [7]. The main reason is that the rectangular tank is closer to the working conditions of the natural river, which is in line with the conditions for the rapid growth of fish, and is easy to construct, low in cost, and high in land utilization [9]. In this paper, the model experiment of a single rectangular aquaculture tank is used to study the characteristics of sewage accumulation to obtain the optimization parameters for aquaculture practice.

2. Materials and methods

2.1. Experimental setup

This experiment was carried out in the rectangular tank of the National engineering research center for marine aquaculture of Zhejiang Ocean University, China. In the model experiment, the physical model tank was built according to the geometric scale of 1:10, the length-width ratio was 2:1, and three kinds of bottoms were designed, flat bottom, and sloped bottom of 1:35 and 1:13, respectively. Fig. 1 shows the top view and front view of the tank. The inner wall of the model tank is 200 cm long, 100 cm wide, 60 cm deep, and the base is 30 cm high. The culture tank and the return tank are separated by a wall, and a 0.1 m × 0.1 m square sewage collection area is opened in the middle of the bottom of the partition wall. The bottom and walls of the sink are painted with aquaculture paint. The experimental water was tap water with a water depth of 0.5 m.

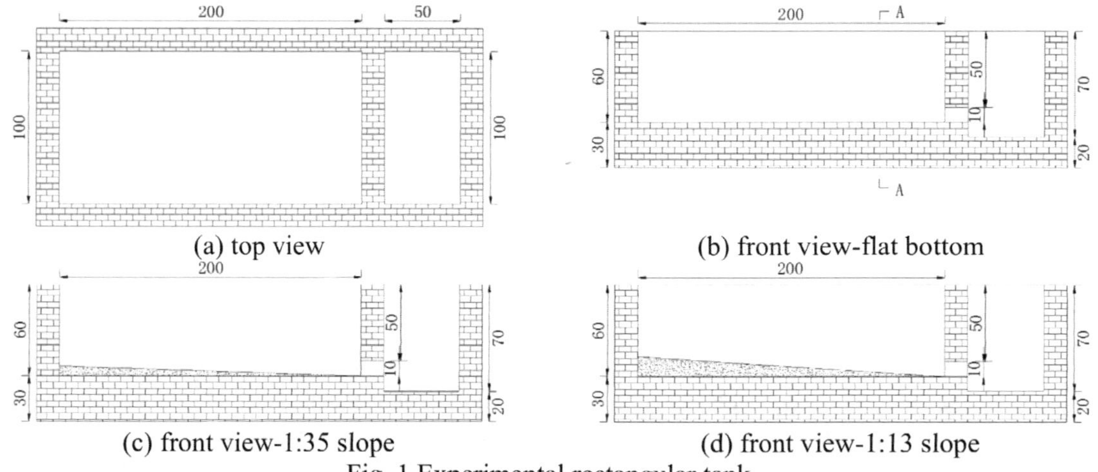

(a) top view (b) front view-flat bottom

(c) front view-1:35 slope (d) front view-1:13 slope

Fig. 1 Experimental rectangular tank

Surface jetting-driven mode is used in this experiment to circulate the water body. The model diagram and structural diagram of the jetting device are shown in Figure 2. The jetting device includes four parts: water pump, water inlet device, jetting device, and bracket. HZS-280 self-priming circulating water pump (Sensen Group Co., Ltd.) is used, and the flow rate is adjusted by adjusting the valve to change the flow rate. The water inlet device and the jet device both are installed 4 cm below the water surface. The main body of the bracket is a plywood board, 120 cm long, 15 cm wide, and 2 cm high. The length of a single jetting tube is 20.5 cm, and the adjacent spacing of four jetting tubes is 5 cm. Each jetting tube is provided with an equal number (12) of jetting holes. The diameter of the jetting holes is 0.3 cm, and the horizontal spacing between adjacent holes is 0.5 cm, the outlet of the jet hole is smooth, which is beneficial for the water flow to be ejected in the same direction.

Fig. 2 Jetting-diven device. 1: inlet device; 2: jetting device; 3: bracket; 4: water pump

The experimental procedure is described below. Firstly, the shrimp No. 2 feed (NINGBO TECH-BANK Co., Ltd.) was weighted and then scattered into the tank evenly. Secondly, the camera fixed directly above the tank was adjusted that the camera range and camera height could cover the entire bottom of the tank. Then the jetting device was installed on the surface of the water carefully. After that, the water pump was started and the recording of the sewage movement lasted about 20 min when the water flow is stable in the tank. Finally, all the equipment was turned off and removed and the recoding data were used for further analysis.

2.2. Experimental design

In this paper, the effects of jetting velocity, jetting position, sewage weight, and bottom slope on the sewage convergence were studied, respectively. The detailed experimental design is shown in Table 1: (1) The effect of jetting velocity on sewage convergence. The jetting position L1 was fixed (40 cm from the back end of the tank), the cement bottom surface, the sewage weight was 30 g, the bottom slope was 1:35, and the three groups of jet velocity were 21.2 cm/s, 28.0 cm/s and 37.4 cm/s, respectively. (2) The influence of the jetting position on the sewage convergence. The jetting velocity was 28.0 cm/s, the cement bottom surface, the sewage weight was 30 g, and the bottom surface slope was 1:35. The three groups of jetting positions were 40 cm, 80 cm, and 120 cm from the back end of the tank, denoted as L1, L2, and L3. (3) The effect of sewage weight on the sewage convergence. The fixed jetting position L1, the jetting velocity of 21.2 cm/s, the bottom surface of the plywood board, the bottom slope is 1:35, and the three groups of sewage masses are 30 g, 60 g and 90 g, respectively. (4) The influence of the slope of the bottom on the sewage convergence. The jetting position was L1, the jetting velocity was 21.2 cm/s, plywood bottom, sewage weight was 30 g. The three groups of bottom slopes are 1:35, 1:13 and flat bottom, respectively.

Table 1 Experiment design

Case	Jetting velocity cm/s	Jetting position	Bottom material	Sewage weight g	Bottom slope	Note
1	21.2					
2	28.0	L1	cement	30	1:35	Effect of jetting velocity
3	37.4					
4		L1				
5	28.0	L2	cement	30	1:35	Effect of jetting position
6		L3				

7				30		
8	21.2	L1	plywood	60	1:35	Effect of sewage weight
9				90		
10					1:35	
11	21.2	L1	plywood	30	1:13	Effect of bottom slope
12					flat	

2.3. *Data analysis*
This paper mainly analyses the final stabilized photos, and uses the rgb2gray function to convert the colour image into a grayscale image, and then the grayscale image is binarized to facilitate the identification of sewage. According to the coordinates of the sewage and the coordinates of the tank, the ratio of the distance between the sewage and the front of the tank is calculated, called the relative distance. The length of the tank is denoted as L, the coordinates of sewage A is (x, y), and the distance from the sewage to the back end of the tank is denoted as A, then the relative distance of sewage A is a/L, namely (L-x)/L. Finally, Histogram function was used to show the results, and the sewage convergence was analysed by comparing the relative distance of all the sewages. The smaller value of the relative distance indicates better sewage convergence.

3. Results

3.1. *Effect of jetting velocity on the sewage convergence*
Fig 3 shows the results of sewage distribution under the velocity of 21.2 cm/s (left panel), 28.0 cm/s (middle panel), 37.4 cm/s (right panel). When the flow velocity is 21.2 cm/s, the sewage is more dispersed, and the farthest relative distance is 0.3; when the flow velocity is 28.0 cm/s, the farthest relative distance also reaches 0.3, but the amount of sewage between 0.25 and 0.3 is less than that at the flow velocity of 21.2 cm/s. When the flow velocity increased to 37.4 cm/s, the farthest relative distance was significantly less than 0.3, and the sewage was concentrated between 0.1 and 0.2. It shows that with the increase of flow rate, the sewage convergence is better.

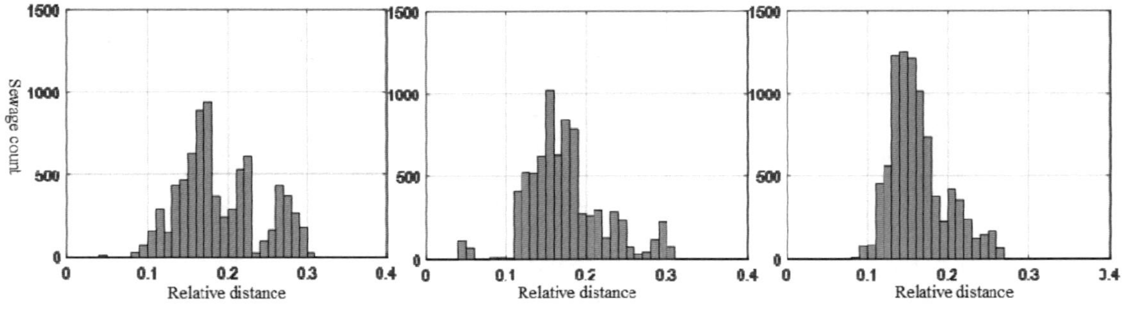

Fig.3 Effect of jetting velocity on the sewage convergence

3.2. *Effect of jetting position on the sewage convergence*
The influence of the jetting positions of L1 (left panel), L2 (middle panel), L3 (right panel) on the sewage convergence is shown in Fig. 4. It can be found that when the jetting position is L1, the

sewage is distributed within 0.3 of the back end of the tank, and the mainly distributed between 0.1 and 0.2; when the jet position is L2, the sewage is obviously far from the back end of the tank, and the farthest relative distance reaches 0.4. When the jet position is adjusted to L3, the farthest relative distance of sewage exceeds 0.5, which is much larger than that of L1 and L2, and the convergence is not good. The results show that the jetting position has a great influence on the sewage convergence.

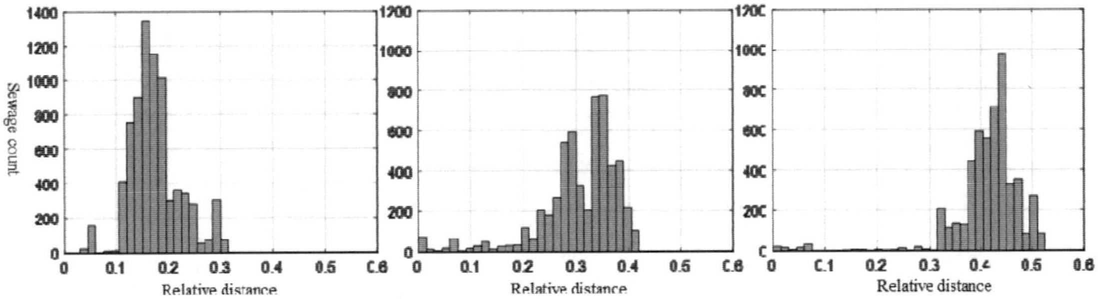

Fig.4 Effect of jetting position on the sewage convergence

3.3. *Effect of sewage weight on the sewage convergence*

Fig. 5 shows the effect of sewage weight 30 g (left panel), 60 g (middle panel), 90 g (right panel) on the sewage convergence. When the sewage weight is 30 g, the farthest relative distance is 0.22. When the weight increases to 60 g and 90 g, the farthest relative distance reaches 0.3. Due to the increase in the amount of sewage, it will accumulate in the process of aggregation, so when the sewage weight is 90 g, the amount of sewage between 0.2 and 0.3 is much more than of 60 g. It can be seen that the sewage weight has a certain influence on the convergence, and the more the sewage, the more difficult to gather.

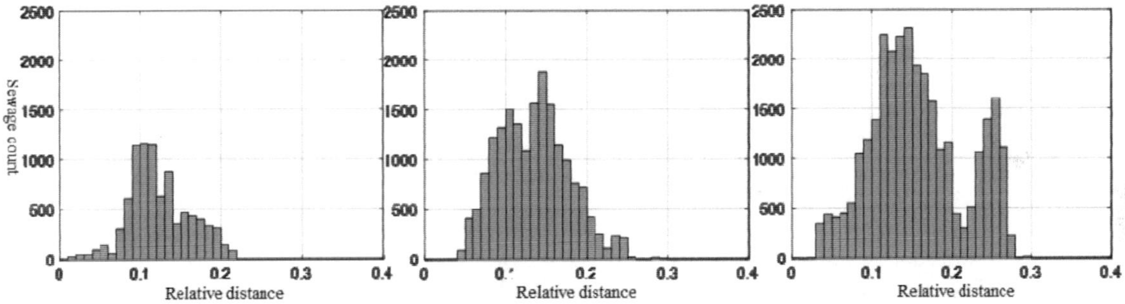

Fig.5 Effect of sewage weight on the sewage convergence

3.4. *Effect of bottom slope on the sewage convergence*

The slope of the bottom of flat bottom (left panel), 1:35 (middle panel), 1:13 (right panel) on the sewage convergence is shown in Fig. 6. It can be seen from the figure that when the bottom of the pool is flat, there is sewage residue in the front half of the bottom of the tank, and very few are farthest from the back end of the tank. The distance is up to 0.9, but most of the swage is distributed between 0-0.2. When the bottom of the tank has a small slope of 1:35, no swage remains in the front half, and the sewage is concentrated between 0 and 0.2. When the slope is adjusted to a large slope of 1:13, there is sewage residue in the front half, and the residual amount is more than that of flat bottom. There is sewage residue in the front half, which may be due to the existence of a vortex at the position of the sewage residue. Overall, the sewage is concentrated between 0 and 0.2 under the three working conditions. It can be seen that the slope of the bottom of the tank has little effect on the accumulation of sewage.

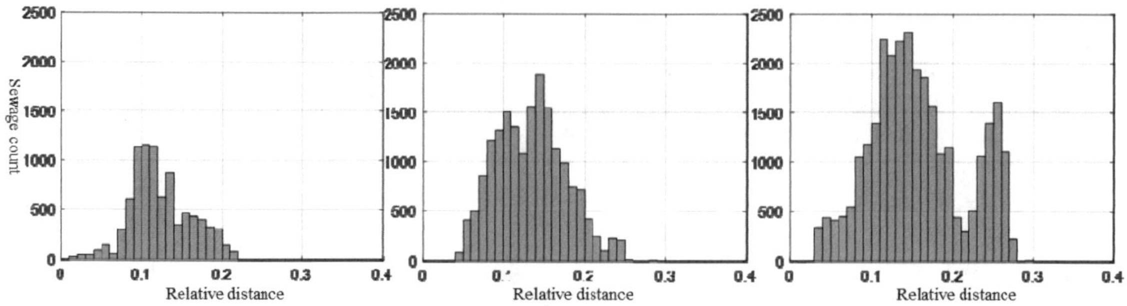

Fig.6 Effect of bottom slope on the sewage convergence

4. Conclusion

In this paper, the characteristics of sewage convergence in rectangular culture tank under different jetting velocity, jetting position, sewage weight and bottom slope were studied. The sewage convergence in rectangular culture tank is obviously affected by jetting velocity, jetting position, sewage weight and bottom slope. The research results found that the higher the jetting velocity, the better the sewage convergence; the closer the jetting position is to the drain hole, the better the sewage convergence; the smaller sewage weight, the better the sewage convergence; when the small slope is 1:35, the sewage convergence is better. It is recommended to adjust the slope of the bottom of the tank to some degree, control the breeding density and feeding amount when constructing the tank. The jetting-driven device should be installed as close to the drain hole for large-size sewage. In order to obtain a better performance of sewage convergence, the flow velocity should be increased as much as possible under the condition of ensuring the healthy growth of the breeding objects.

Acknowledgments

This work was financially supported by the National Innovation and Entrepreneurship Training Program for College Students (202010340011).

References

[1] Timmons M B, Summerfelt S T, Vinci B J. Review of circular tank technology and management[J]. Aquacultural engineering, 1998, 18(1): 51-69.

[2] Summerfelt, Steven, T, et al. Survey of large circular and octagonal tanks operated at Norwegian commercial smolt and post-smolt sites. [J]. Aquacultural Engineering, 2016, 74:105-110.

[3] Oca J, Masalo I. Flow pattern in aquaculture circular tanks: Influence of flow rate, water depth, and water inlet & outlet features [J]. Aquacultural Engineering, 2013:65–72.

[4] Davidson, J., Summerfelt, S. Solids flushing, mixing, and water velocity profiles within large (10 and 150 m3) circular 'Cornell-type' dual-drain tanks. Aquacultural Engineering, 2004, 32(1), 245-271.

[5] JMR Gorle, B.F Terjesen, S.T Summerfelt. Hydrodynamics of Atlantic salmon culture tank: Effect of inlet nozzle angle on the velocity field[J]. Computers and Electronics in Agriculture, 2019, 158.

[6] Oca, J., Masaló, I. Design criteria for rotating flow cells in rectangular aquaculture tanks. Aquacultural Engineering, 2007a ,36(1), 36-44.

[7] Oca J, Masalo I, Reig L. Comparative analysis of flow patterns in aquaculture rectangular tanks with different water inlet characteristics. [J]. Aquacultural Engineering, 2004, 31(3–4):221-236.

Journal of Physics: Conference Series **2271** (2022) 012005 doi:10.1088/1742-6596/2271/1/012005

Flood Inundation Analysis of Xun River Based on 1D and 2D Hydrodynamic Coupling Model

Long Zhao[1], Guoqing Sang [1,*]

[1] Water Conservancy and Environment, University of Jinan, Jinan, 250024, China

[*]Corresponding author's e-mail: sangguoqing@163.com

Abstract. The Xun River is prone to flooding due to its tortuous course and low flood control standards. In this paper, the 1D and 2D hydrodynamic coupling model is used to simulate the flood situation of the Xun River in different return periods, obtain the initial inundation time, maximum inundation depth, maximum inundation area and other information of the flood, and comprehensively analyse the changes in the flood area of different water depth and different townships. The results show that with the increase of return period, the flood inundation range and inundation depth also increase, and the initial inundation time of the flood decreases; the proportion of shallow water areas of 0.05-0.5 meters decreases, and the proportion of deep water areas exceeding 3 meters increases; compared with the flood situation in different townships, Zhonglou Town is more seriously affected, and the area and population of various assets affected by the flood in Zhonglou Town are greater than that of Huangdun Town. This research provides guiding significance for the prevention and control of flood disasters in the Xun River and the rescue and disaster relief.

1. Introduction

Floods are disasters caused by the soaring water level of rivers, lakes and reservoirs, the overflow or collapse of levees, and the entry of water. In addition to causing major disasters to agriculture, floods also cause losses of industry and even life and property, and are one of the ten major natural disasters that threaten the survival of mankind[1-3]. At present, many scholars have studied the flooding situation in different areas in different ways. Yu C[4] proposed a flood dynamic evolution algorithm based on spatial partitions by DEM and simulated the dynamic change process of flood evolution. Dong B L[5] used the physical model to analyse the impact of building density and green belt layout on flood evolution. Yu F Q[6] coupled hydrological model and 2D hydrodynamic model simulated the evolution of floods in the Meixi River basin of Quanzhou. Liu C K[7] used the OpenSceneGraph framework to build a 3DGIS scene, using the flood in the Three Gorges Reservoir area as an example to carry out inundation simulation and loss assessment. In this paper, the 1D and 2D hydrodynamic coupling model of the Xun River basin is constructed, and the flood inundation under different return periods is simulated, which is of great significance for the flood prevention and rescue of the Xun River.

2. Model Principles

2.1 Principles of 1D hydrodynamic model

Conservation equation:

$$\frac{\partial Q}{\partial x}+\frac{\partial A}{\partial t}=q \tag{1}$$

Momentum equation:

$$\frac{\partial Q}{\partial t}+\frac{\partial}{\partial x}\left(\alpha\frac{Q^2}{A}\right)+gA\frac{\partial h}{\partial x}+g\frac{Q|Q|}{AC^2R}=q \tag{2}$$

Where: A represents flow area; Q is discharge through A; t and x represent time and distance; q is later inflow/outflow; h is free surface elevation; g is gravitational acceleration; C is the Chezy resistance coefficient; R is hydraulic radius; and α is momentum distribution coefficient.

2.2 Principles of 2D hydrodynamic model

Continuous equations:

$$\frac{\partial s}{\partial t}+\frac{\partial (ud)}{\partial x}+\frac{\partial (vd)}{\partial y}=0 \tag{3}$$

Momentum equation:

$$\frac{\partial u}{\partial t}+u\frac{\partial u}{\partial x}+v\frac{\partial u}{\partial y}+g\frac{\partial s}{\partial x}+\frac{gu\sqrt{u^2+v^2}}{C^2d}+E\left(\frac{\partial^2 u}{\partial x^2}+\frac{\partial^2 u}{\partial y^2}\right)=0 \tag{4}$$

$$\frac{\partial v}{\partial t}+v\frac{\partial v}{\partial y}+u\frac{\partial v}{\partial x}+g\frac{\partial s}{\partial y}+\frac{gv\sqrt{u^2+v^2}}{C^2d}+E\left(\frac{\partial^2 v}{\partial x^2}+\frac{\partial^2 v}{\partial y^2}\right)=0 \tag{5}$$

Where: s is the free surface elevation, u and v are the depth averaged velocities in the x and y directions, d is the water depth, C is the Chezy resistance coefficient, and E is an eddy viscosity coefficient.

3. Profile of the Study Area

The research area of this paper is the Xun River basin between the South Chenjiagou Reservoir and the Doushan Reservoir. The length of the Xun River is 32 km, and the total area of the basin is 331.13 km^2. The distribution of watersheds is shown in Figure 1. The Xun River basin has an average annual runoff depth of 300 mm. Interannual intra-annual variations in runoff are closely correlated with precipitation. The runoff from July to October accounts for about 87% of the year, and August is the largest, accounting for 56% of the year. The Xun River was historically a river with frequent flooding. In recent years, although it has been partially treated, due to many reasons such as funds, technology, construction, and management, the flood control standards of the river are not high, and flood disasters often occur.

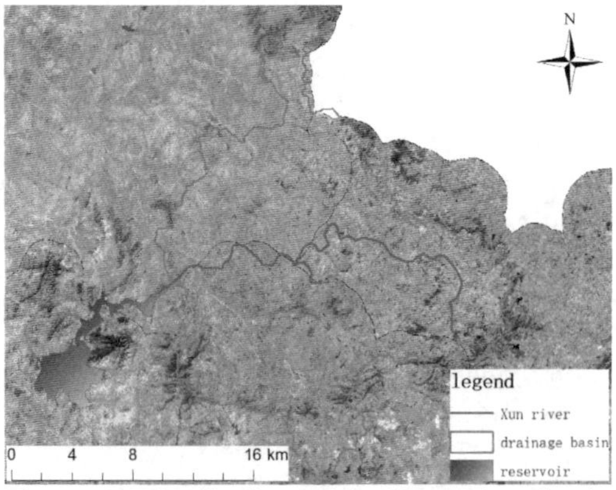

Figure 1. Distribution of the Xun River basin

4. Analysis of results

4.1 Flooding element analysis

The inundation range of different return periods is shown in Figure 2, and the initial inundation time of the flood, the maximum inundation depth, and the maximum inundation area are shown in Table 1. With the increase of the return period, the maximum flood area also increases accordingly, and the maximum inundation depth gradually increases. The initial inundation time of the 200-year flood is 1 hour earlier than the initial inundation time of the 5-year flood, and the increase in the return period, the earlier the initial inundation time, which has a greater guiding effect on the prevention of flood disasters.

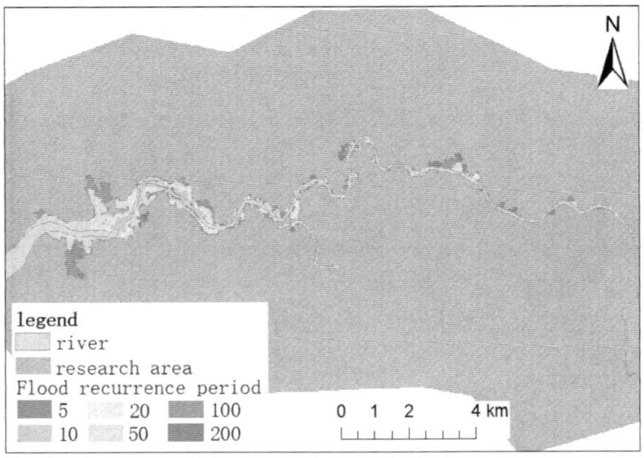

Figure2. Flood inundation ranges at different return periods

Table 1. Flood inundation data at different return periods

Return period(years)	Initial inundation time(h)	Maximum inundation depth(m)	Maximum inundation area(km^2)
5	11.33	1.95	0.13
10	11	2.96	0.9
20	10.83	4.45	1.82
50	10.67	5.72	3.97
100	10.5	6.19	5.64
200	10.33	6.36	6.61

4.2 Comparative analysis of inundated areas in different water depths

The water depth divides the inundated area into five grades: 0.05~0.5m, 0.5~1.0m, 1.0~2.0m, 2.0~3.0m and >3m, and the proportion of inundated area at all levels is plotted in Figure 3. It can be seen from the figure that when the return period is 5 years, the area of inundated water depth of 0.05~0.5m accounts for 79.6%, and there is no inundated water depth greater than 3m area, and when the return period is 200 years, the area with inundated water depth of 0.05~0.5m accounts for 44.6%, and the area with inundated water depth greater than 3m accounts for 4.2%, which indicates that with the increase of the return period, the proportion of inundated water depth of 0.05~0.5m is decreasing. The proportion of inundated water depths greater than 3m is increasing, and the flood hazard is increasing.

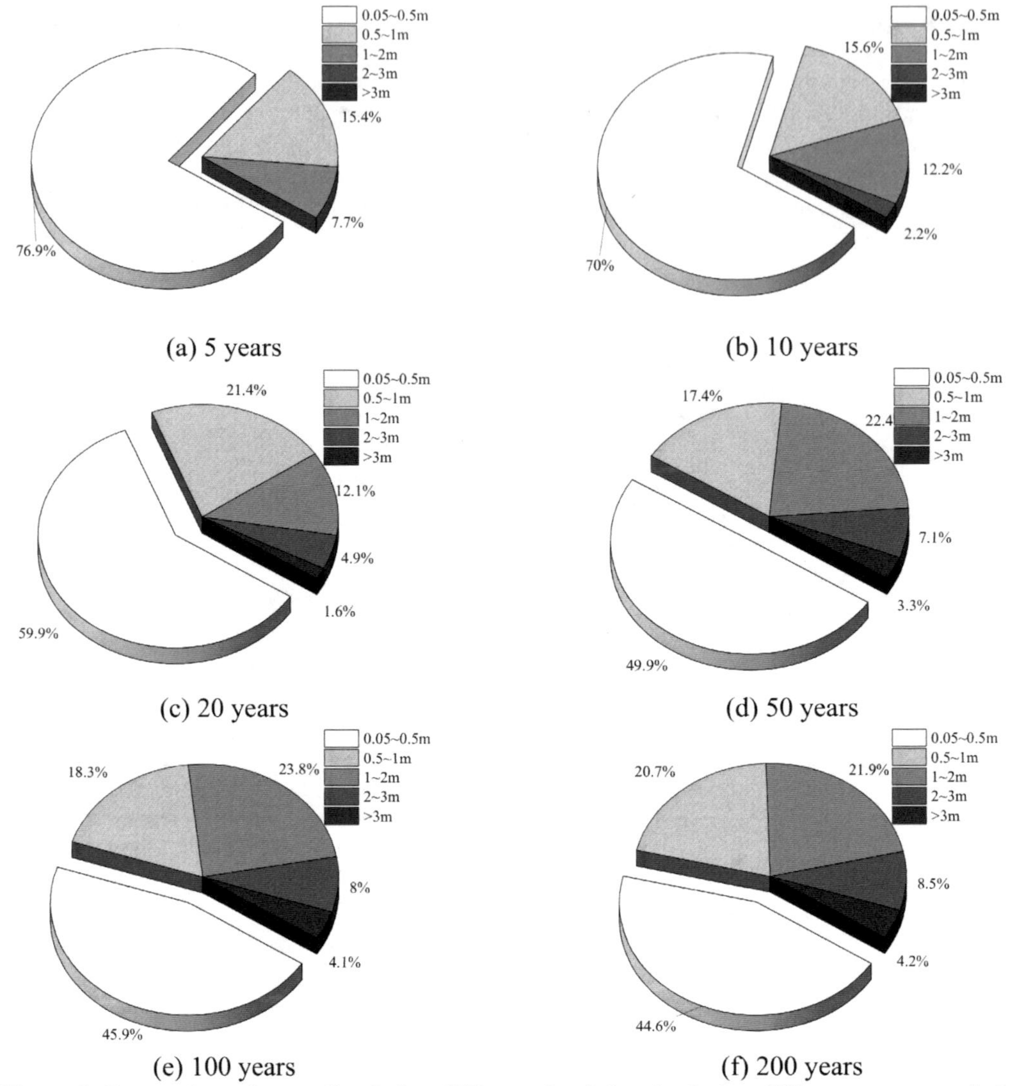

Figure 3. Proportion of areas flooded at different depth levels during different return periods

4.3 Comparative analysis of flood inundated areas of various townships

The flood affected areas in the research area mainly involve two townships, Huangdun Town and Zhonglou Town, the columnar statistical charts are shown in Figures 4. It can be seen from the figure that with the increase of the return period, the total area of flooding, the area of inundated cultivated land, the area of inundated houses and the affects the number of people are gradually increasing, and the impact of flooding on the two townships is increasing. When the return period is 5 years, the total flood area of Huangdun Town and Zhonglou Town is $0.04km^2$ and $0.08\ km^2$ respectively, and the area of inundated cultivated land is 24.0 mu and 46.3 mu respectively. When the return period is 200 years, the flood has the most serious impact on the two towns, among which the total flood area of Huangdun Town is $2.30km^2$, the area of inundated cultivated land is 1248.6 mu, the area of inundated houses is $9183\ m^2$, and the population is 110 people; the total area of flooding in Zhonglou Town is $4.31km^2$, the area of inundated cultivated land is 2164.0 mu, the area of inundated houses is $49600\ m^2$, and the population is 595 people. At this time, the socio-economic impact of the flood on the two towns is very serious, so the relevant departments should be prepared in advance for flood prevention and disaster reduction. Comparing the flood impact of Huangdun Town and Zhonglou Town, it can be seen that

Zhonglou Town is greatly affected by the flood, and when the return period is 200 years, the number of people affected by the flood in Zhonglou Town is more than 5 times the number of people affected by the flood in Huangdun Town. This is closely related to the geographical relationship and economic distribution of Zhonglou Town. Zhonglou Town is located in the lower reaches of the river, the catchment area is large, there are more residents on both sides of the river, and it is more likely to flood.

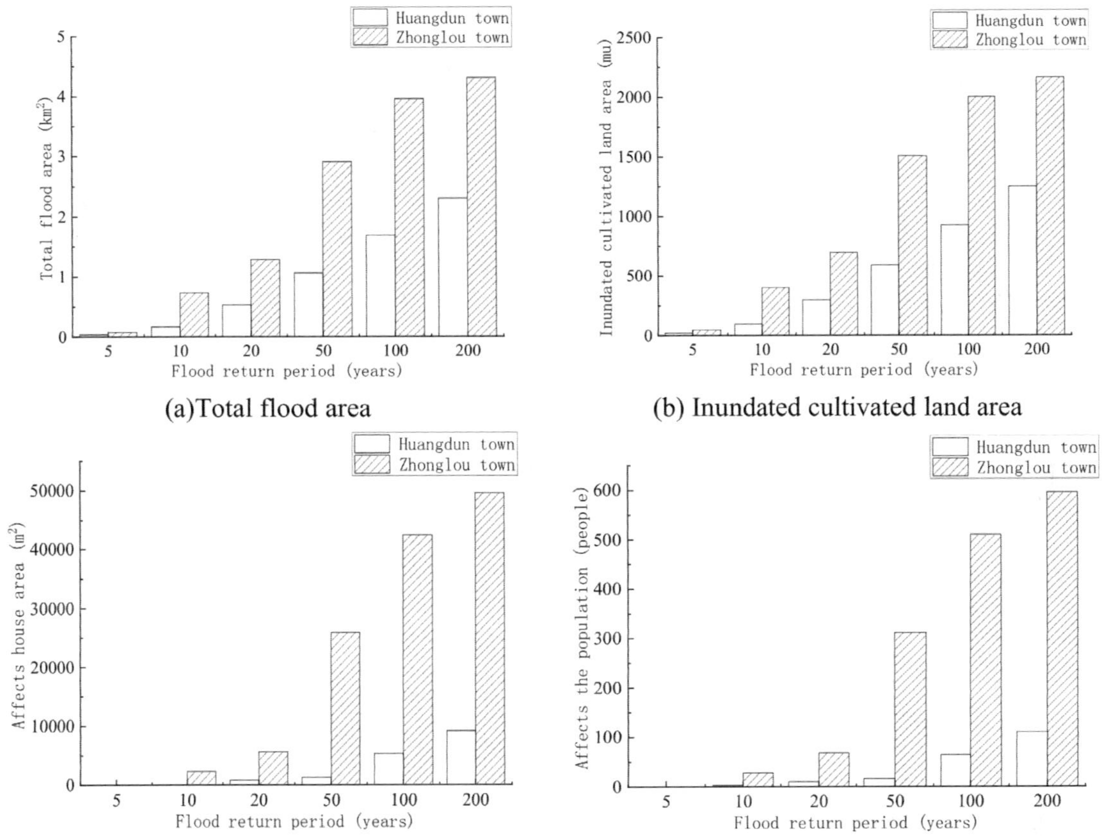

(a)Total flood area (b) Inundated cultivated land area

(c) Affects house area (d) Affects the population

Figure 4. Inundation of Zhonglou Town and Huangdun Town in different return periods

5. Conclusions

In this paper, a 1D and 2D hydrodynamic coupling model is constructed for the Xun River basin, the flood simulation at different return periods is realized, the information such as the initial inundation time, the maximum inundation depth and the maximum area of inundation are obtained, and the inundation area of different water depth and the inundation of each township are analyzed, and the following conclusions are drawn:

1. With the increase of the return period, the maximum flood area and the maximum flood depth continue to increase, and the initial inundation time becomes smaller, which has a greater guiding effect on the prevention of flood disasters. When flooding comes, the population of the threatened area should be transferred in advance to reduce losses.

2. With the increase of the return period, while the flood area expands, the proportion of the flood depth area of 0.05~0.5m is rapidly decreasing, the flood depth is increasing as a whole, and the flood risk is constantly increasing.

3. Compared with the flooding situation of Zhonglou Town and Huangdun Town, under different reproduction periods, the inundation area of various assets in Zhonglou Town is always greater than that

of Huangdun Town, and the river section of Zhonglou Town should be dredged in time and the river embankment should be strengthened to reduce the loss caused by flood disasters.

Acknowledgments
Natural Science Foundation of Shandong Province (ZR2019BEE064).

References

[1] Sun Z L, Zhu X F, Pan Y Z, et al. (2017) Flood Risk Analysis: Progress, Challenges and Prospect . J. Journal of Catastrophology, 32:125-130+136.
[2] Hirabayashi Y, Mahendran R, Koiral S, et al. (2013) Global Flood Risk Under Climate Change. J. Nature Climate Change, 3: 816-821.
[3] Fang J, Li M J, Wang J A, et al. (2015) Assessment and mapping of global fluvial flood risk. J. Journal of Natural Disasters, 24: 1-8.
[4] Yu C, Hu D Y, Zhang Y N, et al. (2020) Research and Application of Flood Dynamic Evolution Algorithm by DEM Spatial Partition. J. Science of Surveying and Mapping, 45: 181-189.
[5] Dong B L, Xia J Q, Chen J H. (2020) Experimental Investigation of Flood Inundation Over Typical Urban Streets. J. Journal of Hydroelectric Engineering, 39: 99-108.
[6] Yu F Q, Yu J S, Jiang W W, et al. (2019) Flood Inundation Simulation Based on Coupled Hydrologic-Hydrodynamic Model. J. South-to-North Water Transfers and Water Science & Technology, 17: 37-43
[7] Liu C K, Ma R, Zhang L, et al. (2020) Research on Flood Simulation and Loss Evaluation Based on 3DGIS. J. Water Resources and Hydropower Engineering, 51: 204-209

Journal of Physics: Conference Series **2271** (2022) 012006 doi:10.1088/1742-6596/2271/1/012006

Research on water heat coupling model of riparian zone based on soil thermal conductivity model

Chen Ma[1,a], Jie Ren[1,2*], Juan Dai[3], Zhaoyang Ma[1] and Fan Wang[1]

[1] State Key Laboratory of Eco-hydraulics in Northwest Arid Region, Xi´an University of Technology, Xi´an, shaanxi, 710048, China

[2] State Key Laboratory of Hydraulics and Mountain River Engineering, Sichuan University, Chengdu, Sichuan, 610065, China

[3] Changjiang River Scientific Research Institute, Wuhan, Hubei, 430010, China

[a] email: 2190421250@stu.xaut.edu.cn

[*] Corresponding author's e-mail: renjie@xaut.edu.cn

Abstract: In recent years, the research on the process of water and heat transfer in the riparian zone has become more and more extensive, a more accurate riparian water and heat coupling model is helpful for us to understand the law of water and heat transfer in the riparian zone. The hydro thermal coupling model of riparian zone is constructed based on different soil thermal conductivity models, the simulation effects of different models are verified and compared by using the experimental data, the parameter sensitivity analysis of the built model is carried out to determine the main parameters affecting the output of model temperature field, and the influence of revetment structure on the process of hydro thermal transport in riparian zone is simulated. The results show that: 1. the Hydrothermal Coupling Model of riparian zone based on Johansen thermal conductivity model has the best simulation effect and high reliability. 2. The permeability coefficient (k_s) and porosity (n) have the most significant influence on the temperature simulation results, and the parameter sensitivity is the highest. 3. The revetment structure hinders the infiltration of surface water into the riparian zone in the horizontal direction, reduces the infiltration water level and temperature fluctuation in the riparian zone.

1. Introduction

Riparian subsurface layer refers to the area where the saturated sedimentary layer extends from the riverbed to the banks on both sides. Groundwater and surface water interact at this time, accompanied by material exchange and energy transportation, which is of great significance for riparian ecological protection[1]. In recent years, with the maturity and advantages of temperature tracing methods, many scholars have gradually shifted their research methods on the riparian subsurface flow zone from the traditional hydraulic method and ion tracing method to the temperature tracing method. For example, Liu Dongsheng et al. carried out the field measured of riparian temperature and water level in the lower reaches of Xin'an River, and summarized the characteristics of subsurface flow exchange and the distribution law of temperature field in winter and summer[2]. Ji Yuyu et al. conducted field tests on the beach of Lancang River, by recording the data of water level and water temperature, they analyzed the response relationship between water temperature and water level, and calculated the lateral undercurrent exchange capacity of the beach[3]. In addition to the above studies, most scholars focus on the experimental means of subsurface flow exchange in the riparian zone, and there are relatively few studies on the hydro thermal coupling model.

Soil thermal conductivity is an important parameter in the study of saturated unsaturated soil Hydrothermal Coupling numerical model. Its value is affected by soil texture, moisture content and other factors. The hydro thermal coupling model constructed in the research at this stage does not consider the influence of heterogeneous heat transfer of soil, and takes the thermal conductivity of soil as a fixed value. For example, Molina girald et al. pointed out that heterogeneous heat transfer has a significant impact on the temperature distribution in the riparian zone by constructing the hydro thermal coupling model of riparian zone[4]. Ren et al. studied the variation characteristics of beach temperature field under the condition of water level change by analyzing the field monitoring data of Dongting Lake and carrying out numerical simulation[5]. This is bound to affect the accuracy of the model results. The uncertainty of soil thermal conductivity is very important in the study of saturated unsaturated hydrothermal transport in riparian zone.

Therefore, in this paper, the soil thermal conductivity model is considered into the riparian hydro thermal coupling model, the simulation effects of riparian hydro thermal coupling model based on different thermal conductivity models are compared and analyzed through indoor test data, and the sensitivity analysis is carried out to determine the main factors affecting the temperature field of the model riparian zone, so as to reduce the workload of model correction, in order to provide reference for relevant research.

2. Research methods

2.1. Laboratory experiment

The indoor test is carried out in a two-dimensional sand tank (as shown in figure 1). The test conditions are: water level 0.7m, water temperature 20 ~ 26°C. The sand tank is composed of left and right water tanks, middle sand tank, water level and water temperature control system and data acquisition device. The main body of the sand trough is 2m long and 0.5m wide. It is filled with sand to simulate the riparian zone. There are 0.15m long and 0.5m wide water tanks on the left and right sides. Perforated baffles are set between the water tank and the sand tank, and the baffles are covered with gauze to prevent sand from passing through. The left sink is provided with water inlet, with the height of 0.5m, 0.6m and 0.7m respectively, and the right sink is provided with water outlet. The water temperature changes sinusoidal through the temperature control system. A temperature measuring point is arranged on the front of the sand tank, and a pressure measuring pipe is arranged on the back. There are 15 temperature measuring points in total, 5 rows are set horizontally, the spacing of each row is 0.12m, and the lowest layer is 0.05m from the bottom. Three columns are set vertically, and the horizontal distance between the first column and the other two columns is 0.35m and 0.65m. The temperature sensor is connected to the computer through the recorder to collect temperature data.

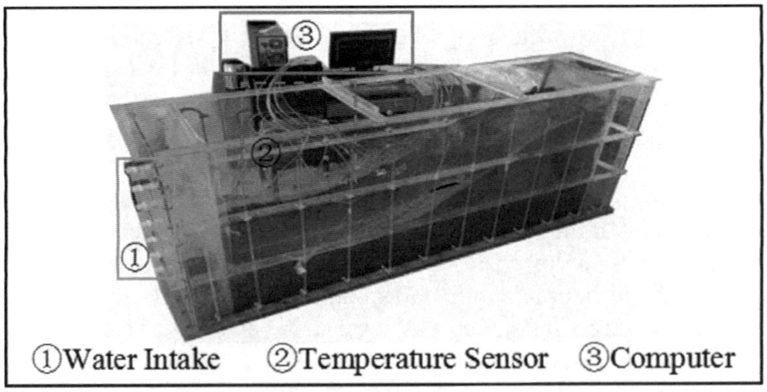

Figure 1. Diagram of indoor test device

2.2. Research methods

The hydro thermal coupling model of riparian zone is constructed based on COMSOL Multiphysics finite element software (figure 2). The geometric horizontal length of the model is 2m, the vertical height is 1m and the slope is 57°. For the seepage field of hydro thermal coupling model in riparian zone, ab and bc are impermeable boundary, ae is infiltration head boundary, ed is no flow boundary and bc is impermeable boundary. For temperature field, the boundaries ab and bc are adiabatic boundaries, the ae section is the water temperature boundary of the infiltration water, the ed and dc sections are the air temperature boundary, and the temperature boundary is set to 18°C.

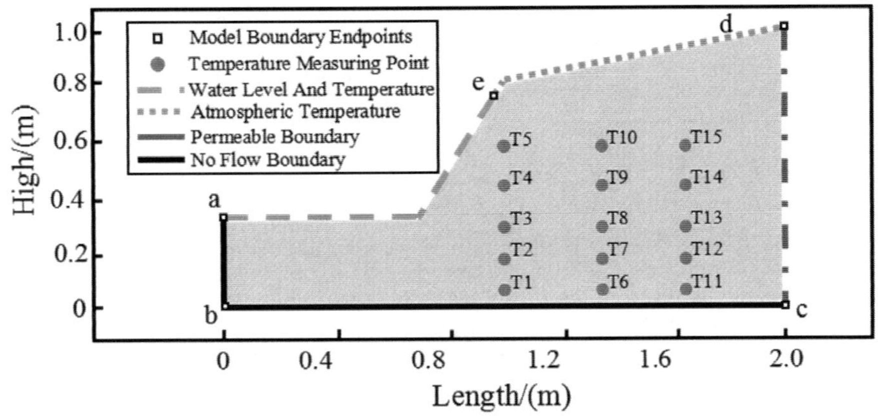

Figure 2. Model setting and boundary conditions

The groundwater module of the hydro thermal coupling model is expressed by Richards equation of saturated unsaturated pore flow:

$$Q = \rho\left(\frac{C_m}{\rho g} + S_e S\right)\frac{\partial p}{\partial t} + \nabla\rho\left(-\frac{k_s k_r}{\mu}\nabla(p + \rho g z)\right) \tag{1}$$

Where: Q is the flow source sink term; ρ is water density; C_m is water capacity; g is the acceleration of gravity; S_e is relative saturation; S is the water storage coefficient; P is the pressure; ∇ is laplace operator; θ is moisture content; k_s is the permeability coefficient; k_r is the relative permeability; μ is the coefficient of viscous motion; z is the vertical elevation.

The soil moisture characteristics are described by Van Genuchten curve.

The temperature field of the model is described by the heat convection heat transfer equation:

$$c\rho\frac{\partial T}{\partial t} = \nabla(\lambda\nabla T) - Cp\nabla(vT) + Q \tag{2}$$

Where: Q is the temperature source sink term; T is the fluid temperature; c is the specific heat capacity; v is the flow rate; ρ is density; λ is the equivalent thermal conductivity; C is the specific heat capacity of water, p is the water density.

The thermal conductivity model can be divided into empirical model, semi empirical model and theoretical model. This paper adopts five empirical models and two theoretical models, including Johansen, Campbell, Lu, Tong, Xiong, Carson, Tarnawski model and three soil effective thermal conductivity models built in COMSOL Multiphysics finite element software[6,7].

The model calculation parameters are shown in table 1.

Table 1. Calculation parameters of hydro thermal coupling model in riparian zone

Parameter	Value	Parameter	Value
Permeability coefficient k_s (m/s)	5.2×10^{-5}	Density of soil ρ_s (kg/m^3)	1600
Porosity n	0.43	Specific heat capacity of water c_w (J/(m$^3\cdot$°C))	4.2×10^{6}
Saturated moisture content θ_s (m^3/m^3)	0.39	Density of water ρ_w (kg/m^3)	1000
Residual moisture content θ_r (m^3/m^3)	0.07	Thermal conductivity of water λ_w (W/(m·°C))	0.58
VG model parameters α(1/m)	11.9	Transverse thermal dispersion α_r (m)	0.01
VG model parameters n_v	2.5	Longitudinal thermal dispersion α_L (m)	0.01

3. Results and discussion

3.1. Model verification and comparative analysis

See figure 9 for the comparison between simulated and measured temperature values at different monitoring points. It can be seen from the figure that the closer to the river, the smaller the error between the simulated value and the measured value, and the higher the simulation accuracy. Among them, the simulation effect of T5 point is the best and T15 point is the worst. The farther the temperature measuring point is from the river, the later the influence of surface water on the temperature field distribution in the riparian zone, the obvious phase lag, and the increase of infiltration distance accelerates the dissipation of infiltration water temperature, and the temperature fluctuation in the riparian zone decreases. Among the 10 models, the simulated values of Tarnawski model, Tong model and three built-in models have large errors with the measured values, and the simulation effect is poor. The simulation values of the other five models are closer to the measured values, and the simulation accuracy is higher.

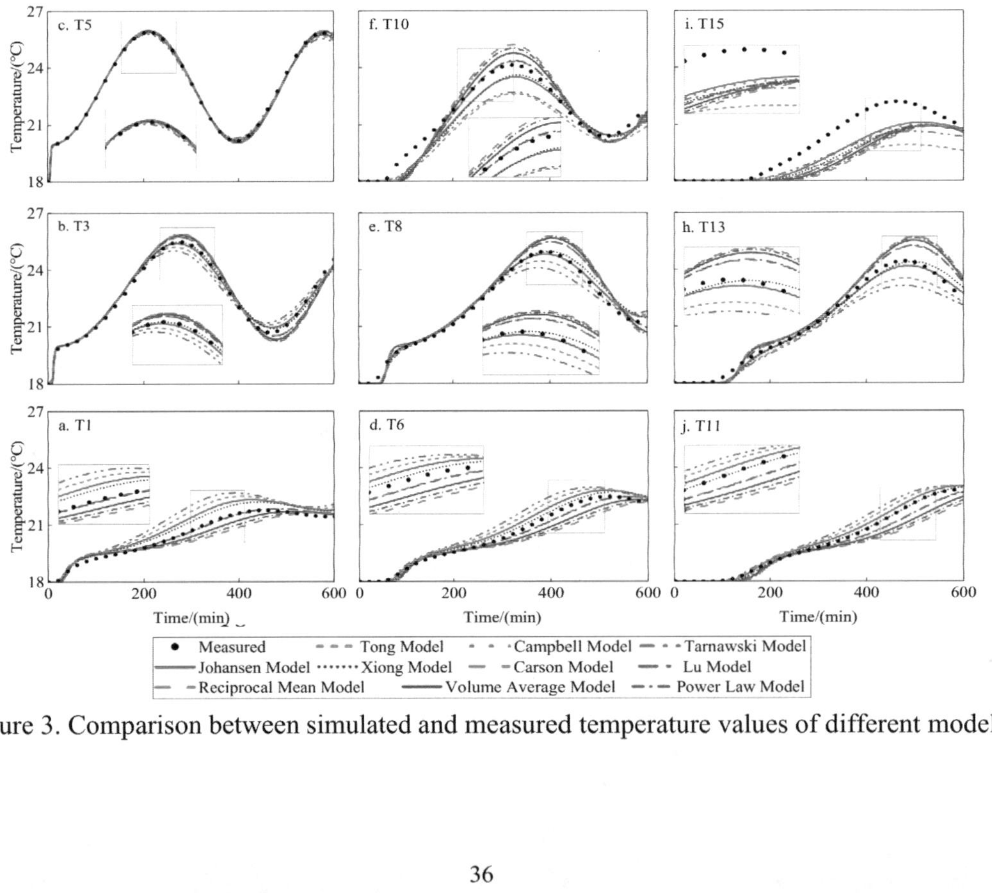

Figure 3. Comparison between simulated and measured temperature values of different models

In order to more intuitively show the simulation effect of water heat coupling models in different riparian zones, root mean square error (RMSE) and relative error (Re) are used to evaluate the simulation accuracy of each model. The results are shown in figure 4. It can be seen from figure 4 that the fluctuation range of evaluation indexes of Johansen model is the smallest, in which RMSE is 0.071°C ~ 0.847°C, Re is 0.003 ~ 0.042, and the mean values of RMSE and Re are 0.327°C and 0.016 respectively, which are lower than other models, indicating that the model has the best simulation effect and can accurately reflect the internal heat transfer process of riparian zone, which is consistent with the conclusion obtained by Zhang Wenbing through comparative analysis combined with field tests[8]. The mean values of RMSE of Campbell model, Carson model, Lu model and Xiong model are 0.330°C, 0.366°C, 0.370°C and 0.381°C respectively, and the mean values of Re are 0.016, 0.017, 0.018 and 0.016 respectively. The simulation effect is slightly worse than that of Johansen model, but the difference is small. The overall simulation effect is good. The simulation effect of Tong model and Tarnawski model is poor, because Tarnawski model and Tong model are developed for specific soil, and the soil thermal conductivity model is not suitable for this study. Overall, the simulation effect of the riparian zone water heat coupling model based on Johansen model is the best.

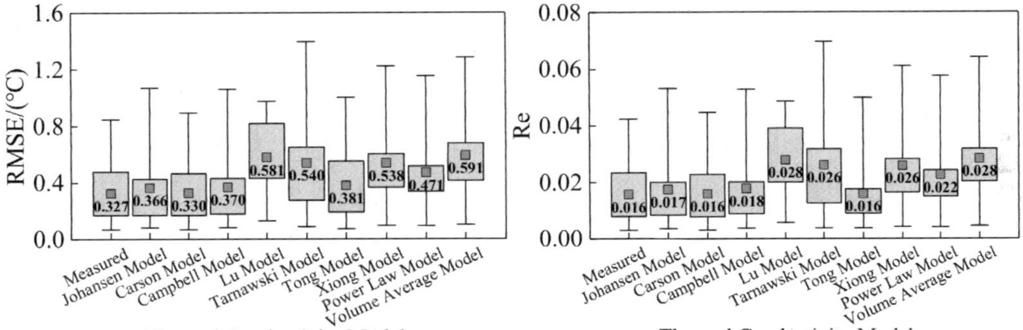

Figure 4. Overall analysis and comparison of evaluation indexes of different models

3.2. Parameter sensitivity analysis

The parameter sensitivity analysis is carried out by orthogonal test to determine the main parameters affecting the temperature field output of the model. On the basis of considering the original parameters of the model, change the input parameter values according to the four levels of ± 5% and ± 10%, and calculate the temperature variation amplitude. Six model input parameters obtained from indoor test are selected: permeability coefficient (k_s), soil porosity (n) and Van Genuchten parameters α and n_v, saturated moisture content(θ_s) and residual moisture content(θ_r), the parameter sensitivity of the riparian zone hydro thermal coupling model based on Johansen model is analyzed.

The sum of squares of range and deviation is analyzed for the orthogonal test results. The results are shown in figure 5. According to the analysis of figure 5, the sensitivity of the model to parameters is: permeability coefficient (k_s) > soil porosity (n) > residual water content(θ_r) > vangenuchten parameter α> saturated moisture content(θ_s) > vangenuchten parameter n_v. permeability coefficient (k_s) and soil porosity (n) have significant effects on the change of model temperature field, while other parameters have no significant effect on the change of model temperature field, and the sensitivity of the model is low.

The orthogonal test results are analyzed, and the results are shown in figure 6. It can be seen from figure 6 that the variation amplitude of temperature simulation value of test models in groups 1~5, 6~10, 16~20 and 21~25 shows a negative increase. The permeability coefficient (k_s) of each five groups is the same, and the soil porosity (n) shows a gradual increase trend. Therefore, when the permeability coefficient (k_s) is the same, the temperature variation amplitude is most affected by the soil porosity (n) and shows an obvious negative correlation. There is no obvious linear correlation between the temperature variation of the simulation values of the 11~15 groups of tests. The reason for this phenomenon is caused by the interaction of multiple parameters, and its influence on the temperature simulation results cannot be ignored.

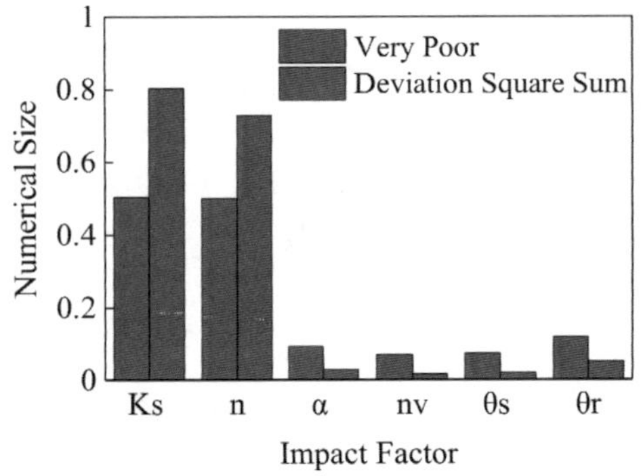

Figure 5. Sensitivity index results of various parameters

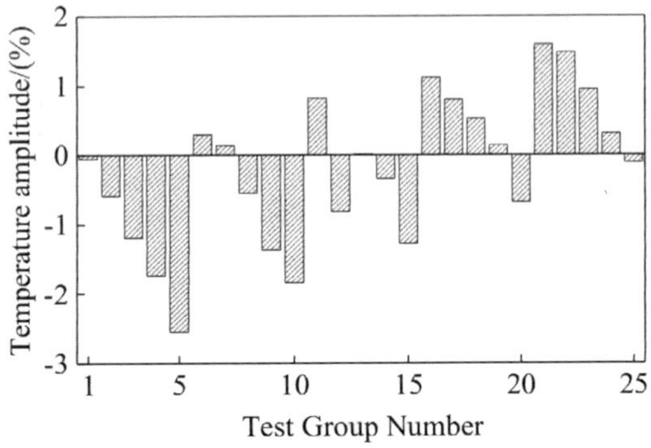

Figure 6. Temperature variation of each group

3.3. *Simulation of water and heat transfer process in riparian zone by revetment structure*

In order to study the influence of bank protection projects with different permeability on the water and heat transfer in the riparian zone, the riparian zone water and heat coupling model based on Johansen model is selected for research. On the basis of setting the non-revetment structure as the control group, three revetments with different permeability coefficients are set with reference to relevant literature[9]: revetment 1 ($k_{s1} = 5.2\times10^{-7}$m/s), revetment 2 ($k_{s2} = 5.2\times10^{-8}$m/s), revetment 3 ($k_{s3} = 5.2\times10^{-9}$m/s).

3.3.1. *Influence of revetment structure on surface water infiltration*

figure 7(a) shows the coastal variation curve of infiltration water level, and (b) shows the variation curve of infiltration water level with time at T6 measuring point. It can be seen from figure 7(a) that the smaller the permeability coefficient of the revetment, the lower the infiltration water level, the stronger the blocking effect on the infiltration of surface water. With the increase of infiltration distance, the blocking effect weakens and the water level gap gradually decreases. It can be seen from figure 7(b) that the revetment affects the infiltration rate of surface water. The time lag of the water level at T6 measuring point is obvious. When there is no revetment, the water level at T6 measuring point basically reaches a stable state in 300min, and the water level rise is not obvious in 300min~600min. When there is revetment, the water level at T6 measuring point is rising in the whole simulation time. The above conclusions are consistent with the analysis of Hu Shuchen[9].

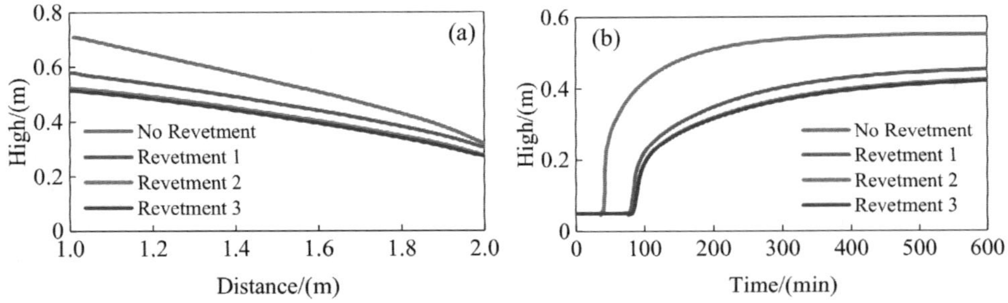

Figure 7. Variation curve of infiltration water level: (a) variation of water level and coastal distance. (b) water level changes with time

3.3.2. *Influence of revetment structure on temperature field*

Figure 8 shows the comparison of temperature simulation values of 6 measuring points under different permeability coefficients of revetment structure. It can be seen from figure 8 that the revetment structure reduces the internal temperature amplitude of the riparian zone, with obvious phase lag, which is directly proportional to the offshore distance. Compared with the simulation value of revetment 1, the temperature simulation values of revetment 2 and revetment 3 at each measuring point have little difference. When the permeability is small to a certain critical value, the influence of revetment permeability on the change of river bank temperature field is no longer obvious.

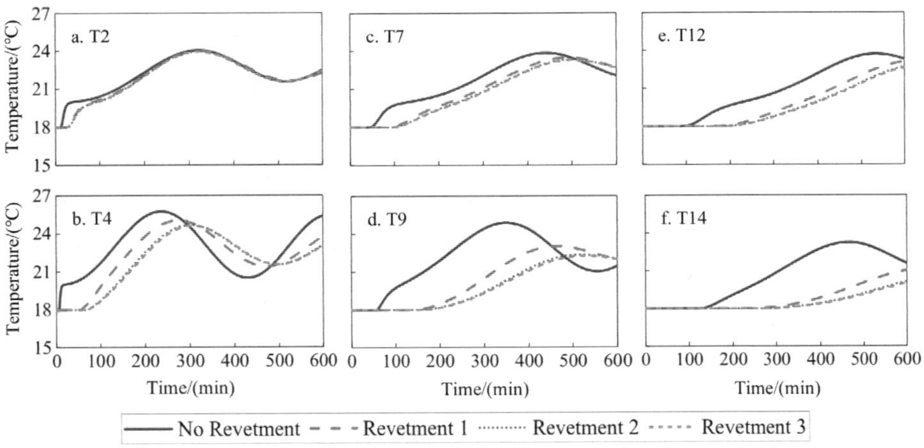

Figure 8. Comparison of temperature simulation values of revetment structure under different permeability coefficients

Figure 9 shows the simulated temperature field distribution in the riparian zone. It can be seen from figure 9 that with the infiltration of surface water, the temperature field in the riparian zone is distributed in high temperature zone, medium temperature zone and low temperature zone in the horizontal direction. With the increase of infiltration time, the range of high temperature zone and medium temperature zone gradually expands and diffuses to the interior of the riparian zone. Because the revetment structure hinders the infiltration of surface water, the range of high temperature zone and medium temperature zone in the riparian zone is smaller than that without revetment. Comparing the temperature cloud images of the same infiltration time of 100 min, 200 min and 300 min, the revetment has a great influence on the temperature field distribution of the riparian zone in the horizontal direction, but not in the vertical direction. The lower the permeability of the revetment, the closer the diffusion distance between the high-temperature zone and the medium temperature zone in the horizontal direction, and the smaller the range.

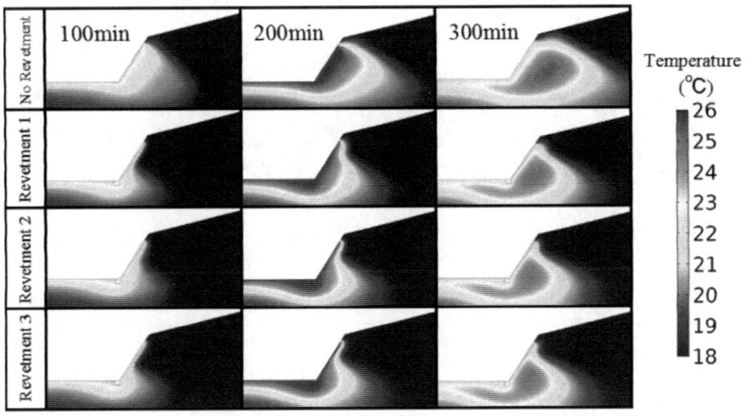

Figure 9. Variation of temperature field in riparian zone under different permeability coefficients of revetment structure

4. Conclusion

The water heat coupling model of riparian zone based on Johansen model has the best simulation effect, and the simulation effect of other models is poor, but they are better than the built-in model of software. Permeability coefficient (k_s) and porosity (n) have the most significant influence on the change of temperature field of the model, and the influence of other parameters is not significant, but the influence can not be ignored. The revetment structure hinders the infiltration of surface water. The smaller the permeability of revetment, the greater the impact on the seepage field and temperature field of riparian zone.

Acknowledgments

This work was supported by the Scientific Research Program Funded by Shaanxi Provincial Education Department (Program No. 20JS102), and the Shaanxi Provincial Innovative Research Team (Program No. 2022TD-01), and the Open Research Fund Program of the State Key Laboratory of Hydraulics and Mountain River Engineering, Sichuan University (Program No. SKHL2106) and the Open Research Fund Program of State Key Laboratory of Eco-hydraulics in Northwest Arid Region, Xi'an University of Technology (Program No. 2021KFKT-3).

Reference:

[1] Cai, Y., Shi, T., Yao, J.L., Liu, S.G., Ruan, X.K., Xu, J. (2021) Advance in field monitoring research of hydrodynamic process of hyporheic exchange in rivers. J. Advances in Water Science, 32: 638–648.
[2] Liu, D.S., Zhao, J., Lv, H. (2017) Comparative analysis of water-heat exchange characteristics of the riparian zone downstream of dam in different seasons. J. Advances in Water Science, 28: 124–132.
[3] Ji, Y.Y., Chen, Q.W., Shi, W.Q., Yi, Q.T., Lin, Y.Q. (2018). Influence of reservoir operation on water and heat exchange in the Mandan's island. J. Advances in Water Science, 29: 73–79.
[4] Molina-Giraldo, N., Bayer, P., Blum, P., Cirpka, O.A. (2018). Propagation of seasonal temperature signals into an aquifer upon bank infiltration. J. Groundwater, 49: 491–502.
[5] Ren, J., Chen, B., Li, Y.L., Dai, J., Men, L.L. (2020). Dynamic Variation Characteristics of the Riparian Zone Temperature Distribution under Fluctuating Water Levels: Field Experiments and Numerical Modeling. J. Journal of Coastal Research, 36: 1178–1196.
[6] Xiong, K., Gui, J.P., Xu, H.C., Jin, H. (2021). Evaluation and Improvement of Thermal Conductivity Model of Room Temperature Soil. J. Water Saving Irrigation, 86–91.
[7] Wang, J.M., He, H.L., Li, M., Dyck, M., Si, B., Lv, J.L. (2020). A review and evaluation of thermal conductivity models of saturated soils. J. Archives of Agronomy and Soil Science, 67: 974–986.
[8] Zhang, W.B., Shen, Z.Z., Ren, J., Xu, L.Q., Zhou, C.C., Yang, J.M. (2022). Hydrothermal coupling model of a riparian zone considering heterogeneous heat transfer of soil. J. Advances in Water Science, 33: 123–134.
[9] Hu, S.H., Zhu, X.L., Jiang, Q.H., Zhang, Z.T., Jin, G.Q. (2020). Lateral hyprohiec exchange process in stratified soil. J. Water Resources and Power, 38: 29–32+82.

Study on flow field characteristics of box artificial reef

Ping Zhu[1], Yuchi Hao[1*], Zhenxing Wei[2], Chaozhe Yuan[1], Sheng Wang[1]

[1]CCCC National Engineering Research Center of Dredging Technology and Equipment Co., Ltd, Shanghai 200082, China

[2]Shanghai Waterway Engineering Design and Consulting Co., Ltd, Shanghai 200120, China

[*]Corresponding author's e-mail: haoyuchi@cccc-drc.com

Abstract: Based on Zhuhai Miaowan artificial reef demonstration project, this paper studied the flow field characteristics of single box artificial reef and reef group under different spacing by using computational fluid dynamics (CFD) method. The results show that the box artificial reef has a significant effect on flow field. The maximum upwelling height near the reef is about 1.4 times the reef height, and the length of the back vortex is twice the reef size. In the reef group arrangement, the flow field distribution plays the best performance at the 5m spacing arrangement. The research results can provide guidance for reef structure optimization and practical arrangement in site.

1. Introduction

In recent years, due to the influence of many factors, such as excessive marine fishing and environmental pollution, marine fishery resources have declined seriously, which has affected the protection and sustainable utilization of marine living resources in China. In order to improve the current situation of the depletion of offshore marine fishery resources, marine pasture projects have been carried out in coastal areas. The artificial reef, playing an important role in the construction of marine pasture projects, is an artificial structure arranged on the seabed to improve the habitat environment of aquatic organisms, provide a place for aquatic organisms to grow and develop, and then achieve the effect of protecting and proliferating resources.

The flow field characteristic of artificial reef is the decisive factor for giving play to the function of artificial reef. The flow field distribution near the reef is changed due to the influence of the reef wall and its blocking effect, after the reef is put on the seabed. Upwelling is formed near the upstream the reef, which promotes the water exchange between upper and lower layers and transports the bottom sediments and nutrients over the reef. Because of the upwelling, the marine primary productivity and bait effect are enhanced greatly, leading to a significant agglomeration effect on fish. In addition, the vortex zone with low flow velocity is formed downstream of the reef, providing shelter, breeding and habitat for fish. Therefore, the study on the flow field characteristics of artificial reef is of great significance for the structure design and layout optimization of artificial reef.

Physical model experiments and CFD method are mainly used to study the flow field characteristics of artificial reef. Liu et al[1] analyzed the upwelling intensity of artificial reef under different flow conditions through flume experiments. Jiang et al[2] used CFD method to analyze the hydrodynamic characteristics around the cube monomer reef and observed the characteristics of the flow field of the reef through the physical model experiments, which verified the robust of CFD method. Tang et al[3] compared the flow field characteristics of different types of artificial reefs by numerical simulation. Through CFD simulation, Wang[4] concluded that the flow field distribution of the shutter type perforated artificial reef was most suitable for benthos. It can be seen that CFD method is relatively mature in studying the flow field of artificial reef, and has been widely recognized.

The flow field characteristics of artificial reef are closely related to the reef structure, field layout and hydrodynamic conditions. Taking the box artificial reef as the research object, based on the typical flow condition of Zhuhai Miaowan artificial reef demonstration project, this paper used CFD method to study the flow field characteristics of single reef and reef groups under different layout spacing, providing guidance for the optimal layout of artificial reef on site.

2. Numerical model

2.1 governing equations

In the numerical model, the fluid is considered as incompressible viscous fluid, and the governing equations, including continuity equation and momentum equation are applied.

(1) Continuity equation

$$V_F\frac{\partial\rho}{\partial t}+\frac{\partial}{\partial x}(\rho uA_x)+\frac{\partial}{\partial y}(\rho vA_y)+\frac{\partial}{\partial z}(\rho wA_z)=0 \tag{1}$$

Where V_F indicates the fraction of open volume to flow; ρ indicates fluid density; velocity components of (u,v,w) are in (x,y,z) direction; Ax indicates the fraction of open surface at x direction; Ay and Az also indicates the fraction of surface at y and z direction, respectively.

(2) Momentum equation

$$\begin{aligned}\frac{\partial u}{\partial t}+\frac{1}{V_F}\left(uA_x\frac{\partial u}{\partial x}+vA_y\frac{\partial u}{\partial y}+wA_z\frac{\partial u}{\partial z}\right)&=-\frac{1}{\rho}\frac{\partial p}{\partial x}+f_x\\ \frac{\partial v}{\partial t}+\frac{1}{V_F}\left(uA_x\frac{\partial v}{\partial x}+vA_y\frac{\partial v}{\partial y}+wA_z\frac{\partial v}{\partial z}\right)&=-\frac{1}{\rho}\frac{\partial p}{\partial y}+f_y\\ \frac{\partial w}{\partial t}+\frac{1}{V_F}\left(uA_x\frac{\partial w}{\partial x}+vA_y\frac{\partial w}{\partial y}+wA_z\frac{\partial w}{\partial z}\right)&=-\frac{1}{\rho}\frac{\partial p}{\partial z}+G+f_z\end{aligned} \tag{2}$$

Where G indicates mass acceleration, and (fx,fy,fz) indicates viscosity accelerations.

2.2 Geometric model

The artificial reef studied in this paper belongs to the restoration and proliferation reef, which is aimed at improving the bait and fish breeding place in the reef area. At the same time, it can provide sufficient shelter space for young fish. The schematic diagram of artificial reef structure is shown in fig.1. The reef specification is 3.5m × 3.5m × 3.5m. Concrete panels are used around and on the top of the reef, in order to increase the surface area and internal complexity of the reef.

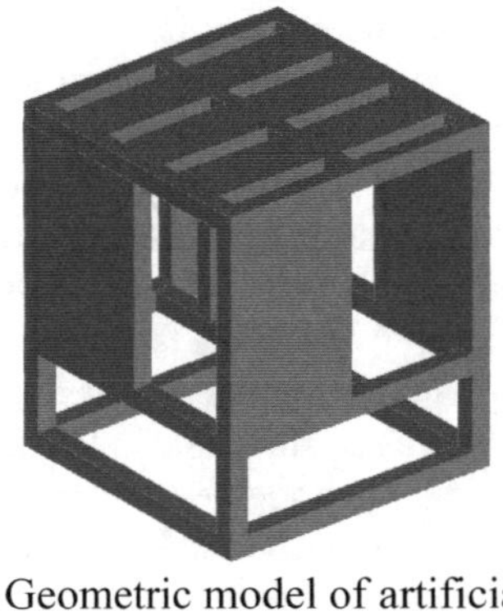

Fig.1 Geometric model of artificial reef

2.3 simulation setup

(1) Computational conditions

Based on the real marine hydrodynamic characteristics of the project site, the typical flow velocity is selected to carry out CFD simulation. The specific computational conditions are shown in Table 1. The flow field characteristic of single artificial reef is analyzed in case 1 and the other cases study the flow field of 9 reefs with 3×3 spatial arrangement under different spacing(2m, 5m, 8m).

Table 1 Calculation Condition

Case	Arrangement	Spacing（m）	Velocity(m/s)
1	single	/	0.52
2	3×3	2	0.52
3	3×3	5	0.52
4	3×3	8	0.52

(2) Computational domain and boundary conditions

The quality of simulation results is closely related to the comparison between reef size and computational domain[5]. Combined with site conditions and computational principles, the computational domain of 35m×30m×30m for single artificial reef is shown in Fig.2. The upstream length of the reef is 10.5m, while the downstream length is 21m.

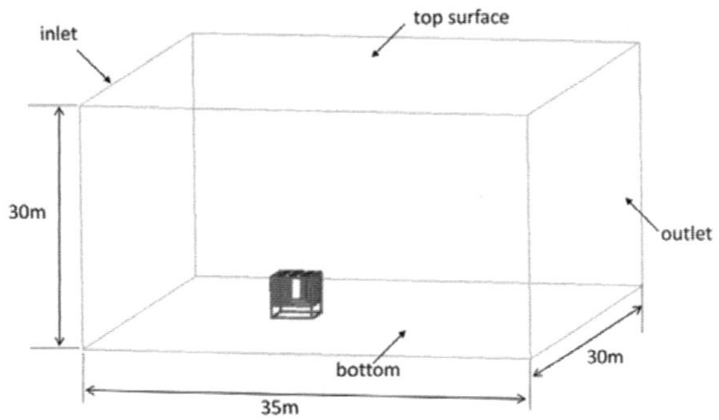

Fig.2 Computational domain and boundaries

The computational grid of artificial reef adopts structured grid. In order to improve the calculation efficiency and ensure the calculation accuracy, the computational grid is refined near the reef structure.

The boundary conditions are described as follows. Both the inlet and outlet are set to the velocity boundary conditions regarding the practical flow condition, with the typical velocity of 0.52m/s. The top surface is defined as air pressure boundary while the bottom and lateral boundary are both specified as wall boundary conditions.

3. results and discussions

The flow field characteristics of artificial reef mainly takes the upwelling on the upstream and the back vortex generated on the back surface of reef as the key evaluation factors. Therefore, based on the longitudinal profile of flow field at the center of the reef and the transverse profile of flow field at 2m depth, this paper analyzed the characteristics of upwelling and back vortex near the reef, and then evaluated the flow field characteristics of the box artificial reef.

3.1 flow field characteristics of single artificial reef

The flow field of the single reef is shown in Fig.3. The upwelling, which transports the nutrients on the seabed to the top of the reef, is significant near the upstream of the reef. The maximum height of upwelling is about 4.8m, 1.4 times the reef height. In the downstream area of the reef, due to the blocking effect of the reef, extensive eddy structures are generated, where the flow velocity is slow. According to the streamlines around the single reef , as shown in Fig.4, the maximum width of the back vortex of the reef is about 6m, and the length of the back vortex is about twice the length of the reef.

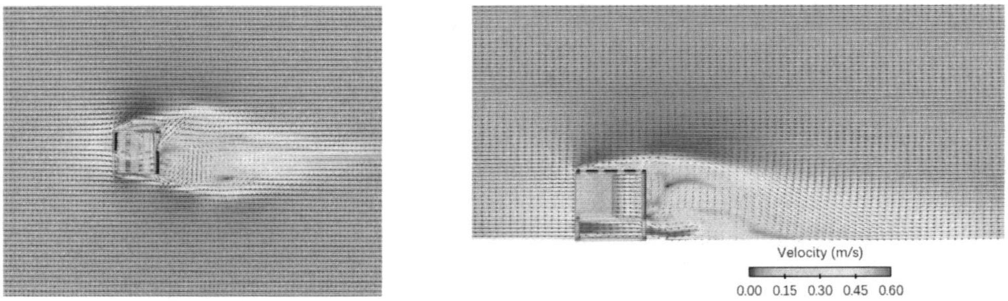

(a) transverse profile (b) longitudinal profile

Fig.3 flow field of single reef

(a) transverse profile (b) longitudinal profile

Fig.4 streamlines of single reef

Compared with the cube artificial reef[6], The box artificial reef, leading to a larger vortex area, is more suitable for fish proliferation.

3.2 flow field characteristics of artificial reef group

Fig.5 shows the flow field characteristics of reefs under three different spacing arrangements. It can be seen that the layout spacing has a significant impact on the flow field around the reef. When the distance between adjacent reefs is 2m, the maximum height of upwelling reaches about 5.5m, and the eddy structures inside and between reefs are significant. Due to the small spacing, the flow field area, disturbing by the reef groups, is slightly small, about 17m×22m in the horizontal range.

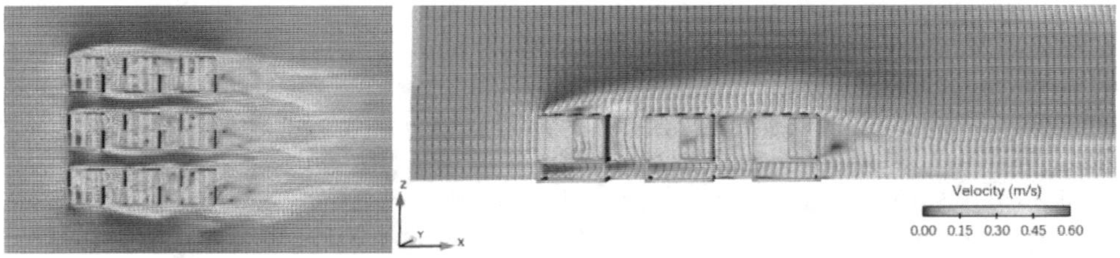

(a) 2m spacing

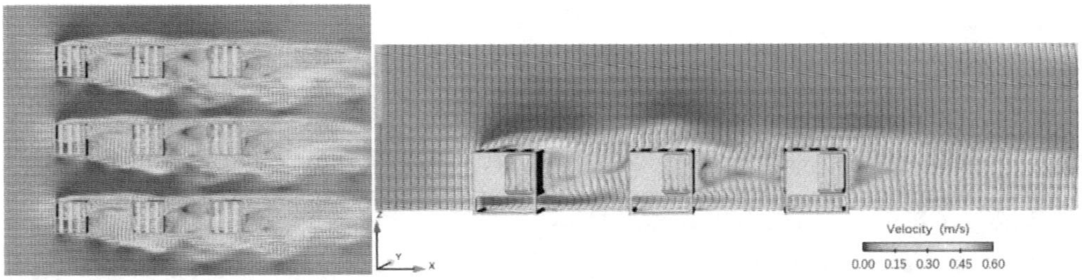

(b) 5m spacing

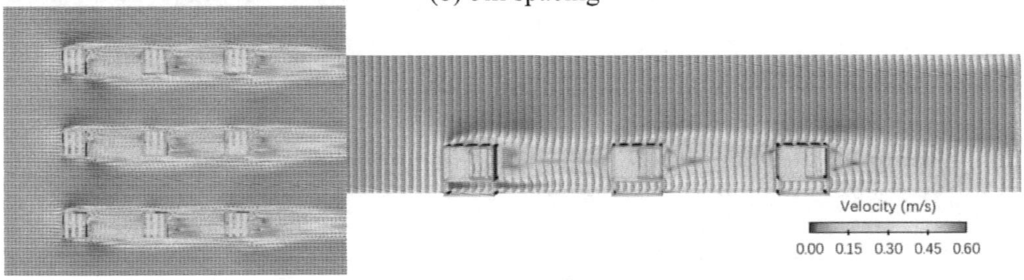

(c) 8m spacing

Fig.5 flow field of reef group at different spacing

When the spacing is set to 5m, the strength of upwelling is still strong, and the upwelling height is about 5.2m. Compared to the reef arrangement with a spacing of 2m, the eddy current range of the reef with 5m spacing in the horizontal direction further develops on both sides of the reef. The eddy current structures of reef group are widely distributed. About 23m×30m area in the horizontal range, the artificial reef group under 5m spacing plays an important role in flow field distribution.

When the arrangement spacing is 8m, the upwelling is fully developed in the front row of artificial reef group only. Near the last two rows of reefs, the upwelling and vortex effects are weakened to some extent. The development of eddy current structure on both sides of the reef is weak as well. The width of eddy current is similar to that of artificial reef. Under this spacing arrangement, the reef group has insufficient regulation of the overall flow field.

By comparing the flow field characteristics of artificial reefs with different spacing arrangements, the optimal spacing between reefs is determined. when the reef spacing is 2m, the artificial reef group has a significant regulation effect on the flow field around the reef, while the influence area is slightly small. When the reef spacing is 8m, the influence range of flow field is large. However, the artificial reef group has a weak regulation effect on the flow field, due to the large spacing between reefs, When the spacing is set to 5m, the strength of upwelling and back vortex is significant around the artificial reef group, which has a large influence area on the flow field as well. Therefore, in the field project of Miaowan artificial reef, it is most appropriate to set the layout spacing between reefs at about 5m, optimizing the flow field characteristics of artificial reef group.

4. conclusion

Based on the practical engineering, this paper studied the flow field characteristics of box artificial reef by CFD method, and mainly analyzed the flow field of single reef as well as the reef group with different spacing arrangement. The conclusions are as follows:

(1) The box reef plays a significant role in flow field. The maximum upwelling height is close to 1.4 times the reef height. The back vortex of reef is fully developed, with the length of back vortex reaching twice the reef size.

(2) The 5m spacing is most suitable for box reef arrangement in site. Under this arrangement spacing, the strength of upwelling and back vortex is significant and the influence range of flow field is relatively wide as well.

Reference

[1]Liu S B, Liu D L, Liu W W, et al. (2007) The hydrodynamic theory, ecological effect and risk assessment of the artificial fish-reef. Proceedings of the 26th International Conference on Offshore Mechanics and Arctic Engineering.

[2]Jiang Z Y, Liang Z L, Tang Y L, et al. (2010) Numerical simulation and experimental study of the hydrodynamics of a modeled reef located within a current. Chinese Journal of Oceanology and Limnology, 28(2) : 267-273.

[3]Tang Y L, Long X Y, Wang X X, et al. (2017) Comparative analysis on flow field effect of general artificial reefs in China. Transactions of the Chinese Society of Agricultural Engineering, 33(8): 97-103.

[4]Wang X, Li X Y, Yue J, et al. (2021) Numerical study on flow field characteristics of a novice open-aperture square artificial reef with shutter. Journal of Water Resources & Water Engineering, 32(5): 158-165.

[5]Zheng Y X, Guan C T, Song X F, et al. (2012) Numerical simulation on flow field around star artificial reefs. Transactions of the Chinese Society of Agricultural Engineering, 28(19): 185-193.

[6]Liu Y, Zhao Y P, Cui Yong, et al. (2012) Experimental study of the flow field around cube artificial reef

Analysis on the Influence of Flood Drainage of Cao'e River Sluice on the Hydrodynamics of Waiguantang River

Junfeng Yu[1], Weijin Chen[2,a], Zhihao Fang[3,4], Fuqing Bai[4], Dongfeng Li[4], Zihao Li[4] ,Yishan Chen[4],Feng Xie[4], Mei Chen[4]

[1]Shaoxing Water Conservancy and Hydropower Project Quality and Safety management Center, Shaoxing, 31200, China

[2]Shaoxing No.1 Water Conservancy Ecological Construction Co., Ltd, Shaoxing, 31200, China

[3]Zhejiang Ocean University, School of Naval Architecture and Maritime, Zhoushan 216022， China

[4]Key Laboratory for Technology in Rural Water Management of Zhejiang Province Zhejiang University of Water Resources and Electric Power, Hangzhou 310018，China

[a]Coresponding authors: chenweijin2011@126.com

Abstract: In order to fully study the impact of Cao'e River Sluice drainage on the external pipe pond and improve the water quality of Shaoxing plain river network, this paper establishes a two-dimensional hydrodynamic model, analyzes the rivers and lakes in Shaoxing plain, and uses the established model to simulate the initial conditions. Through the comparative analysis of the information of the two-dimensional numerical hydrodynamic model and the condition of the river network, we can draw the following conclusions: under normal circumstances, the decline rate of the water surface elevation in Waiguantang river is faster than that in the river network, the overall water surface elevation is relatively stable, and the water surface elevation decreases with the increase of water source distance. The variation range of velocity from the source is large, which is different from the uniform change of water surface elevation. Therefore, the occurrence of flood disaster can be reduced through accurate data analysis.

1. Introduction

Waiguantang river is located in the south of Chengnan Street and Donghu Street in Yuecheng District, adjacent to hills and relatively high terrain, as shown in Fig. 1 below. The drainage of Waiguantang river network is of great significance to the social and economic development of Shaoxing. Cao'e River sluice is an important sluice to control the drainage of Waiguantang river network. When the gate is opened at different times during flood drainage, the water level elevation drop values in different township streets in Waiguantang river network are different. According to the scatter analysis of the changes of water surface elevation and velocity with time near Waiguantang river Chengnan Street and Donghu Street, the scatter diagram can clearly see the impact of gate opening on flood drainage of river network. According to the degree of change, township streets can formulate different drainage policies. The research on the impact of gate regulation on river network drainage is of great significance. In this paper, the mathematical model of point river network near Chengnan street and Donghu street in Waiguantang river is established by hydraulic software to study the hydrodynamic force, analyze the main rivers of Shaoxing river network, and analyze the flood control and drainage. The changes of river surface elevation, river velocity, peak value of river velocity, water surface elevation and velocity of rivers near important streets are simulated and calculated, and the impact on the surrounding streets is analyzed.

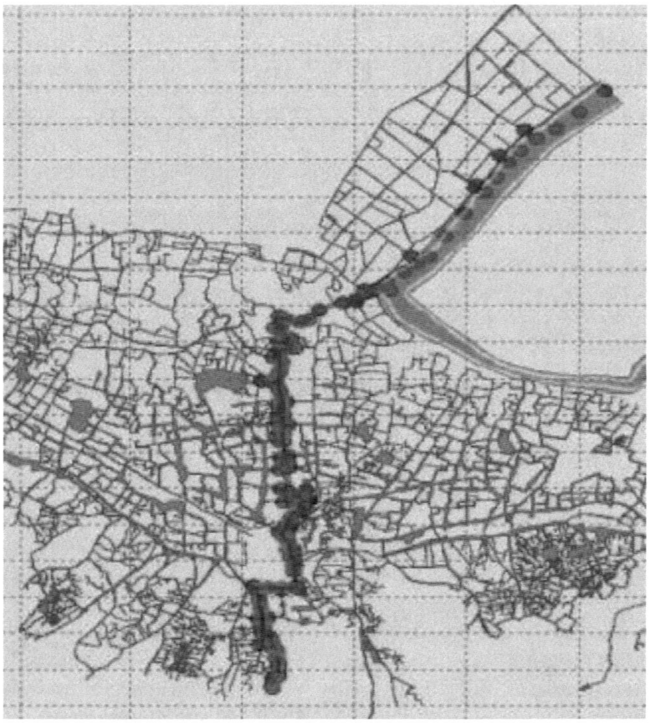

Figure 1 Location map of Waiguantang River

2. Two dimensional hydrodynamic plane mathematical model of river network

2.1. Basic equation of two-dimensional model of water flow in river network

(1) Basic equation

Flow continuity equation:

$$\frac{\partial z}{\partial t}+\frac{\partial (hu)}{\partial x}+\frac{\partial (h\upsilon)}{\partial y}=0 \tag{1}$$

X. Y-direction flow equation:

$$\frac{\partial u}{\partial t}+u\frac{\partial u}{\partial x}+\upsilon\frac{\partial u}{\partial y}-f\upsilon+g\frac{\partial z}{\partial x}+g\frac{u\sqrt{u^2+\upsilon^2}}{c^2h}=\lambda\Delta u \tag{2}$$

$$\frac{\partial u}{\partial t}+u\frac{\partial \upsilon}{\partial x}+\upsilon\frac{\partial \beta\upsilon}{\partial y}+fu+g\frac{\partial z}{\partial y}+g\frac{\upsilon\sqrt{u^2+\upsilon^2}}{c^2h}=\lambda\Delta \upsilon \tag{3}$$

Where: x and y are spatial coordinates; z is the water level; t is the time; u, v is the velocity component in X direction and Y direction respectively; h is the water depth; f is Coriolis force, $f=2W\sin\Phi$; W is the rotation velocity of the earth; Φ is the latitude of the earth; c is Xie Cai coefficient; g is gravitational acceleration.

(2) Definite solution condition

Initial conditions: give the initial terrain, velocity, water depth, etc. of each calculation grid node

Boundary conditions: the flow process at the inlet of a given model; discharge process of given water level at outlet.

2.2. 2D model validation

2.2.1. Study area

Shaoyu plain belongs to the Xiaoshao canal water system of Cao'e River Basin. The North is a plain area, crisscrossed by reticulated rivers and numerous lakes. Shaoxing has the unique river and ocean power of river network flow movement in this area. Relying on this characteristic, the development and utilization of water resources, flood control, drainage and pollution control of the plain river network can be realized. The West of the basin is Dongxiaojiang River; Cao'e River in the East; Kuaiji mountains in the South; In the North are the estuaries of Cao'e River and Qiantang River. The water flows from west to east and from south to north, mainly through Xinsanjiang gate and Mashan gate and then into Cao'e River. The model range and boundary conditions are shown in Fig. 1.

3. Analysis of Cao'e River sluice to Guantang river network

A total of 72 hours from May 1 to May 3, 2019 are selected as the experimental analysis time point. As can be seen from Fig. 2, Waiguantang river is located in the south of Chengnan Street and Donghu Street in Yuecheng District, adjacent to hills and relatively high terrain, so the water surface elevation of Waiguantang river is relatively stable. The closer the water level of each point of the river is to the river water source, the higher the water surface elevation. As can be seen from Fig. 3, with the passage of time, the change range of water surface elevation at each point in the Zhijiang River of Guazhu lake with time is relatively small, and the change rate tends to decrease. After 72 hours, the working conditions are shown in the figure below. The change trend of water surface elevation with time in each time period is relatively stable.

Figure 2 Location distribution map of Shaoxing river network basin and 21 streets

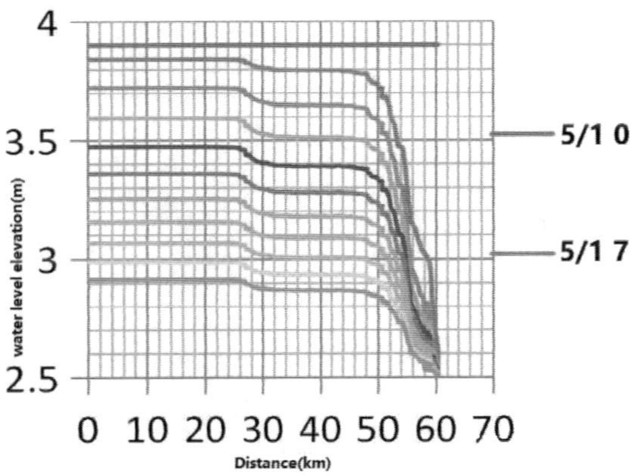

Figure 3 Water surface elevation change map of each point in the Zhijiang River of Guazhu Lake

As can be seen from Fig. 4, the water surface elevation of the points near Chengnan and Donghu streets began to decrease due to the opening of Cao'e River Sluice and Binhai sluice. The water surface elevation finally reached 2.905m and 2.858m after 72 hours from the original 3.9m, and the maximum flow velocity reached 0.000618m/s and 0.0248m/s after about 23 hours from zero, After about 49h, the flow velocity reached 0.0045m/s and 0.0178m/s respectively.

It can be seen from Fig. 4 and Fig. 5 that the difference of flow velocity at the points near Chengnan and Donghu streets is greater than 0 in the first 23 hours and less than 0 in the last 49 hours. That is, the change rate of flow rate is greater than 0 in the first 23 hours, and the flow rate continues to increase until the change rate of flow rate is equal to 0 in the 23rd hour. At this moment, the flow rate reaches the maximum value, and the change rate of flow rate is less than 0 in the last 49 hours, and the flow rate continues to decrease. However, the flow velocity generally shows an increasing trend.

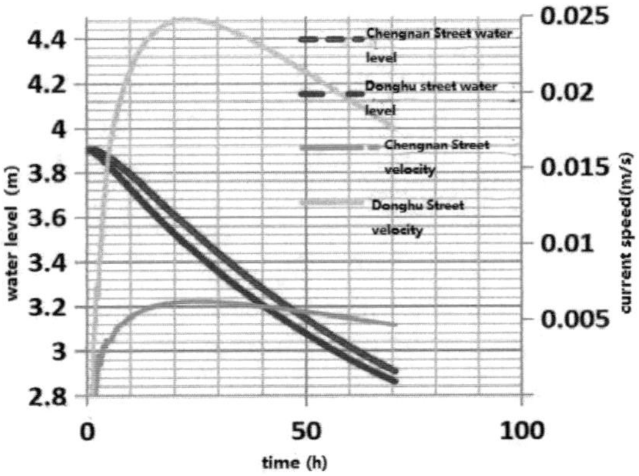

Figure 4 Variation of water level and velocity in Chengnan Street and Donghu Street

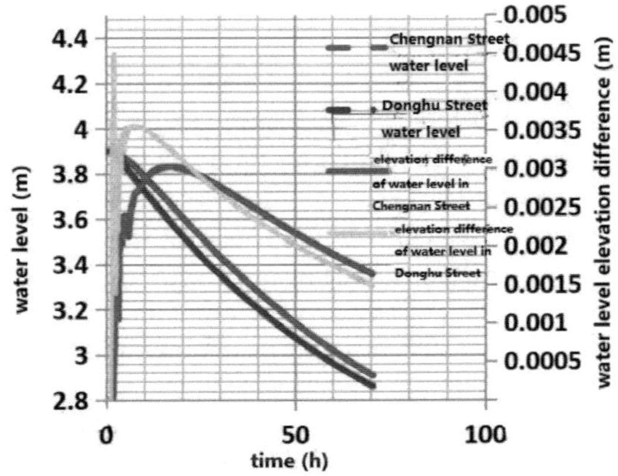

Figure 5 Variation diagram of water level and water level elevation difference in Chengnan street and Donghu Street

The point 22.995 km away from the source of Waiguantang river and with a velocity of 0.607m/s - adjacent to Hewei building is analyzed. The specific location is shown in Fig. 6.

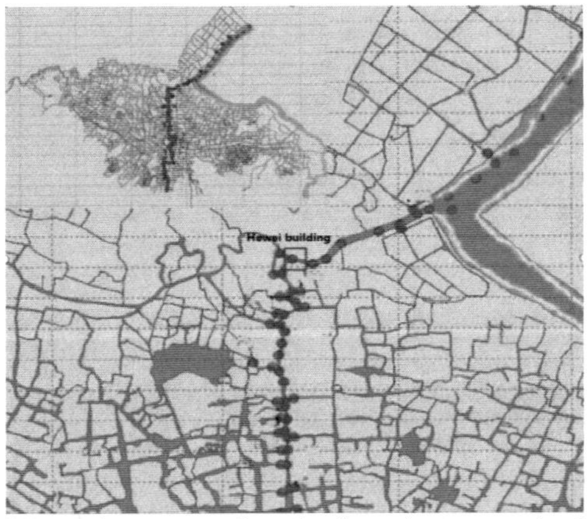

Figure 6 Location map of analysis points near the building

It can be seen from Fig. 6 and Fig. 7 that due to the geographical location in the south, close to the mountains and high terrain, the water level at the location 22.995 km away from the source shows a downward trend. From 0:00 on May 1, 2019 to 22:00 on May 3, 2019, after 72 hours, the water level changes from 3.9m to 2.1455m; The change of the velocity at the peak point of the river velocity is different from the change of the total velocity of the river. After 8h, the velocity reaches the maximum value of 0.616m/s. Therefore, it can be considered that the main factor affecting the river is the cross-sectional area of the river.

It can be seen from Fig. 8 that the water level increase rate changes with time. The water level increase rate rises rapidly within about 2H and reaches the maximum value. The trend changes to a rapid decline within about 5h, and the water level increase rate decreases slowly and tends to zero within 65h; The increase rate of flow rate also reached the maximum value in about 2h, with a steep trend. It decreased rapidly in about 6h, and the change value became 0.303m/s, and then changed slightly and tended to a stable value of 0.006m/s.

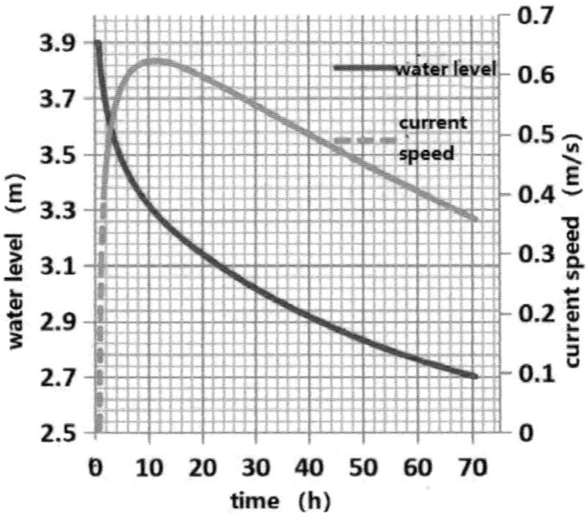

Figure 7 Variation diagram of water level and velocity with time at analysis point

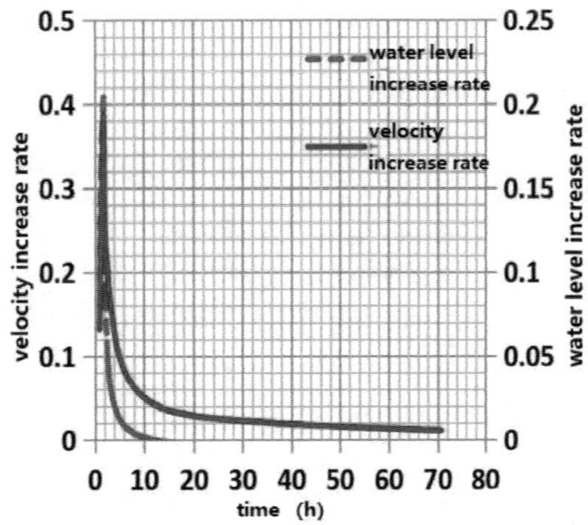

Figure 8 Variation diagram of water level increase rate and velocity increase rate with time

To sum up the data, in order to prevent the residents near the river and the important village - Tiantangdian village from suffering from flood disasters, when the rainstorm is coming, it is necessary to open the Cao'e River Gate and Binhai gate 7h in advance, that is, open the gate at 0:00 on May 1, 2019, and the water level can be controlled. If the gate is opened too early, it will lead to a shortage of water resources, too late will cause flood disasters and unnecessary losses. Therefore, according to the data feedback, the gate will be opened on time, so that the residents near the river and important sub district offices will not be affected by the flood disaster formed after the rainstorm, nor will it cause a waste of water resources.

4. Conclusion

In this paper, the flood discharge capacity of main rivers is studied from the simulated boundary condition and the opening and closing of the gates of Cao'e River gate and Binhai gate at the same time, combined with the hydrodynamic data calculated numerical model. By analyzing the change of water surface elevation and flow velocity with distance at different times and the changes of water surface elevation change rate and velocity change rate with time, so as to analyze when the gate of each main river will be opened in advance at the gate of Cao'e River when the rainstorm is coming without causing flood disaster, and the important streets near the basin and villages near each main river can not be affected by flood disaster.

Acknowledgments

This research was supported by the Joint Funds of the Zhejiang Provincial Natural Science Foundation of China（No.LZJWZ22E090004, No. LZJWZ22C030001）and the Funds of Water Resources of Science and Technology of Zhejiang Provincial Water Resources Department, China(No.RB2115)，the National Key Research and Development Program of China(No.2016YFC0402502)，the National Natural Science Foundation of China(51979249).

References

[1] H.M. Song, Z.N. Wang, W. Gu. (2021) Analysis on the impact of Baimao water intake project in Wuwei County on river flood discharge based on Mike21. Jianghuai water conservancy technology., (01): 10-12

[2] C.D. Qin, J. Qian, M.D. Li, C.Wang. (2021) Numerical simulation of water environment impact of songjiawan reservoir based on Mike21 model. Gansu water resources and hydropower technology.,57 (02): 1-10 + 19

[3] W.J. Yan , X.H. Wang, X.Y. Shi. (2021) Study on the division of drinking water source protection areas of large lake reservoir type based on Mike21 -- Taking Huating Lake Reservoir as an example. Environmental pollution and prevention.,43 (01): 109-114
[4] Y.Chen, Y.C.Zhou. (2020) Effect analysis of waterlogging regulation scheme in sponge city based on Mike model. Smart city and smart architecture (10): 63-65
[5] D.F. Li,Y.J.Cheng, B. Zou, (2004)2D finite element mathematical model of the Yellow River estuarine sediment and its application(I) Advances in Marine Science 01(2004)21-28.
[6] D.F.Li,X.Z.Zhang, Q.L.Han, (2004) 2-D mathematic model of the Yellow River estuarine tidal wave and sediment and its application(II) Advances in Marine Science 03(2004)284-291.

Experimental study on formation characteristics and influence of ice jam in melting period

Yi Sun[1*], Zhigang Gu[1], Jing Ma, Xiaoyan Yi[1]

[1] Yellow River Inst. of Hydraulic Research YRCC, Zhengzhou, Henan, 450003, China

[*]Corresponding author's e-mail: sunyi@hky.yrcc.gov.cn

Abstract: Ice problem widely exists in the north rivers, especially the ice disaster of the water elevation rise triggered by local ice jam in melting period caused by flowing ice. Therefore, it is of great significance to study the formation and influence of ice jam during the melting period. In this paper, the characteristics of ice flow in Yumenkou reach of the Middle Yellow River during the melting period is studied by laboratory physical model tests using paraffin as model ice. The results showed that: The formation of ice jam is controlled by flow rate and ice density during the melting period. When the flow rate is small and the density of flowing ice is high, there are enough ice cubes in a certain space who have enough time to combine with each other to form an ice jam. An ice jam is made up of at least two ice cubes stacked on top of each other vertically whose backwater influence is at least greater than the maximum height of one ice cube. An ice jam is more likely to form in the wider reaches of the river. The reason for that is, when the flow ice density reaches a certain level, the flow velocity is reduced more due to the wider river width, and the ice cubes are more likely to collide, squeeze and combine with each other, which are not easily transported by water with low flow intensity.

1. Introduction

The ice problem of river is widespread over a wide area of the Northern hemisphere, such as Canada, the United States and China, etc, where often suffers from the disaster of ice flood in winter and early spring. China is one of the countries with severe ice disaster in the world[1]. The Yellow River, Heilongjiang river, Songhua River, Nenjiang river and other northern rivers have different degrees of ice problem in winter which often causes disaster. The existence of ice in river reduces the capacity of flow and sediment transport, intensifies the bank erosion, affects channel stability, and is prone to form flood disaster[2-4].

The evolution process of river ice can be roughly divided into ice period, frozen period and thawing period[5]. During the ice period, river ice mainly moves with the flow in the form of ice flowers. Ice jam and ice dam in frozen period and thawing period are the main factors causing disasters. Therefore, the formation and evolution mechanism of ice jam and ice dam and its influence on water flow are concerned by researchers. Ke et al.[6] revealed the formation mechanism of ice jam from the perspective of force analysis of flowing ice and analysed the causes of ice jam formation in the Bayangaole reach of Yellow River by using a large amount of actual data. According to the measured data, Sui and Fang[7] studied the influence of ice jam on flow from the relationship between flow roughness and Froude number and established the corresponding relationship. Wang et al.[8] believed that Froude number was closely related to the formation of ice jam through flume test and pointed out that when FR>0.139, ice jam could not be formed in curve section.

The formation of ice jam changes the boundary condition of the river channel. Therefore, the influence of its thickness and range on water elevation and flow capacity is of great significance to anti-ice work[9-11]. Yang et al.[12] proposed an equation describing the formation and evolution of ice jam according to the mechanism of ice jam formation and evolution. Mao et al.[13] established a comprehensive dynamic mathematical model of ice-water coupling in the formation and evolution of ice jam by applying river dynamics and thermodynamics, and verified it by using the prototype measured data of the curve section of the Yellow River. Meanwhile, they studied and analysed the evolution law of ice jam thickness and water elevation. Zhu et al.[14] studied the ice flow process and water elevation change in the Wuding River Basin. Cai and Lu[15] studied the empirical formula of subglacial flow capacity in frozen reach. Song et al.[16] studied the ice-passing ability of gate pier through laboratory tests at low temperature. Beltaos et al.[17] studied the influence of flow on bridge piers during the river melting period. Yoshikawa et al.[18] studied the flow and transit phenomenon during the melting period and developed a one-dimensional mathematical model to calculate the ability of ice dam generation. Zhang et al.[19] predicted the change of velocity and water elevation caused by ice dam by determining the comprehensive roughness of ice dam.

River ice process is a complex process, which involves many factors such as local climate, river bed geomorphology and hydraulic condition, etc. The formation and extent of ice jam are related to the degree of ice damage. Therefore, this paper focuses on the characteristics and influence of ice jam formation in melting period through physical model test.

2. Materials and Methods

2.1. Prototype overview

The study reach is located in the Yumenkou reach of the Yellow River. The upper reach is a type of canyon reaches, with bed slope of about 3‰~6‰, and the average width of main channel is about 170m. The river widened rapidly to 1000~3000m wide at the downstream of Yumenkou. The ice in Yumenkou reach is divided into upstream coming ice and local shore ice. Therefore, the variation of ice condition in this reach can be divided into four periods, namely, flow ice period, frozen period, thawing period and ice end period[20]. The total time of four periods of the Yumenkou reach generally starts from mid-to-late November every year and lasts 80~130d, with an average of 86d for many years. The largest ice cube is more than 20m in diameter and 2.5m in thickness. The average ice density is 10%~30% (Ice density $D = \frac{f}{F} \times 100\%$, where f is the water surface area occupied by flowing ice; F is the total water area under the same water elevation).

2.2. Physical model overview

The experiment was carried out on the physical model of Yumenkou reach of Yellow River in the test base of Yellow River Institute of Hydraulic Research. The physical model is a fixed bed abnormal model with a plane scale of 250 and a vertical scale of 60. The model range is shown in Figure 1. Topography and river boundary are the topography after flood in 2010. River boundary includes existing river works, roads and Bridges, etc. Paraffin with a small amount of gypsum powder was used as the model ice in the experiment, and its bulk density is 910kg/m^3, which is very close to the bulk density of natural ice, 917kg/m^3, which could meet the requirements of similar floating ice movement.

Considering that the size of the prototype ice is mostly 25~100m^2, the size of the model ice should mostly be in the range of 4~16cm^2 according to the plane scale of the ice model 250. Considering the bonding and breaking of the prototype ice during transportation and the similarity of the model ice transportation under the model water flow condition, the thickness of the model ice should be 0.3~0.5cm. In this way, the volume of model ice cubes should be in the range of 1.2~8.0cm^3. The grain distribution curve of the model ice cube is shown in Figure 2. It can be seen that the percentage of model ice is 20%~80% when the area of model ice is 4~16cm^2, and 15%~85% when the volume of model ice is 1.2~8.0cm^3, namely, the area and volume of model ice are mostly within the required range, and the size of model ice can basically meet the test requirements.

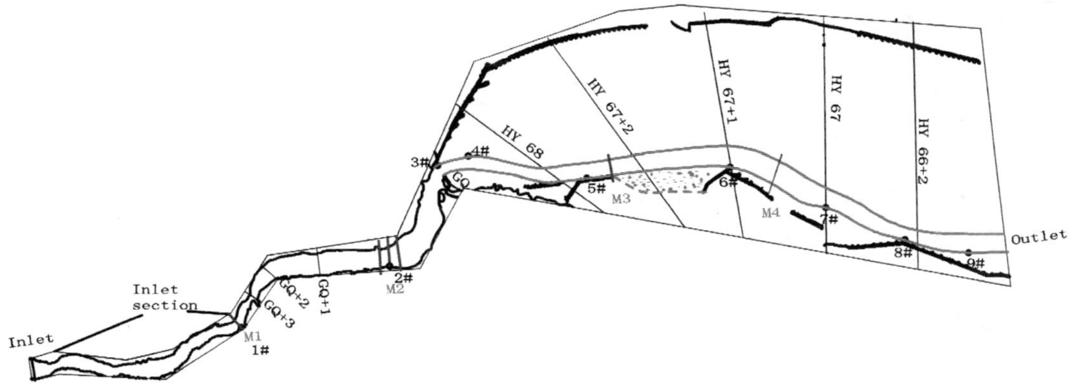

Figure 1. Diagram of the ice model

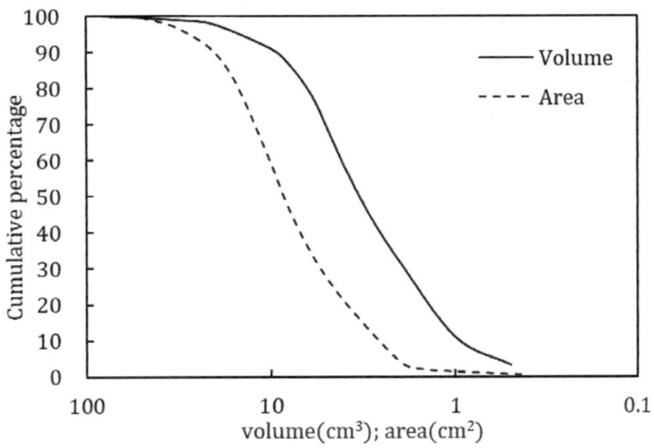

Figure 2. Gradation curves of area and volume of model ice

2.3. *Model verification*

The water elevation of the model was verified by selecting the relatively complete data in melt period of Yumenkou reach in 2011. The initial topography and riverbed boundary are post-flood conditions in 2010. The measured data of flow and ice density at Longmen Station in the current year were selected for the verification test, which are shown in Table 1. On February 25, 2011, ice flow began to appear in the reach with a flow rate of 583m^3/s and a density of 40%. On March 14, the flow rate was 753m^3/s and the density was 3%, then the ice flow basically disappeared.

As can be seen from Figure 3 and Figure 4, the water elevation process along the model is basically consistent with that of the prototype, and the ice density during the discharge process of the model is basically consistent with that of the prototype. The water elevation difference between the prototype and the model along the river is generally within 0.15m, which is in good agreement.

Table 1. Ice model validation of flow and ice density

Date	Flow rate(m3/s)	Ice density(%)
2011-2-25	583	40
2011-2-26	547	30
2011-3-1	500	30
2011-3-2	495	10
2011-3-8	475	5
2011-3-11	644	5
2011-3-14	753	3

Figure 3. Water surface verification of Q=583m^3/s, D=40%

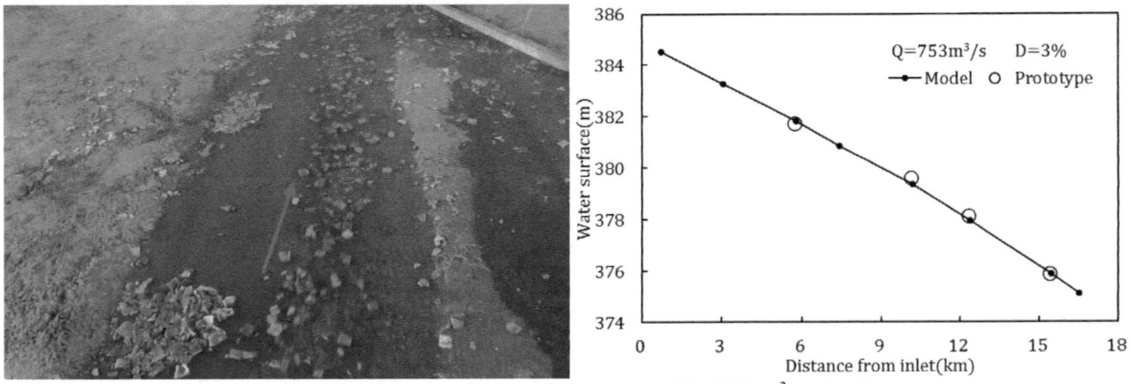

Figure 4. Water surface verification of Q=753m^3/s, D=3%

2.4. Experiment group

The experiment flow rate was 400m^3/s, 600m^3/s and 1000m^3/s from small to large, and the ice density was 20%, 50% and 80%. There were 9 groups of tests in total as shown in Table 2.

Table 2. Test group

Test	**Flow rate(m^3/s)**	**Ice density(%)**
RUN1A	400	20
RUN1B	400	50
RUN1C	400	80
RUN2A	600	20
RUN2B	600	50
RUN2C	600	80
RUN3A	1000	20
RUN3B	1000	50
RUN3C	1000	80

3. Results

3.1. Location of ice jams

During the experiment, the ice jam phenomenon occurred in 4 test groups. The positions of ice block are concentrated near the M2, M3 and M4 sections, especially the number of ice block phenomenon is largest near the M3 section and formed the ice jam, as shown in Table 3. In the total 9 groups of tests, ice jam can be formed only when the flow rate is small (400m^3/s and 600m^3/s) and the ice density is high (50% and 80%), and the positions where ice jam are formed are located in M3 section. There are only ice blocks formed at M2 and M4 sections, and after a certain period of time, the ice blocks are washed broken and transported to the downstream. When the flow rate is large (1000m^3/s) and the ice density is 50% and 80%, ice block can only occur locally in the vicinity, and is quickly washed broken and transported downstream.

Table 3. Statistical table of ice block number in different cross-section

Test	**Flow rate(m^3/s)**	**Ice density(%)**	**M2**	**M3**	**M4**
RUN1B	400	50	2	13	
RUN1C	400	80	2	13	
RUN2B	600	50	1	8	1
RUN2C	600	80	1	7	1

3.2. Variation of water elevation

Figure 5~Figure 8 shows the water surface lines along the channel under different test time of typical test groups. As can be seen from Figure 5 and Figure 6, when the flow rate is 400m^3/s and the ice density is 50%, the ice is blocked in M2 first, and then ice jam is formed in M3, which lasts until the end of the test. When the ice density increases to 80%, the time of ice jam formation is significantly earlier and directly formed in M3. It can be seen from Figure 7 and Figure 8, when the flow increases (600m^3/s), the ice block time of M2 and the final ice jam formation time of M3 are both slightly earlier than that of 400m^3/s. When the ice density increases to 80%, the ice block time of M2 and the final ice jam formation time of M3 is further earlier than that of 50%, but later than that of 400m^3/s. The reason is that the flow strength increases with the increase of the flow rate, and so does its capacity of ice transport.

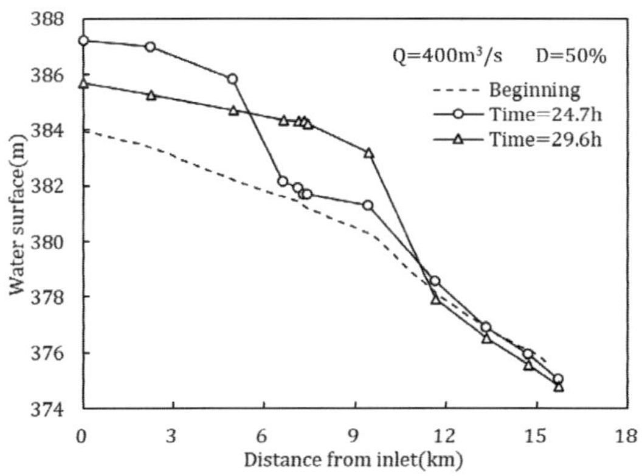

Figure 5. Bed surface topography of M2

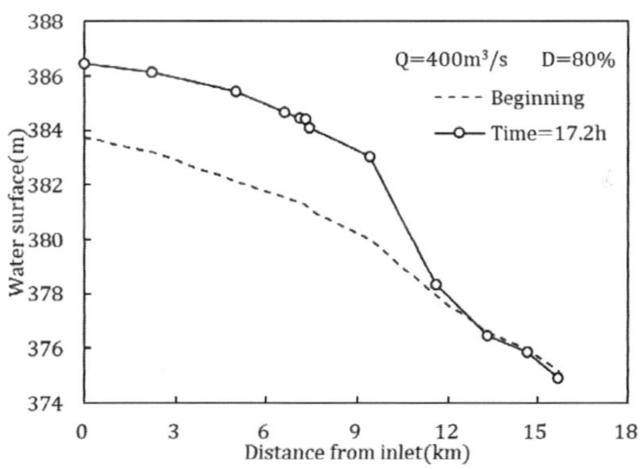

Figure 6. Bed surface topography of M3

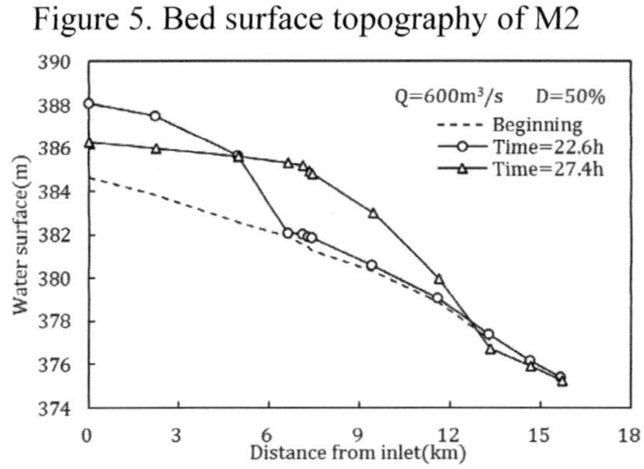

Figure 7. Water surface of RUN2B

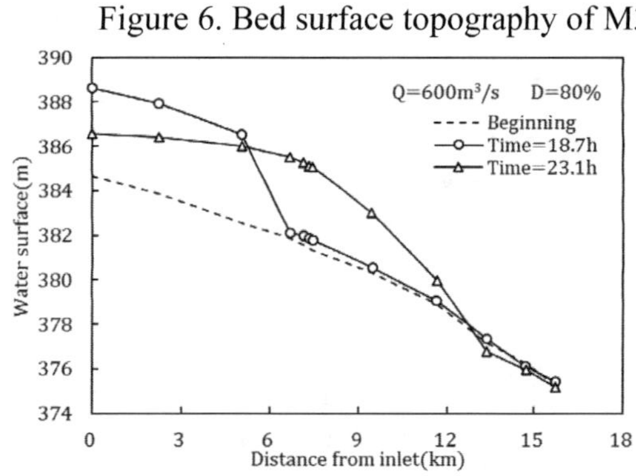

Figure 8. Water surface of RUN2C

3.3. Ice jam impacts
Table 4 shows the water depth and backwater height of each experimental group under the influence of ice. It can be seen from the table that the influence of ice can increase water depth by 0.17~1.35m, which is consistent with the maximum ice thickness of 2.5m observed by hydrology station. The maximum backwater of experiment group formed by ice jam can reach 2.65~3.58m. Obviously, this value is greater than the maximum ice thickness of 2.5m, that is, the ice jam is formed by at least two ice cubes in the vertical direction.

Combined with the topography of M2 and M3 sections (Figure 9 and Figure 10), at M2 section, the river width is relatively narrow and the backwater caused by ice block is high which means the water head between upstream and downstream is large, that leads to the instability of the ice jam and makes it easier to be swept away. At M3 section, the river width is wide. Under the same the flow rate, the velocity is smaller, and the ice stays longer in the local area, which makes it easier to form ice jam. The ice jam backwater is relatively low, and the water head between upstream and downstream is small, so the ice jam is more stable after formation and is not easy to be washed collapsed.

Table 4. Water depth influenced by ice

Test	**Flow rate(m^3/s)**	**Ice density(%)**	**M2 section**		**M3 section**	
			Water depth(m)	**Backwater Height(m)**	**Water depth(m)**	**Backwater Height(m)**
RUN1A	400	20	1.01	0.17	1.65	0.33
RUN1B	400	50	3.49	2.65	4.29	2.98
RUN1C	400	80	4.18	3.33	4.37	3.06
RUN2A	600	20	1.28	0.19	1.80	0.28
RUN2B	600	50	4.22	3.12	5.08	3.57
RUN2C	600	80	4.64	3.55	5.09	3.58
RUN3A	1000	20	2.07	0.16	2.44	0.21
RUN3B	1000	50	2.28	0.36	2.66	0.43
RUN3C	1000	80	3.26	1.35	3.48	1.31

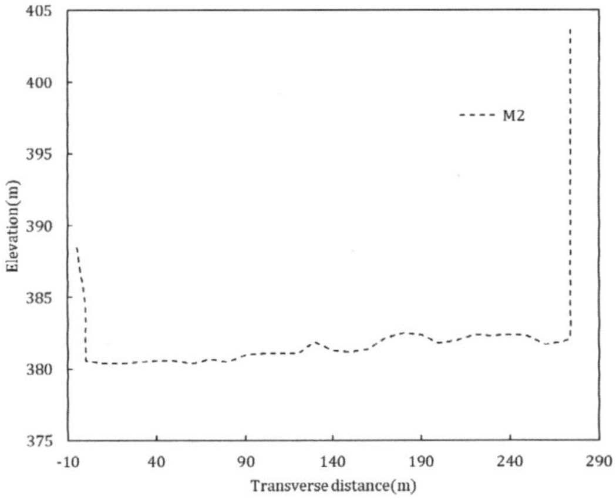

Figure 9. Topography of M2 section

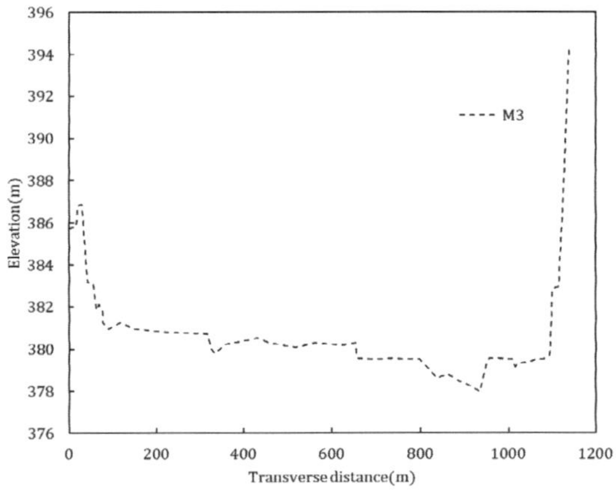

Figure 10. Topography of M3 section

4. Conclusions
Based on laboratory experiments, the formation characteristics and influence of ice jams in the melting period are studied. The results show that:

(1) Low discharge and high ice density are the requirements for the formation of ice jam during the melting period. When the flow rate is large, the strength of the water is large, and the ice is transported to the downstream before it combines with each other. When the flow ice density is low, the ice cubes are too far apart to bond with each other.

(2) Ice jam is formed by at least two ice cubes in vertical direction, and the backwater influence caused by ice jam is at least greater than the maximum height of one ice cube.

(3) In canyon river, ice jams are more likely to form in the wider reaches of the river. The reason is that when the ice density reaches a certain level, the wider river width reduces the flow velocity, and the ice cubes are more likely to collide, squeeze and combine with each other, which is not conducive to the transport under low flow intensity.

Acknowledgments
This research was supported by the National Natural Science Foundation of China (51809107), the Fundamental research Funds of Yellow River Inst. of Hydraulic Research(HKY-JBYW-2020-10) and the National Key R&D Program of China (2018YFC0407801).

References

[1] Wang, E.L., Li, Y.A., Liu, C.Q. (2021) Research progress of ice jam origin and prediction. Hydro Science and Cold Zone Engineering, 4:66-72.
[2] Turcotte, B., Morse, B., Bergeron, N., E., et al. (2011) Sediment transport in ice-affected rivers. Journal of Hydrology, 409: 561-577.
[3] Morse, B., Hicks, F. (2005) Advances in river ice hydrology 1999-2003. Hydrological Processes, 19: 247-263.
[4] Hicks, F. (2009) An overview of river ice problems: CRIPE07 guest editorial. Cold Regions Science and Technology, 55: 175-185.
[5] Ke, S., J. Study on ice condition of Yellow River. Yellow River Water Conservancy Press, Zhengzhou.
[6] Ke, S., J., Lv, G., Y., Ren, Z., Y. (2000) Study on mechanism of ice jam formation in Bayangaole section of Yellow River. Journal of Hydraulic Engineering, 7: 66-69.
[7] Sui, J., Y., Fang, D., X. (1993) Roughness coefficient of river ice jams. Journal of Hydraulic Engineering, 8: 10-18.
[8] Wang, J., Fu, H., Yi, M., K., et al. (2007) Analysis of stages under ice-covered in winter. Advances in Water Science, 18: 102-107.
[9] Prowse, T., D., Culp, J., M. (2003) Ice breakup: a neglected factor in river ecology. Canadian Journal of Civil Engineering, 30: 128-144.
[10] Lindenschmidt, K., Sydor, M., Carson, R., W. (2012) Modelling ice cover formation of a lake–river system with exceptionally high flows (Lake St. Martin and Dauphin River, Manitoba). Cold Regions Science and Technology, 82: 36-48.
[11] Smith, D., G., Pearce, C., M. (2000) river ice and its role in limiting woodland development on a sandy braid-plain, Milk river, Montana. Wetlands, 20: 232-250.
[12] Yang, K., L., Liu, Z., P., Li, G., F., et al. (2002) Simulation of ice jams in river channels. Water Resources and Hydropower Engineering, 33: 40-47.
[13] Mao, Z., Y., Wu, J., J., Zhang, L., et al. (2003) Numerical simulation of river ice jam. Advances in Water Science, 14: 700-705.
[14] Zhu, R., R., Liu, C., M., Li, L., et al. (2008) Mathematical Modeling of River Ice Evolution in the Downstream of Wudinghe Basin, Yellow River. Journal of Glaciology and Geocryology, 30: 520-526.
[15] Cai, L., Lu, D., T. (2002) Mathematical model of reservoir operation for ice-jam flood control. Journal of Hydraulic Engineering, 6: 67-71.
[16] Song, A., Wang, Z., Y., Shi, Q., Z., et al. (2005) Ice model test and hydraulic engineering. Water Resources and Hydropower Engineering, 36: 140-143.

[17] Beltaos,S., Miller, L., Burrell, B., C., et al. (2007) Hydraulic effects of ice breakup on bridges. Canadian Journal of Civil Engineering, 34: 539-548.
[18] Yoshikawa, Y., Watanabe, Y., Hayakawa, H., et al. (2011) A study of ice break and ice flow during river ice breakup. Journal of Japan Society of Civil Engineers, Ser. B1 (Hydraulic Engineering), 67: I_1075-I_1080.
[19] Zhang, X., L., Zhao, B., Wang, M., et al. (2011) Analysis on the influence of ice condition on the reconstruction and extension project of Fugu-Hequ Yellow River Highway Bridge. Water Conservancy Science and Technology and Economy, 17: 37-39.
[20] Hou, Z., J., Sun, Y., Gu, Z., G. (2014) Experimental study on the ice physical modeling with a spanning river bridge, Journal of Sediment Research, 4: 49-54.

Numerical analysis of the hydrodynamic performance of RDT with varying blade numbers

Wen JIANG [1a], Ruixue YU [1b], Liangliang LIU [1c], Ao ZOU [1d], Tao BIAN [123e*]

[1] School of Intelligent Manufacturing, Jianghan University, Wuhan 430056, China
[2] Guangdong HUST Industrial Technology Research Institute, Dongguan 523808, China
[3] School of Artificial Intelligence and Automation, Huazhong University of Science and Technology, Wuhan, Hubei 430074, China

[a]langxin2333@163.com, [b]282527641@qq.com,[c]painless1996@qq.com
[d]949410586@qq.com, [e*]biant@jhun.edu.cn

Abstract—This study investigates the effect of the blade number of rim-driven thruster (RDT) on the hydrodynamic performance of RDT. Numerical investigation is performed to compare the thrust coefficient and efficiency of the RDT with various blade numbers. The results show that the thrust coefficient of the RDT increases and the efficiency of the RDT decreases with the number of blades increasing.

1. Introduction

1.1 Background

In recent years, more and more people turned their attention to the more advanced rim-driven thruster (RDT). RDT could provide many advantages superior to the traditional ship propulsion equipment, including reducing vibration and noise, lightweight and small volume [1].

Ojaghlu et al. [2] presented a method to determine electric motor specifications according to hull and propeller. Yakovlev et al. [3] designed the propeller using the direct blade pitch optimization method, and tested the model in the deep-water towing tank of the Krylov Shipbuilding Research Institute. The efficiency had a good consistency, and the torque had some deviations. Cao et al. [4] predicted the loading distributions for a rim-driven thruster. There was tip joint vortex, root region vortex and no tip leakage vortices in RDT. Cai et al. [5] compared the ducted propeller and RDT blade thickness in the design, compared the hydrodynamic performance, and made the RDT physical object to determine the simulation feasibility. Liu et al. [6] performed numerical investigation to simulate the gap flow between RDT and concluded that friction torque increased with the gap, and friction torque in both axial and radial direction had mutual effects on each other. Liu et al. [7] compared RDT with a Ducted Propeller. Due to the existence of a hub, RDT produced less thrust on the engine, and the overall efficiency of RDT was significantly lower than that of the ducted propeller.

The work of the references in the introduction is summarized as follows:

(1) The differences between CFD simulation and actual experiment are studied. It is feasible to use CFD simulation to analyze the real object.

(2) Foreign countries have studied RDT earlier and have made related products. Domestic analogous products are still in their development, but theoretical study has steadily progressed.

Corresponding research has been carried out on the influencing parameters of the propeller, such as pitch ratio, diameter, blade area ratio and other parameters.

1.2 Research Purpose

In this work, numerical investigations are performed to investigate and compare the hydrodynamic performance of RDT with different blade numbers. The fundamental purposes of this work are as the following:

(1) The CFD simulations are carried out to detect and compare the thrust coefficient and efficiency of the RDT with various blade numbers.

(2) The pressure distribution on the blade surface and the flow field in the wake of the RDT is presented and analyzed in this study.

2. CFD Model Setup and Validation

2.1 Numerical method validation

The numerical simulation of No.19A+Ka4-70 ducted propeller is carried out. The process is geometric modeling, grid independence and boundary conditions. The relevant parameters are that the diameter is 250mm, the blade area ratio is 0.7, the pitch ratio is 1.0, and the speed is 450r / min. The calculation domain is divided into two parts: the stationary domain and the rotating domain. The inlet and outlet boundary conditions are inlet velocity and pressure outlet respectively.

$$J = \frac{V_A}{nD} \quad (1) \quad K_T = \frac{T}{\mathrm{r}\, n^2 D^4} = \frac{T_N + T_P}{\mathrm{r}\, n^2 D^4} \quad (2) \quad K_Q = \frac{Q}{\mathrm{r}\, n^2 D^5} \quad (3) \quad \mathrm{h} = \frac{K_T}{K_Q}\frac{J}{2\mathrm{p}} \quad (4)$$

J is the advance coefficient, K_T is the thrust coefficient, K_Q is the torque coefficient and h is the propulsive efficiency; V_A is the inlet velocity in m / s, T is the overall thrust, T_P is the blade thrust, T_N is the duct thrust in N; Q is the propeller torque in N · m; n is the propeller speed in r/min. The open water performance experiment of No.19A+Ka4-70 ducted propeller is calculated and compared with the experimental data. What is striking in Figure 1 is that the simulation data are in good agreement with the experimental data, indicating the feasibility of the calculation model in this paper.

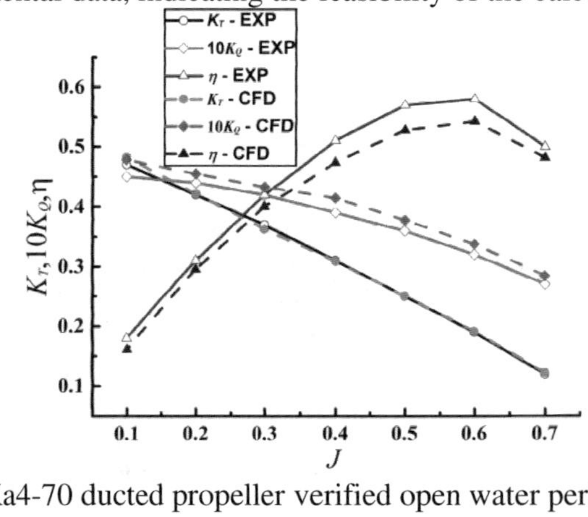

Fig. 1 No.19A+Ka4-70 ducted propeller verified open water performance diagram

2.2 The investigated RDT

The design object is derived from No.19A+Ka4-70 ducted propeller. Because the blade tip, blade root and traditional propeller of RDT are equivalent to being adjusted, the blade thickness is adjusted from 0.2R to 1.0R. When the blade area ratio, pitch ratio and other parameters are sure, the number of blades ranges from 3 to 7, the extension area of individual blades decreases, the axial projection area decreases, and the general extension area is the same. Several RDT models are shown in Figure 2. It can be seen that the projection area of one blade of 3 blades is more significant than that of 7 blades.

(a) 3blade RDT (b) 5blade RDT (c) 7blade RDT
Fig. 2 Partial RDT model

3. Result and discussion

3.1 Open water performance

In order to study the change of blade numbers on open water performance, the changes of K_T and h from 3 blades to 7 blades are analyzed. Figure 3 reveals open water performance comparison, 3blade ~ 7blade represents 3 ~ 7 blades RDT.

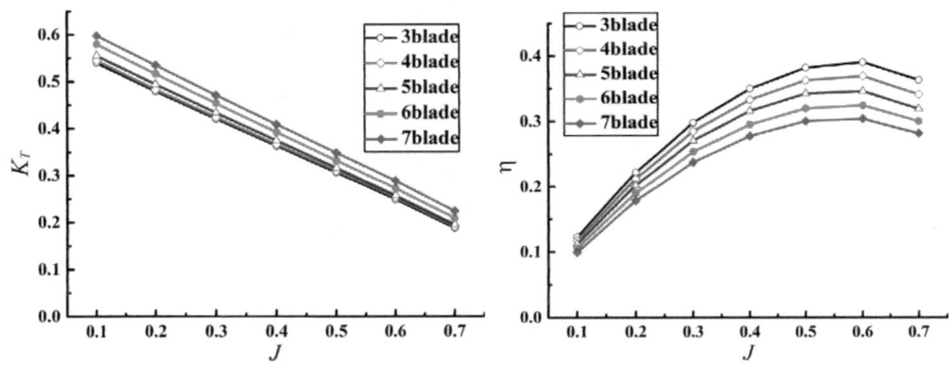

(a)Thrust coefficient (b) Efficiency
Fig. 3 Open water performance comparison

On the premise of a specific blade area ratio and other parameters, according to the analysis in Figure 3 above, for each RDT, with the advance coefficient J from 0.1 to 0.7, the thrust coefficient continually decreases, the efficiency first increases and then decreases, reaching the maximum value at J = 0.6.

The efficiency curve of RDT is similar to that of the traditional propeller. When the inlet velocity is low, the propeller is idling without resistance, therefore the thrust is greatest. On the other hand, the propulsive efficiency is poor. The propeller resistance increases as the inlet flow velocity increases, while the thrust decreases in contrast. The efficiency is at its peak when the inflow velocity reaches a particular value. The thrust is reduced to a minimum as the inlet velocity increases, indicating that this RDT is not suitable for this working situation.

Under the same advance coefficient J, as the number of blades grows from 3 to 7, the thrust coefficient increases successively, and the efficiency decreases.

When there are only several blades, the thrust of the blade itself is small but the efficiency is high, meaning that a single blade can generate thrust in greater space. As the number of blades grows, the space decreases, resulting in reduced efficiency, by contrast, the overall thrust increases. Therefore, under the same condition of small thrust requirements, the RDT with high propeller efficiency and few blades is preferred.

3.2 Analysis of suction surface and pressure surface pressure

Under the identical working conditions, RDT with the number of 7 blades has the largest thrust, whereas the efficiency becomes the largest when the number of blades is 3. Therefore, we select 3 blades and 7 blades as typical representatives for subsequent analysis to analyze the pressure on the suction surface and pressure surface pressure of 3-blade to 7-blade blades. Some contours of pressure are shown in Figure 4 below. The letter Cp stands for pressure coefficient.

The intercepted position is part of the axial screenshot in the middle of the blade in the rotating area, as shown in Figure 4, including half of the blade guide. When the same advance coefficient J = 0.7, for 3 and 7 blades, contours of positive pressure and negative pressure with roughly the same shape will be generated in the rotating area. There are circles of isolines at the blade. The negative pressure is the largest near the blade, and then slowly shrinks, within the values are different. For the 7 blades with high thrust, there will be greater positive pressure and negative pressure, which should also prove the low propulsive efficiency. For 3 blades with low thrust and high propulsive efficiency, the pressure distribution is smaller. The same pressure profile will be generated outside the rotating region.

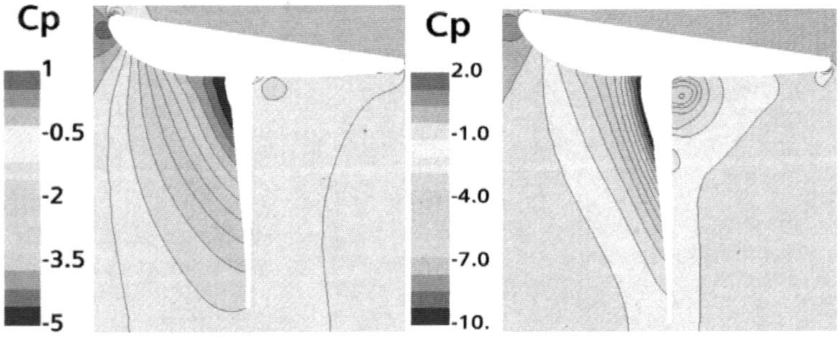

(a)3blade (b)7blade

Fig. 4 Contour map of pressure coefficient

3.3 Flow field velocity analysis of RDT

The pressure distribution has been analyzed. In order to further analyze and study, Figure 5 shows the radial velocity just out of the duct outlet. Only part of the non-rotating area is intercepted. Because there is no change in the non-rotating area, the rotating area is specially analyzed. With the change of the number of blades, the wake velocity map distribution changes. 3 blades apparently correspond to triangles, and 7 blades can apparently observe similar seven angle graphics, forming a regular velocity distribution. The distribution of 7 blades is finer than that of 3 blades. In the rotation area and areas close to the outside, the velocity is greater. The difference in overall velocity is minor. In principle, the outlet velocity of 7 blades is greater, so it can produce greater thrust.

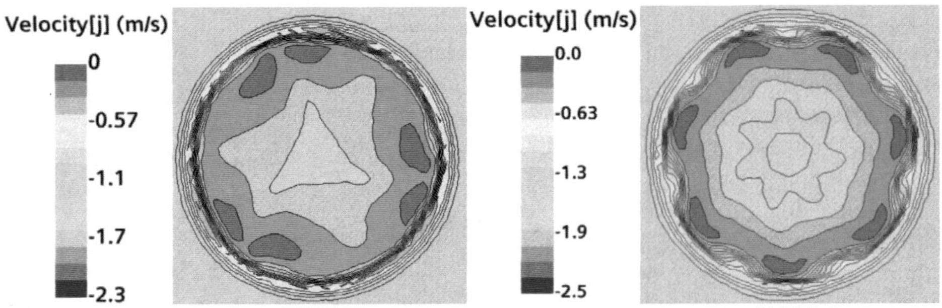

(a)3blade (b)7blade

Fig. 5 Contour map of wake velocity

4. Conclusion

In this paper, based on No.19A+Ka4-70 ducted propeller, the actual motion of RDT is simulated by using the same blade parameters and the number of blades from 3 to 7. The open water performance change and pressure caused by the change of the number of blades are analyzed, and the following conclusions are drawn:

(1) Under the same advance coefficient, the thrust coefficient of the RDT increases and the efficiency of the RDT decreases with the blade number increasing.

(2) Under the same advance coefficient, the pressure distribution changes from scattered to concentrated with the blade number increasing, while the overall pressure and position of the blades do not change much.

(3) The radial velocity and pressure increase with the blade number increasing. The cloud pattern formed is roughly similar. The distribution becomes denser as the number of blades increases.

Acknowledgments
This work was financially supported by the Guangdong Provincial Key Laboratory of Manufacturing Equipment Digitization [Grant No. 2020B1212060014].

References

[1] Yan, X. , Liang, X. , Ouyang, W. , Liu, Z. , Liu, B. , & Lan, J. . (2017). A review of progress and applications of ship shaft-less rim-driven thrusters. Ocean Engineering.
[2] Ojaghlu, P. , & Vahedi, A. . (2018). Specification and design of ring winding axial flux motor for rim-driven thruster of ship electric propulsion. IEEE Transactions on Vehicular Technology.
[3] Yakovlev, A. Y. , Sokolov, M. A. , & Marinich, N. V. . Numerical Design and Experimental Verification of a RIM-Driven Thruster*.
[4] Cao, Q. M. , Hong, F. W. , Tang, D. H. , Fang-Lin, H. U. , & Lin-Zhang, L. U. . (2012). Prediction of loading distribution and hydrodynamic measurements for propeller blades in a rim driven thruster. Journal of Hydrodynamics, 24(1), 8.
[5] Cai, M. , Yang, C. , Wu, S. , Zhu, Y. , & Yu, X. . (2016). Hydrodynamic analysis of a rim-driven thruster based on RANS method. Oceans. IEEE.
[6] Liu, B. , Yan, X. , Ouyang, W. , Lan, J. , & Liang, X. . (2017). Research on regular pattern of gap flow in shaftless rim-driven thruster. 2017 4th International Conference on Transportation Information and Safety (ICTIS).
[7] Liu, B. , & Vanierschot, M. . (2021). Numerical study of the hydrodynamic characteristics comparison between a ducted propeller and a rim-driven thruster. Applied Sciences, 11(11), 4919.

Energy Storage Expansion Planning Method for Active Distribution Network Based on Improved Particle Swarm Optimization Algorithm

Yidan Hu

Nanjing University of Science and Technology ZiJin College, Nanjing Jiangsu 210023, China

*E-mail: wuzhilxy@163.com

Abstract: With the increasing proportion of new energy sources such as photovoltaic and wind power in the regional power grid, the unreliable and random problems of new energy generation are magnified, which will cause damage to the security and stability of the power grid. How to improve the reliability and accuracy of active distribution network energy storage expansion planning and ensure the safe, stable and economic operation of the power grid is the research direction of all countries in the world. In this paper, an improved particle swarm optimization algorithm based on particle swarm optimization for adaptive improvement is proposed. Compared with the traditional particle swarm algorithm, the feasibility and superiority of the algorithm are illustrated. The lower-level programming model is solved by using the interior point method of tracking the center trajectory. The dual gap is improved to speed up the convergence of the algorithm. The improved particle swarm optimization algorithm is combined with the interior point method of tracking the center trajectory to solve the established bi-level programming model. The planning model is solved by using the algorithm. Through the analysis of the planning results, the optimal planning and operation scheme of renewable distributed generation in active distribution network is obtained, and the feasibility of the model is verified. In this paper, the active distribution network has the characteristics of active management, which improves the ability of the distribution network to accept the expansion of energy storage, and becomes an important direction of the development of distribution network.

1. Introduction

According to the long-term and short-term trend of load change in the regional power grid, power grid companies coordinate the output of various types of generating units. Through the overall arrangement of the output coordination between conventional generating units and new energy generating units, while ensuring the spinning reserve of the regional power grid, they also make overall arrangements to optimize the combination of various types of generating units and planned maintenance. While reducing the waste of wind energy and solar energy, it improves the ability of safe and stable operation of the power grid system, improves the utilization rate and effectiveness of all resources including solar energy and wind energy, and seeks the economic and social benefits of new energy grid connection.

The related technologies of active distribution network can improve the acceptance capacity of renewable power generation such as wind turbines and photovoltaics, reduce the network loss of distribution network and improve the power quality of the system. Reference [2] takes the optimal operation efficiency and economy of distributed generation as the objective, and the growth of load as the constraint. In reference [3], an active distribution network planning model is established for the output removal of renewable distributed generation and distribution network lines, and the input and benefit of the model are analyzed. In reference [4], with the goal of system network reconfiguration and distributed generation access capacity, the integrated planning of distributed generation access system and distribution network construction is carried out. Reference [5] considers the time sequence of wind turbine and photovoltaic power output, takes the power supply rate and economy as the objective function, and optimizes the siting and sizing of distributed generation.

There are relatively few studies on the access of renewable distributed generation with fluctuating output power to active distribution network. Therefore, the planning and operation of active distribution network with renewable distributed generation is worthy of further study. In this paper, an improved particle swarm optimization algorithm is proposed, and the improved particle swarm optimization algorithm is used to solve the model to verify the feasibility of the algorithm. The comparison and analysis of the results show the rationality of the active distribution network planning and operation scheme in this paper. This scheme has a certain reference value for improving the stability of the operation of the active distribution network with renewable generation, the reasonable planning of the location and capacity of the distribution network with distributed generation, and reducing the cost.

The main innovations of this paper are:

(1) A multi-objective programming model with minimum cost and minimum network loss is proposed;

(2) The adaptive improved particle swarm optimization algorithm and the interior point method of tracking the center trajectory are used to solve the problem.

2. Particle swarm optimization

PSO simulates the behavior of birds searching for food. If a bird in the search range represents the possible solution of each problem, it is regarded as a "particle" of the problem to be solved. It is assumed that there are M particles in the defined range, and these particles form a group in an N-dimensional search space. Each particle has its position x_i and velocity v_i, and $i = 1, 2, \cdots, M$, in which the position of each particle x_i is a possible solution. The "flight" direction and distance of each particle are determined by the velocity v_i. The fitness function is used to calculate the fitness of each particle, and the objective function of the planning model determines the fitness of the particle. PSO uses the fitness value of the particle to judge the quality of the solution and replace it accordingly, and obtains the optimal solution of the optimization problem through repeated iterations [6]. The update formula of the particle swarm speed v_i is shown in formula (1), and the update formula of the position x_i is shown in formula (2):

$$v_{id} = \omega \times v_{id} + c_1 r_1 (p_{id} - x_{id}) + c_2 r_2 (p_{gd} - x_{id}) \tag{1}$$

$$x_{id} = x_{id} + v_{id} \tag{2}$$

Where:

ω——Inertia weight;

c_1 and c_2——Learning factor;

p_{id}——Optimal location for individual particles;

p_{gd}——Global particle optimum position;

r_1 and r_2——Random number in the range $0 \sim 1$.

The specific flow of the basic particle swarm algorithm is as follows:

(1) Initialize the population size N, the position x_i and the speed v_i of each particle of the particle swarm;

(2) Calculate the fitness value $\mathrm{Fit}[i]$ of each particle;

(3) Compare that fitness value $\mathrm{Fit}[i]$ of each particle with the individual extreme value $p_{best}(i)$. If $\mathrm{Fit}[i] > p_{best}(i)$, replace $p_{best}(i)$ with $\mathrm{Fit}[i]$;

(4) Compare that fitness value $p_{best}(i)$ of each particle with the global extreme value $g_{best}(i)$. If $\mathrm{Fit}[i] > g_{best}(i)$, replace $g_{best}(i)$ with $\mathrm{Fit}[i]$;

(5) Update the position x_i and velocity v_i of the particle using the equations (1) and (2);

(6) If the obtained result meets the set error range or the algorithm reaches the maximum cycle number, end the search. If the condition is not satisfied, return to (2) to continue the iterative calculation.

PSO does not need to set too many parameters, which is one of its greatest advantages [7].

However, the parameters in PSO determine the convergence of the algorithm. The most important parameter is the inertia weight ω. Increasing the value ω can improve the global search performance of PSO, and decreasing the value ω can improve the local search performance of PSO. Reasonable determination of the value of inertia weight can avoid the algorithm falling into local optimum, improve the search performance of the algorithm, and achieve better convergence effect by improving the inertia weight.

3. Improved particle swarm optimization

To improve the local search capability of PSO and improve the convergence performance of PSO, the inertial weight ω is adaptively improved, as shown in Equation (3) :

$$(\omega_{\max} - \omega_{\min})\exp\left(-\frac{1}{1+\frac{\text{Maxter-Iter}}{\text{MaxIter}}}\right) + \omega_{\min} \tag{3}$$

Where $\omega_{\max}$ and $\omega_{\min}$ are the maximum and minimum inertia weights. Maxter is the maximum number of iterations, and Iter is the current number of iterations.

The PSO learning factor also has a certain impact on the convergence and algorithm optimization, so the learning factors c_1 and c_2 are adaptively improved according to the formula (4).

$$C_1 = C_2 = C_{\max} - \frac{(C_{\max} - C_{\min}) \times \text{Iter}}{\text{MaxIter}} \tag{4}$$

Where $C_{\max}$ and $C_{\min}$ are the maximum and minimum values of the learning factor.

To study the influence of the parameters in PSO on the convergence performance of the algorithm, and to verify the superiority and feasibility of the adaptive improved PSO, the following scheme is formulated: the inertia weight of PSO in scheme 1 is 0. 8, and the inertia weight of PSO in scheme 2 is 0. 6. The adaptive improved PSO is used to solve the same objective function in scheme 3. The convergence performance of each scheme is verified by giving a linear function and finding its minimum value. The iteration curves of the particle swarm algorithm for different schemes are shown in Figure 1.

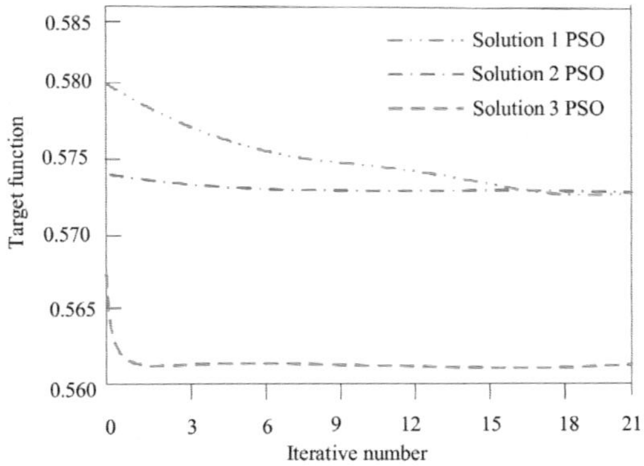

Fig. 1 Comparison of particle swarm optimization algorithms with different schemes

From the analysis of Figure 1, it can be seen from the comparison of the curves of scheme 1 and scheme 2 that the inertia weight ω of the particle swarm has a great impact on the convergence accuracy of the algorithm, especially on the convergence speed of the algorithm. By comparing the third scheme with the other curves, it is found that the first scheme PSO converges in the eighth generation, the second scheme PSO converges in the sixth generation, and the improved PSO converges immediately in the third generation. The value obtained by the improved PSO is better than that obtained by the unimproved PSO. It shows that the adaptive improved PSO has the characteristics of fast convergence speed and good optimization accuracy.

4. Experimental analysis

Based on the actual wind speed data and light intensity in a certain area, the simulation analysis is carried out, and the power generation of wind power generation and photovoltaic power generation is simulated according to the probability density function. Taking the distribution network node as the simulation example, the system voltage level is 12.66 kV. The system total active load is 3.715 MW, and the system total reactive load is 2.30 MVAR [8]. The node diagram of the distribution network is shown in Figure 2.

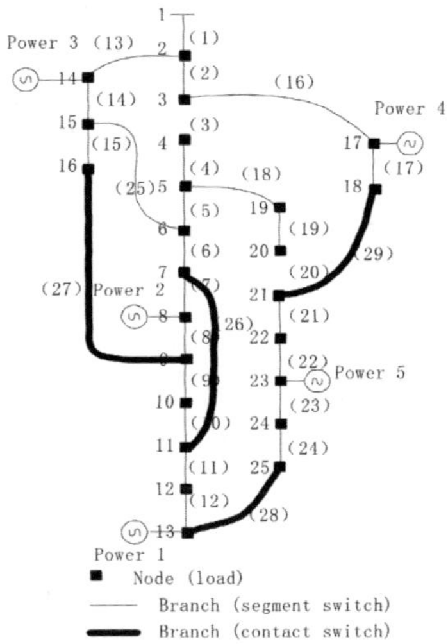

Fig. 2 Distribution network node diagram

According to the wind speed data and Weibull two-parameter distribution formula, the shape parameters k and c are 2.28 and 8.87. The rated light intensity of photovoltaic power generation is 1000 W/m2, and the shape parameters of Beta distribution are $\alpha = 0.89$, $\beta = 0.83$. The economic service life of distributed generation is 20 years. The discount rate is 10%, and the electricity price is 0.5 yuan/kWh. The node voltage range is set to be 0.95 to 1.05 p. u.; and the upper limit of branch transmission power is 4.5MW. The total installed capacity of distributed generation shall not exceed 25% of the total load of the network.

Randomly select a group of nodes as the position of distributed generation connected to the distribution network system, and solve the objective function by using the traditional particle swarm algorithm and the improved particle swarm algorithm respectively. The iterative curve of the objective function is shown in Figure 3.

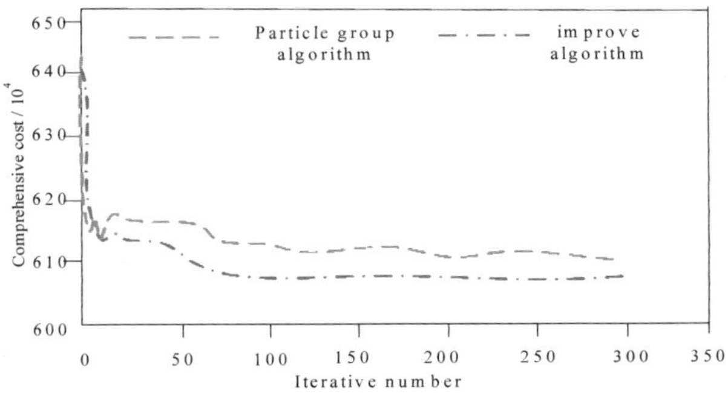

Fig. 3 Convergence curves of improved and unimproved PSO

In Figure 3, the particle swarm algorithm converges at the 238th generation, with an annual comprehensive cost of 6.08 million yuan; while the improved particle swarm algorithm converges at the 63rd generation, with an annual comprehensive cost of 6.04 million yuan. It not only has faster convergence, but also has a better objective function value than the value obtained by the traditional particle swarm algorithm, which shows the effectiveness of the algorithm [9].

The improved adaptive particle swarm optimization algorithm is used to solve the planning model, and the results are shown in Table 1.

Table 1 Results of distributed generation planning schemes under different conditions

Comparison item (ten thousand yuan)	Plan one	Plan two	Plan three	Plan four
Distributed generation installation costs	-	146.35	150.7	152.52
Distributed generation maintenance costs	-	52.92	54.97	56.32
Purchase electricity from higher authority	1114.5	623.1	601.5	585.3
Network loss cost	50.67	30.27	21.28	19.17
Government environmental subsidy	-	196.56	205.2	211.68
Annual comprehensive cost	1165.17	656.08	623.25	601.63

The total amount of renewable distributed power and reactive power compensation allowed by active distribution network under each scheme are shown in Table 2.

Table 2 Installation capacity of renewable distributed power generation and reactive power compensation under different schemes

Compare item	Plan one (MW)	Plan two (MW)	Plan three (MW)	Plan four (MW)
Photovoltaic installed capacity	0	0.7	0.68	0.63
Fan installation capacity	0	1.12	1.22	1.33
Reactive power compensation capacity	0	0.06	0.95	0.12
Total installed capacity	0	1.82	1.90	1.96

Through analysis and comparison, the corresponding problems are explained. Compared with Scheme 1 and Scheme 2, the annual comprehensive cost of the system connected to distributed generation is much less than that of the system not connected to distributed generation. This is because distributed generation generates electricity as a generator, the cost of purchasing electricity from the superior is greatly reduced, and the access of distributed generation will reduce the network loss. The corresponding network loss cost is also reduced [10]. Because of the reasonable access of distributed generation, the network loss of the system is reduced, and the cost of network loss is also relatively reduced.

5. Conclusion

In this paper, the active power output of PV generation and wind power generation with uncertainty in the expansion planning of active distribution network energy storage is simulated. The corresponding power generation is obtained as the basis for solving the planning model. The improved particle swarm optimization algorithm and the traditional particle swarm optimization algorithm are used to solve the same objective function respectively, and the simulation results are analyzed to verify the effectiveness of the proposed comprehensive ratio formula and the feasibility of the improved algorithm. It improves the ability of distribution network to accept distributed generation, and verifies that active distribution network is an important direction of distribution network development.

In the future work, it will be considered as a dynamic load in the actual distribution system, and the corresponding dynamic change will occur with the change of voltage. The access of distributed energy is not only a simple digital change, but also a more complex situation. Therefore, further consideration is required.

References

[1] Ming Wu, Lingfeng Kou, Xiaogang Hou, Yu Ji, Bin Xu, Hongjun Gao. A bi-level robust planning model for active distribution networks considering uncertainties of renewable energies[J]. International Journal of Electrical Power and Energy Systems,2019,105.

[2] A. Garry, F. Cadoux, M.-C. Alvarez-Herault, N. Hadjsaid. Risk aversion model of distribution network planning rules considering distributed generation curtailment[J]. International Journal of Electrical Power and Energy Systems,2018,99.

[3] Cong Zhang, Haoyong Chen, Honwing Ngan, Zipeng Liang, Manlan Guo, Dong Hua. Solution of

reactive power optimisation including interval uncertainty using genetic algorithm[J]. IET Generation, Transmission & Distribution,2017,11(15).

[4] Sarangthem Sanajaoba Singh, Eugene Fernandez. Reliability assessment of a remote hybrid renewable energy system using Monte Carlo simulation[J]. Int. J. of Renewable Energy Technology,2018,9(3).

[5] Lvbin Peng, Bo Hu, Kaigui Xie, Heng-Ming Tai, Kaveh Ashenayi. Analytical model for fast reliability evaluation of composite generation and transmission system based on sequential Monte Carlo simulation[J]. International Journal of Electrical Power and Energy Systems,2019,109.

[6] Mughees Neelam, Mohsin Syed Ali, Mughees Abdullah, Mughees Anam. Deep sequence to sequenceBi-LSTM neural networks for day-ahead peak load forecasting[J]. Expert Systems With Applications,2021,175.

[7] Eskandari Hosein, Imani Maryam, Moghaddam Mohsen Parsa. Convolutional and recurrent neural network based model for short-term load forecasting[J]. Electric Power Systems Research, 2021, 195.

[8] Li Lechen, Meinrenken Christoph J., Modi Vijay, Culligan Patricia J. Short-term apartment-level loadforecasting using a modified neural network with selected auto-regressive features[J]. AppliedEnergy,2021,287.

[9] Song Jiancai, Zhang Liyi, Xue Guixiang, Ma Yun Peng, Gao Shan, Jiang QingLing. Predicting hourly heating load in a district heating system based on a hybrid CNN-LSTM model[J]. Energy & Buildings, 2021, 243.

[10] Shafiei Chafi Zahra, Afrakhte Hossein, Golilarz Noorbakhsh Amiri. Short-Term Load Forecasting Using Neural Network and Particle Swarm Optimization (PSO) Algorithm[J]. Mathematical Problems in Engineering,2021,2021.

Research on active reconfiguration strategy of distribution network for real-time risk

Yuan Li[1,2a*], **Yonghua Chen**[1,2b], **Bingwen Wang**[1,2c]

[1]State Grid Electric Power Research Institute Co., Ltd., Nanjing 315211, China

[2]NARI Technology Co., Ltd., Nanjing 315211, China

[a*]liyuan7@sgepri.sgcc.com.cn, [b]chenyonghua1@sgepri.sgcc.com.cn, [c]wangbingwen@sgepri.sgcc.com.cn

Abstract: Based on the rapid development of smart distribution network, an active reconfiguration strategy for real-time risk prevention is proposed. Firstly, the operation risk assessment model of distribution network is constructed. Based on the distribution internet of things, the risk source information is collected in real time, and the system risk value under the current operation mode is evaluated from the dual dimensions of distribution network equipment operation reliability and failure consequences. On this basis, the proactive network reconstruction strategy and implementation scheme for risk prevention are proposed. When the risk value of the system exceeds the safety warning line, the network reconstruction of the distribution network is carried out with the minimum risk value and the minimum switching operation times as multiple objectives, so that the distribution network can recover the safe and reliable operation state in time. Finally, the feasibility and effectiveness of the active reconfiguration strategy of distribution network under the risk state proposed in this paper are demonstrated by taking the IEEE 33 node distribution system as an example.

1. Introduction

Distribution network is an important link between power generation and transmission system and end users. As the "last kilometer" of power system, its safe and stable operation directly affects the reliable power consumption of thousands of households. However, the structure of distribution network is complex and the operating conditions are changeable. More than 80% of system failures and blackouts occur in distribution network every year, resulting in huge economic losses[1]. With the continuous improvement of distribution automation system and the large-scale construction of ubiquitous power Internet of things, it is possible to holographic perceive the real-time operation risk of main equipment of distribution system. Using the risk source data to quantify the system operation risk and timely take active prevention and control measures for the situation exceeding the safety threshold can reduce the probability and impact range of system power failure accident from the source, which has strong practical research significance.

The premise of implementing effective active reconfiguration measures is to establish a reasonable distribution network risk assessment system. At present, domestic and foreign scholars' research on distribution network risk assessment is mainly divided into the following two categories: one is to serve the planning and design of system grid structure and maintenance tasks with the goal of operation economic index or equipment health index, and its risk assessment cycle is long, unable to provide guidance for real-time security defense control of distribution system[2-4]. The second is the real-time risk assessment of load aiming at formulating the expected accident set, that is to predict the accident under the normal operation of the distribution network and formulate the load recovery and transfer scheme after high-risk faults in advance[5-7]. The purpose of risk assessment is to improve the speed of power supply restoration of distribution network load after fault, rather than pre-warning and eliminating potential risks. Aiming at the high-risk working conditions before the accident, this paper realizes the goal of real-time risk prevention and control of distribution network by implementing active reconstruction measures before the fault, balancing the out of limit condition of equipment, transferring important loads in advance, and reducing the impact range of high-probability accidents.

In this paper, an active reconfiguration strategy for distribution network risk prevention is proposed, which aims to change the system operation topology through network active reconfiguration measures in the process of distribution network risk increase, deterioration and evolution, reduce the fault risk and avoid the occurrence of malignant power outage. Firstly, this paper establishes the real-time risk assessment model of distribution network, selects the risk index representing the operational reliability of distribution network equipment and the severity of outage consequences, and gives the corresponding weight value to the index by using analytic hierarchy process to determine the real-time operation risk value of the system. When the risk value is higher than the safety warning line, take reducing the risk value and minimizing the number of network reconfiguration switch operations as the objective function to carry out active reconfiguration, so as to achieve the goal of improving the operation safety and stability of distribution network by changing the network topology in the risk state.

2. Distribution network operation risk assessment

The risk operation of distribution network is characterized by the high probability of system failure and serious consequences. Therefore, the comprehensive quantitative analysis of the system risk operation degree is carried out from two aspects: the operational reliability of distribution equipment and the consequences caused by equipment failure, so as to obtain the real-time risk value of the system, and provide a controllable objective function for the next active reconstruction strategy.

2.1. Equipment operation reliability diagnosis

The risk factors affecting equipment failure and shutdown are mainly divided into internal risk sources (operating conditions and poor historical records) and external risk sources (bad weather and construction damage). Quantitative analysis is carried out to diagnose the real-time operation reliability of the equipment.

(1) Under operating conditions, the physical performance of the main distribution equipment will be damaged to a certain extent under overload or voltage out of limit, which may seriously lead to equipment protection tripping and load loss. With the double accumulation of electrical quantity out of limit degree and out of limit time, the failure probability of equipment increases exponentially. Therefore, the risk value of equipment overload and voltage out of limit is defined as:

$$L_A = \begin{cases} 0 & I \le 0.8 \\ e^{(I-0.8)} - 1 & I > 0.8 \end{cases} \tag{1}$$

$$R_A = \int_t^{t+\Delta t} L_A dt \tag{2}$$

Where: I is the unit value of current flowing through the equipment; L_A is the instantaneous risk value of overload at time t. integrate L_A within the evaluation period Δt to obtain the cumulative risk value R_A of overload.

$$L_V = \begin{cases} 0 & |V-1| \le 3\% \\ e^{|V-1|} - 1 & |V-1| > 3\% \end{cases} \tag{3}$$

$$R_V = \int_t^{t+\Delta t} L_V dt \tag{4}$$

Similarly, where: V is the unit value of node voltage; L_V is the instantaneous risk value of voltage out of limit at time t. The cumulative risk value R_V of voltage out of limit can be obtained by integrating L_V within the evaluation period Δt.

(2) Historical bad records refer to the aging degree of equipment operation, historical fault and out of limit operation records and defect records of the same batch of equipment. The above data will affect the current health status of the equipment, and then bring certain risks to the real-time operation reliability of the equipment. Based on the statistical data of the whole operation process of the equipment, the risk value of equipment failure under the influence of historical adverse records can be obtained:

$$R_H = k_1\left(e^{A\frac{t}{T}} - 1\right) \times k_2 \sum_j E_j \Delta T_j \times k_3 P_x \tag{5}$$

Where: R_H is the risk value of equipment failure under the influence of historical records, which is composed of aging degree, out of limit records and defect records of equipment in the same batch. k_1, k_2 and k_3 are the proportional factors of each part respectively; A is equipment aging parameter; t is the duration of equipment operation; T is the rated service life of the equipment; E_j is the severity of the equipment exceeding the limit for the j-th time, and ΔT_j is the duration of equipment exceeding the limit in history. P_x is the confidence level of equipment operation reliability under the influence of familial physical defects of the same batch of equipment.

(3) Construction damage: man-made construction will cause probabilistic damage events to the nearby distribution equipment and affect the reliable operation of the equipment. Therefore, the corresponding risk weight RC shall be selected for the distribution equipment within the construction range according to different construction intensity.

(4) In bad weather, most of the main electrical equipment of the distribution network are exposed outdoors and are vulnerable to meteorological disasters such as lightning, icing, rainfall and typhoon. Therefore, the corresponding risk weight values are set for the main electrical equipment of the distribution network within the range of bad weather, as shown in Table 1.

Tab.1 Risk Weights in Severe Weather

Type	Risk Weight (R_W)
Normal	1.0
Thunderstorm & Gale	1.0-2.0
High Temperature	1.0-1.2
Dense Fog	1.0-1.2
Freeze	1.0-1.5

To sum up, the real-time operation conditions, historical records, bad weather, construction damage and other external factors of distribution network equipment will have a certain impact on the operational reliability of the equipment. The internal risk value (operating conditions and historical records) is accumulated as the failure probability of the equipment itself, which increases exponentially under the influence of external risk sources (bad weather and construction damage). Therefore, the operational reliability of the equipment is defined as:

$$R_P = \left(R_A + R_V + R_H\right) \times R_W \times R_C \tag{6}$$

Where: R_P is the operational reliability of the equipment; R_A and R_V are the cumulative value of the risk of exceeding the limit of electrical quantity of equipment; R_W is the risk weight value of bad weather; R_C is the risk weight value under the influence of construction damage.

2.2. Equipment failure consequence impact assessment

In the distribution network, the consequences of equipment shutdown are closely related to the network topology location of the equipment. In the radial network, once the equipment is out of operation, the power quality of the upstream load of the feeder where the equipment is located will be affected to a certain extent through fault location, fault isolation, power transfer and other treatment measures, while the downstream load of the fault will have different degrees of short-term power loss. Therefore, the evaluation of the consequences of equipment failure mainly refers to the downstream load of the feeder and the weight of the importance of the load.

According to the different requirements of power users for power supply reliability, the weight table of load importance is set, as shown in Table 2.

Tab. 2 Weight Table of Load Importance

Load Level	Weight Coefficient (c)
Primary Load	1.5
Secondary Load	1.3
Tertiary Load	1.0

The higher the proportion of short-time power loss of downstream important loads caused by equipment shutdown, the more serious the fault consequences caused by equipment shutdown. According to the load level, the consequences of power failure accident of class I load are the most serious, followed by class II load and class III load. Define the failure consequence value of the equipment as:

$$R_F = \frac{\sum_{i=1}^{M} P_i \times c_i}{\sum_{t=1}^{N} P_t \times c_t} \tag{7}$$

Where: R_F is the failure consequence evaluation value of the equipment; P_i is the active power of the i-th load downstream of the equipment; c_i is the weight coefficient of this load; M is the total downstream load; P_j is the active power of the j-th load on the feeder where the equipment is located; c_j is the weight coefficient of this load; N is the total number of loads on the feeder where the equipment is located.

2.3. System risk assessment model

Considering the multi-faceted and complex characteristics of distribution network risk, the "triage" of each main equipment in the system is carried out first. Generally, risk assessment mainly refers to the possibility of each risk and the severity of its consequences. Therefore, by evaluating the operational reliability of the equipment and its failure consequences, the product of the two is defined as the overall risk value of the equipment. When the operational reliability of the equipment is low and the failure consequences are serious, the risk value of the equipment will increase significantly. The system risk assessment model is constructed, as shown in Figure 1. The "consultation" of the risk state of the distribution network system is realized through the accumulation of the risk value of each equipment, as shown in equation 8:

$$VAR = \sum_{i=1}^{Q} \left(R_{P,i} \times R_{F,i} \right) \tag{8}$$

Where: VAR is the risk value of distribution system; $R_{P,i}$ is the operational reliability of the i-th equipment in the system; $R_{F,i}$ is the failure consequence evaluation value of the i-th equipment in the system; Q is the total number of devices in the system.

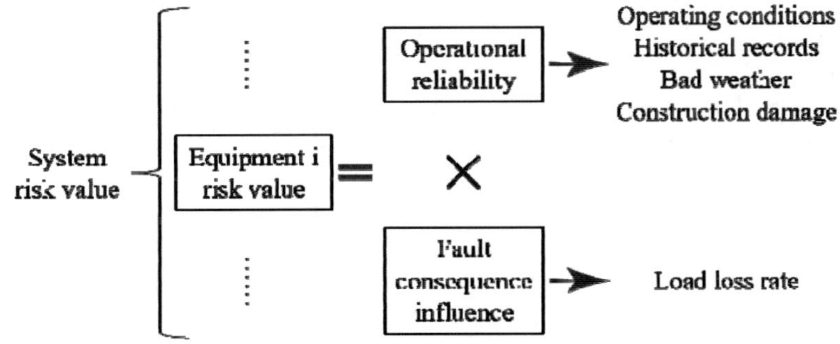

Fig.1 Model of Risk Assessment System

3. Distribution network risk prevention and control model

This paper takes active reconfiguration as the main means of risk prevention and control of distribution network. Its mathematical model and implementation process are introduced in detail below.

3.1. Objective function

According to the above, considering that the number of equipment switching actions should be minimized and the operation reliability should be improved, the minimum real-time risk value and the minimum number of switching operations of the distribution system are analyzed and calculated as a multi-objective function.

$$\min VAR \tag{9}$$

$$\min \sum_{k=1}^{N_S} |x_i - x_{i0}| \tag{10}$$

Where: VAR is the risk value of distribution system; N_S is the total number of operable switches in the distribution network; x_i is the state of switch i after reconstruction; x_{i0} is the state of switch i before reconstruction. The state of the switch is represented by binary "0" and "1" ("0" means the switch is open and "1" means the switch is closed). The objective of equation (10) is to minimize the number of switching operations in the process of distribution network reconfiguration.

3.2. Constraints

(1) Power flow constraint: the distribution network reconstruction scheme shall meet the power flow constraint equation.

(2) The node voltage constraint is $U_{i,\min} \le U_i \le U_{i,\max}$, where $U_{i,\min}$ 、 $U_{i,\max}$ are the upper and lower limit voltages of node i respectively.

(3) The branch capacity constraint is $i_l \le i_{l,\max}$, where i_l, $i_{l,\max}$ are the current flowing through each branch and the maximum allowable current carrying capacity respectively.

(4) Network topology constraints: the reconstructed distribution network structure shall meet radial constraints.

3.3. Optimization algorithm

Multi objective immune particle swarm optimization (IPSO) is applied for optimization and reconstruction. The algorithm flow is shown in Figure 2. The particle in the algorithm is composed of the opening and closing states of all switches in the system, which represents the result of network reconstruction. The active reconstruction strategy aiming at risk prevention and control, namely, on the premise of ensuring the radial operation of distribution network, generates several non-dominated solutions to reduce the risk value of the system as the solution set of the active reconstruction scheme through iterative optimization of particle swarm optimization.

The immune idea is introduced to improve the population diversity and convergence speed of the algorithm. In the iterative process of particle swarm optimization, the particles with lower concentration are selected according to the concentration probability selection mechanism to suppress the particles with higher concentration, so as to ensure the diversity of particle swarm and the global search ability of the algorithm. In the process of particle screening, the candidate solution with large affinity with the optimal solution is retained to reduce the number of iterations and improve the speed of the algorithm. The position update formula of particles is:

$$v_{id}^{t+1} = \omega v_{id}^{t} + cr_d\left(P_{id}^{t} - x_{id}^{t}\right) \tag{11}$$

$$x_{id}^{t+1} = x_{id}^{t} + Round\left\lfloor v_{id}^{t} \right\rfloor \tag{12}$$

$$A_{k,j} = \frac{1}{1+Z_{k,j}} \tag{13}$$

Where: v_{id}^{t+1} is the d-dimensional velocity of particle i; ω is the inertia factor; c is the learning factor; r_d is the random number between intervals [0,1]; p_{id}^{t} is the learning object of the d-th dimension of particle i; x_{id}^{t} is the position value of the d-th dimension of particle i; $Round$ indicates rounding calculation. $A_{k,j}$ is the affinity between antibody k and antigen j; $Z_{k,j}$ is the Euclidean distance between antibody k and antigen j.

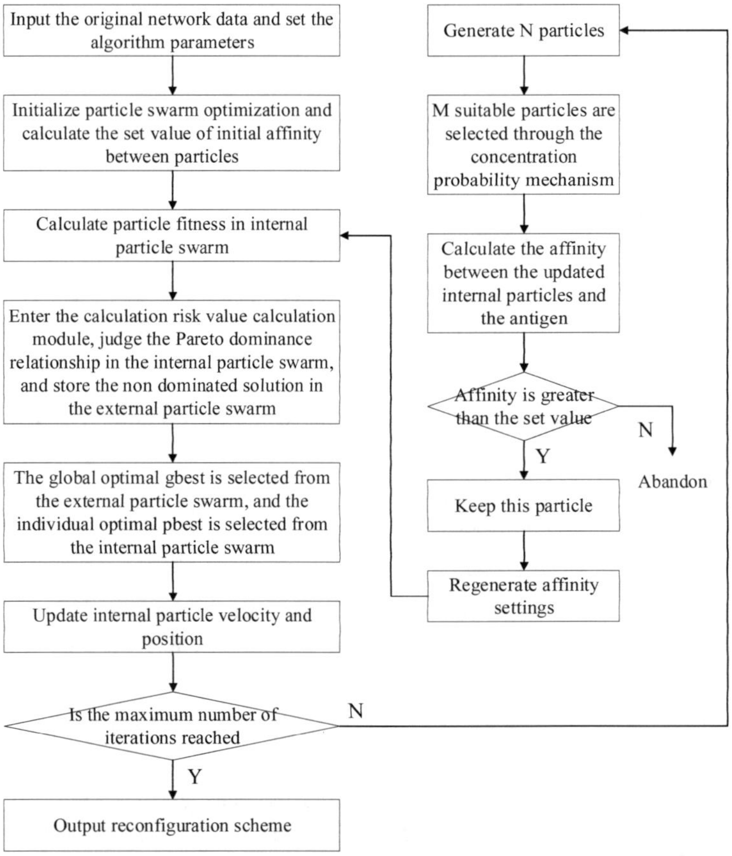

Fig.2 Multi-objective Immune Particle Swarm Optimization Strategy

4. Distribution network risk prevention and control system

Based on the active reconfiguration strategy under the risk state of distribution network, the risk prevention and control system of distribution network is proposed, as shown in Figure 3. In the process of real-time operation of the distribution network, the distribution network cyber physical system is used to carry out real-time situation awareness feedback on the electrical quantity operation level, load intensity, real-time external operation environment and network topology under the current operation mode. After data analysis in the risk assessment module, the real-time risk value VAR of the current system is obtained, and then logical judgment is made. If the VAR is lower than the preset safety warning line $VARset$, return to the risk calculation sub module of the current distribution network; if it is higher than $VARset$, the active reconfiguration is carried out with the minimum risk value and the minimum number of switching operations as the multi-objective. Upload the optimized reconstruction scheme to the distribution network management system (DMS), and use the advanced distribution automation system for network reconstruction after being approved by the decision-makers.

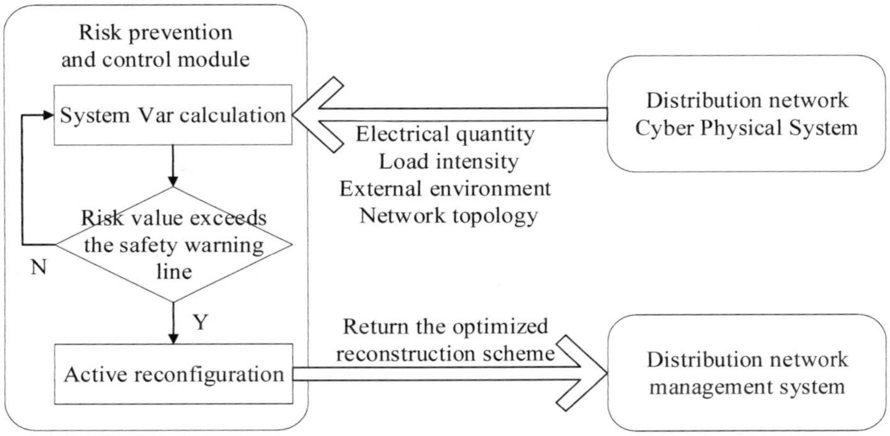

Fig.3 Diagram of Risk Prevention and Control System

The proposed risk prevention and control system can be used for on-line safety monitoring of distribution network, guide the prevention and control reconstruction strategy under risk state, and reduce the probability and loss of accidents. It can also be used as an off-line computing platform. According to the historical data stored in the Cyber Physical System and the influence of construction plan, the off-line computing system can operate in a safe and reliable manner, and cooperate with the adjustment of the short-term operation mode of the distribution network.

5. Simulation Verification

Taking IEEE 33 bus distribution network system as an example, the effectiveness of the proposed model and algorithm is verified. As shown in Figure 4, the system has 37 branches, 33 nodes and 5 tie switches. The rated voltage is 12.66kV. The load grade of each node is shown in Table 3. According to the feedback information uploaded by the urban micro weather station, the power distribution range affected by bad weather (thunderstorm and gale) is divided in the figure.

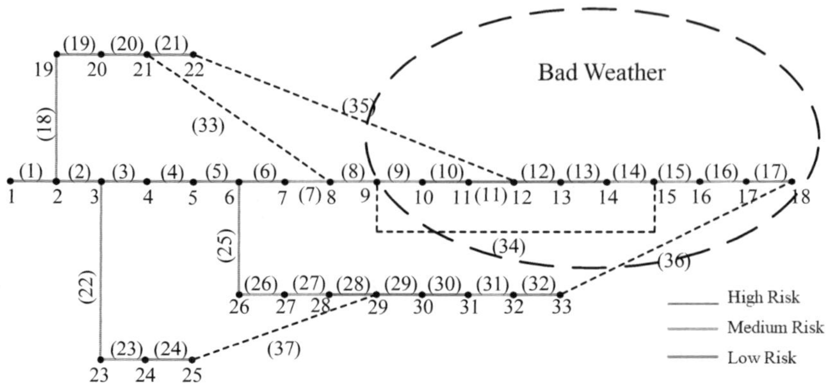

Fig.4 IEEE 33 Bus System with its Topological Graph

Tab.3 Node Load Level

Load Level	Load Node
Primary Load	4,16,17,29,30
Secondary Load	5,9,12,13,14,15,22,24,25
Tertiary Load	Other Nodes

The risk assessment system described in this paper is used to evaluate and calculate the risk of the distribution system in the initial operation state. The current system risk value is 9.772. As the equipment 9~18 is affected by bad weather and carries a lot of downstream load, the risk value of the system increases. Table 4 shows the calculation results of typical high-risk equipment.

Tab.4 Risk Calculation Results under Initial Operation State

Equipment	Equipment Reliability Diagnosis					Failure Consequence	Value at
	R_A	R_V	R_H	R_W	R_C	Assessment R_F	Risk VAR_i
6	0	2.53	0	1	1	0.714	1.806
12	0	3.74	0.2	2	1	0.299	2.356
17	0.12	4.45	0	2	1	0.060	0.548
20	0.79	0	0	1	1	0.391	0.309
23	2.25	0	0.2	1	1	0.696	1.702
26	1.52	0.263	0.2	1	1	0.637	1.263
30	3.398	0.400	0	1	1	0.311	1.181

Taking the minimum risk value and the minimum switching operation times of distribution system as the optimization objective, the multi-objective immune particle swarm optimization algorithm is used to optimize the active reconfiguration scheme. The population size of the algorithm is 30 and the external particle swarm optimization size is 10; $w=0.5$, $wdamp=0.09$, $v_{\max}=4$, $v_{\min}=-4$, and the maximum number of iterations is 50. The Pareto optimal solution set obtained by algorithm iteration is shown in Table 5.

Tab.5 Optimal Solution Set

Reconstruction Scheme	Disconnect Switch	System Risk Value	Switch Operation Times
Original Network	33,34,35,36,37	9.772	Nothing
1	8,14,33,36,37	5.634	4
2	9,14,28,33,36	2.527	6
3	10,28,32,33,34	3.876	6
4	7,10,14,36,37	2.143	6
5	7,9,14,32,37	2.728	8

Because this paper uses multi-objective particle swarm optimization algorithm, it will produce multiple groups of non dominated solutions to form the optimal frontier. From the optimization results, the five reconfiguration schemes reduce the risk value of the system with as few switching actions as possible. The subsequent decision-makers can choose the preferred reconstruction scheme according to the actual situation. This paper takes scheme 2 with the most obvious reduction of risk value as an example for discussion and analysis. Its topology is shown in Figure 5, and the risk assessment results of corresponding equipment are shown in Table 6.

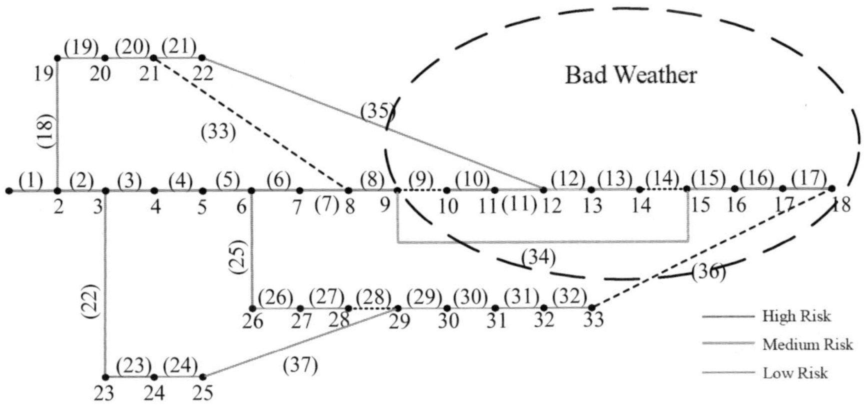

Fig.5 Network Topology Diagram of Reconfiguration Scheme 2

Tab.6 Risk Calculation Results after Network Reconfiguration

Equipment	Equipment Reliability Diagnosis					Failure Consequence Assessment R_F	Value at Risk VAR_i
	R_A	R_V	R_H	R_W	R_C		
6	0	0	0	1	1	0.516	0
12	0	0	0.2	2	1	0.257	0.103
17	0	2.21	0	2	1	0.078	0.344
20	0	0	0	1	1	0.652	0
23	0	0	0.2	1	1	0.831	0.166
26	0	0	0.2	1	1	0.197	0.039
30	1.819	2.76	0	1	1	0.253	1.15

Compared with the network structure improved by the reconstruction method of scheme 2, it is specifically reflected in two aspects:

(1) This reduces the impact of low reliability devices. As can be seen from Fig.4, the reconstruction scheme reduces the load carried downstream by opening switches 9 and 14 to place high-risk equipment at the end of the power supply path, thus reducing the impact range of short-time power loss of load caused by equipment outage.

(2) This ensures the reliability of power supply paths for important loads. In the reconstruction solution, turn on switch 37 and turn on switch 28 to place highly reliable devices at important positions in the power supply path, that is, to shoulder a relatively large downstream load. In this way, the reliability of the power supply path of important loads is improved and the power failure probability of important loads is reduced.

Through example analysis, the power distribution network based on the proposed risk assessment system of quantitative risk, with rather low reliability equipment downstream load and improve the reliability of important load power supply path as a means of control, risk assessment and the active reconstruction control works in harmony, to achieve the target of reducing the risk of power system blackouts.

6. Conclusion

In this paper, an active reconfiguration strategy of distribution network for risk prevention and control is proposed. Firstly, considering the equipment reliability and network risk value, a real-time risk model of distribution network is constructed to effectively evaluate the risk level of the current operating status of distribution network; when the system is in a high-risk status, the multi-objective immune particle swarm optimization algorithm is used to actively search for a more reasonable network topology and improve the operating status of the equipment. Simulation tests show that the proposed strategy can effectively reduce the accident risk of the system and avoid the occurrence of faults. With the construction and development of the distribution internet of things, the active reconfiguration strategy can be integrated into the distribution management system. The bidirectional flow function of the distribution internet of things information and control can control the system risk in real time and reduce the probability and loss of accidents, which has certain practical value.

Acknowledgments

This work was supported by the Science and Technology Projects of SGCC [grant numbers 5400-202117140A-0-0-00].

References

[1] S. Miletić, D. Karavidović. Multi-objective maintenance optimization[C] Electricity Distribution (CIRED 2013), 22nd International Conference and Exhibition on. IET, 2013.
[2] Shen X , Li R . The Plan for Distribution Network Maintenance Based on psobc[J]. Applied Mechanics and Materials, 2014, 521:494-497.
[3] JORGE M B, HÉCTOR V O, MIGUEL L G, et al. Multi fault service restoration in distribution networks considering the operating mode of distributed generation[J]. Electric Power Systems Research, 2014, 116(11):67-76.
[4] DE CASTRO L N, VON ZUBEN F J. Learning and optimization using the clonal selection principle[J]. IEEE Transactions on Evolutionary Computation, 2002, 6(3):239-251.
[5] Zhang, R.F., Li, X., Jiang, T. et al. (2021) Review on Resilience Assessment and Enhancement of Urban Integrated Energy System. J. Journal of Global Energy Interconnection., 4(02): 122- 132.
[6] HUANG Junhui, HAN Jun. A diagnostic method for distribution networks based on power supply safety standards[J]. Protection and Control of Modern Power Systems, 2016, 1(1): 63-70.
[7] HE Jinghan, LIU Lin, DING Fanfan, et al. A new coordinated backup protection scheme for distribution network containing distributed generation[J]. Protection and Control of Modern Power Systems, 2017, 2(2): 102-110.

Journal of Physics: Conference Series **2271** (2022) 012013 doi:10.1088/1742-6596/2271/1/012013

Study of Power Flow Optimisation for Distributed Energy Systems

TianYi Shi[1*]

[1]Pittsburgh College, Sichuan University, Chengdu 610225, Sichuan, China

[*]Corresponding author's e-mail: 2019141520046@stu.scu.edu.cn

Abstract. This paper takes energy saving as the reference point. The distributed energy system is based on the principle of energy "stepped utilization", where the power generation system is built at the end of the customer to meet the customer's cooling and heating load, and the energy of different temperatures is utilized according to the temperature matching principle to improve the utilization rate of primary energy. The system is optimised from the perspectives of reliability and economy of operation of the multi-energy complementary distributed energy system, and is analysed using an improved particle swarm optimisation algorithm. The system is optimised to meet the needs of customers in terms of electricity, cooling and heating, while optimising the reliability of the energy system and optimising the environmental and economic aspects of the distributed energy system to maximise the energy efficiency and economic benefits of the system.

1. Introduction

In today's increasingly contradictory economic development and environmental pollution, distributed energy technologies based on clean and renewable energy sources are the key to solving China's energy problems. Distributed energy systems combine different types of energy modules, energy storage devices and customer loads to provide an integrated use of energy [1]. However, the coupling of individual energy sources in a multi-energy complementary distributed energy system makes leads to a complex management of the energy system, therefore, the control of multi-energy complementary distributed energy systems is a key issue in the study of this system [2]. The optimisation of distributed energy systems aims to configure an optimal energy supply solution for the current environment and to coordinate the energy interactions among the modules and the system. This paper investigates the scheduling of energy within a distributed energy system based on a model predictive control approach to configure an optimal energy supply scheme for the system.

2. Mathematical model for the optimization objectives of distributed energy systems

2.1. Optimisation of the objective function

By reasonably optimising the power output of each module, the wind, CHP and storage system is able to meet the power requirements as far as possible, reducing the s with the grid and ensuring that the overall operating costs of the distributed energy system are minimised. The control model for the multi-energy complementary distributed energy system developed in this paper is divided into two parts.
The system is internally designed to meet the load requirements as far as possible.

$$F_1 = minJ = \sum_{t}^{T}(P_{Lj} - P_{Wj} - P_{CHPj} - P_{SOCj})^2 \tag{1}$$

Minimal internal operating costs of the system, including depreciation of equipment, fuel costs for natural gas consumed by the CHP system, and costs of operating and maintaining equipment purchased from and sold to the grid [4]. The cost of fuel for the natural gas consumed by the CHP system, the cost of operating and maintaining the equipment, and the cost of purchasing power from and selling power to the grid. The costs incurred for the interaction of electricity with the grid. All costs in this chapter are expressed in terms of the unit time utility cost of each module.All costs in this chapter are valued in terms of the unit cost of power per unit of time for each module, i.e. (Yuan/KW (MW)).

$$F_2 = minC = C_{dp} + C_{gas} + C_{OMC} + C_{bg} - C_{sg} \quad (2)$$

In equation1, F1Indicates the least amount of electrical energy interacting with the grid.

In equation 2, F2 indicates the minimum cost of operating a distributed energy system.

2.2. Load power balance constraint within distributed energy systems (equation constraint).

Electrical power balance constraint.

$$P_{grid} + P_W + P_{CHP} + P_{SOC} = P_L \quad (3)$$

The P_{grid} ,P_W, P_{CHP}, P_{SOC} and P_L represent the power interacting with the grid, the output power of the wind turbine, the output power of the microcomputer, the charging and discharging power of the battery and the load load, respectively.

2.3. Operational constraints of the various devices within the distributed energy system.

Output power constraints for power modules.

$$\begin{cases} P_W^{min} \le P_W \le P_W^{max} \\ P_{CHP}^{min} \le P_{CHP} \le P_{CHP}^{max} \end{cases} \quad (4)$$

P_W^{min} indicates the minimum effective power that the turbine starts to output when the cut-in wind speed causes the turbine to start, and P_W^{max} indicates the maximum effective power that the turbine outputs before it stops running. P_{CHP}^{min} indicates the minimum effective power that the micro gas turbine outputs when it starts, and P_{CHP}^{max} indicates the rated power that the micro gas turbine outputs.

Battery charge state constraint and charge/discharge constraint.

Charge state constraint：

$$SOC_{min} \le \mathrm{SOC} \le SOC_{max} \quad (5)$$

As an energy storage device in a distributed energy system, the battery is mainly involved in the regulation of electrical energy and must therefore work reasonably well under certain constraints in the battery operation process.

Charging and discharging constraints：

$$0 \le P_c(t) \le P_c^{max}(t) = min[I_c^{max} U_{bat} \frac{E_\eta(SOC_{max}(1-\eta)SOC(t-1))}{\eta_c \Delta t}] \quad (6)$$

Eq. 6 is the charging power constraint for the batter Principles and forms of standard particle swarm algorithms

Each particle has a corresponding fitness function value fitness, which is solved by the objective function (fitness function), fitness = f(x), which is used to measure the quality of the solution during the search for an optimum [5]. The solution x of the objective function is also the current position of the particle. In the particle swarm optimisation process, we use g_{best} to record the individual optimal position of each particle in the population during the iterative process, and at the same time, in each iteration there is a best position among all the individual optimal positions, i.e. the global optimal position, which is denoted as g_{best} Let the population size of the particle swarm be N, i.e. the population contains N particles, there are also N P_{best}. If these N particles are searched in a D-dimensional vector space. The particle swarm optimisation algorithm finds the optimal solution by iterating continuously. In each iteration of the search, the particles in the population update their flight speed and position according to the two current values (P_{best}, g_{best}).

2.4. Standard particle swarm optimization algorithm flow

Particle swarm optimization is actually the problem of solving the function optimum by continuous exhaustive enumeration, the particle swarm algorithm.

The specific steps are: Initialisation-Calculate the fitness value-Iterative update-Termination condition.

2.5. Problems with the standard particle swarm algorithm

As can be seen from the PSO standard form formula, the PSO search process can be roughly depicted in Figure 1-2 below.

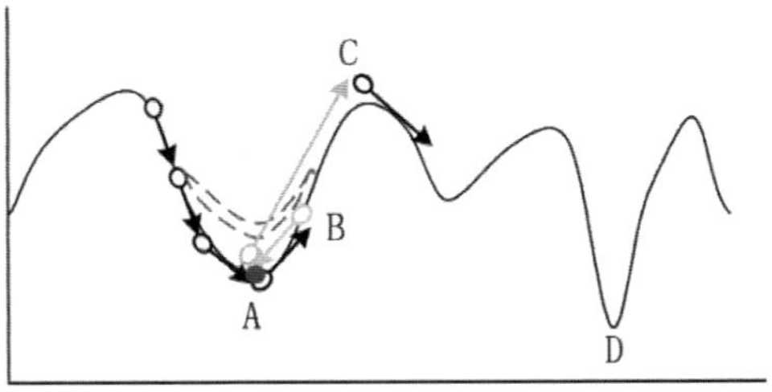

Figure 1. Illustration of the PSO search process

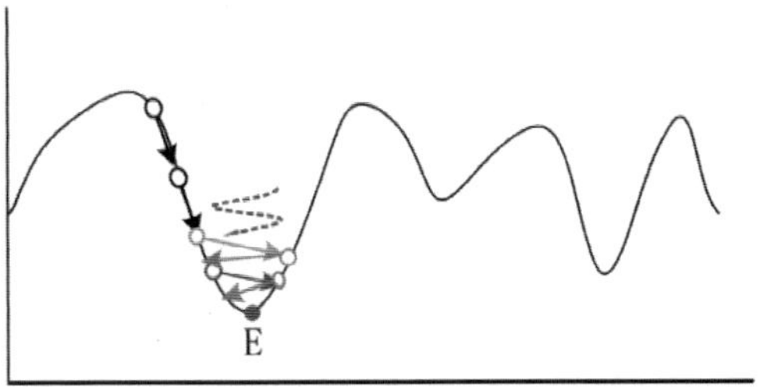

Figure 2. Optimal solution to the oscillation phenomenon

The circles in the diagram indicate the position through which the particle passes in each search, the arrow indicates the direction of the search, and the length of the arrow indicates the speed of each iteration of the particle update. The particle keeps updating its speed and position along the direction of the smaller target value.

When the particle updates to position B, it finds that the value of the fitness function at position A is smaller than that at position B, so it changes its update direction towards A. If at the beginning of the

If at the beginning of the initialisation there are no particles around position D, then eventually all particles will converge towards A. At this point, point A becomes the optimal solution to the objective function, but in fact the optimal solution is point D. This is a case where the algorithm falls into a local optimum [6]. Due to the large step size of each iteration, the situation in the diagram will occur, where the particles will go back and forth to find the optimal solution near point E, but never reach the position of point E. The phenomenon of oscillation occurs. If we reduce the value of ω and keep shortening the step length of the particle, the oscillation of the particle near the optimal solution will be relieved, and the particle will gradually approach the point E and finally reach the location of the optimal solution point E.

2.6. Improved particle swarm optimization algorithm

In this paper, the improved particle swarm algorithms are collectively referred to as Bayesian-Adaptive Weighted Particle Swarm Optimisation algorithms.

Adaptive weighting method: The adaptive weighting method used in this paper is a non-linear and dynamic way of setting inertia weights, which mainly changes according to the change of the fitness value obtained from each iteration, and adjusts the size of the inertia weight value according to the size of the fitness value.

Bayesian Optimization: Bayesian optimisation involves fitting a model to find the hyperparameter configuration that maximises (minimises) the acquisition function, evaluating that configuration and repeating the process.

2.7. Configuration of algorithm parameters

Each particle has a corresponding fitness function value which is derived from the objective function (fitness function), i.e. fitness = f(x), which is used to measure the quality of the solution during the search (Figure 3).

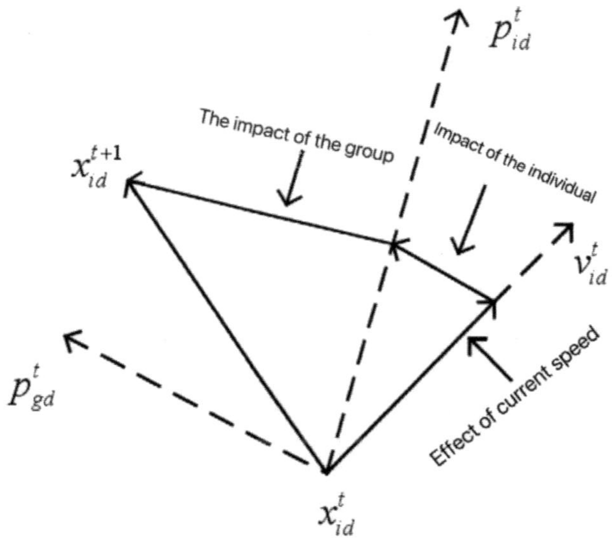

Figure 3. How the particle positions are updated

2.8. Analysis of optimization results

In order to verify the effectiveness of the Improved Particle Swarm Algorithm (Bayesian-AWPSO), several hyperparameter variables mentioned in the above section were optimised by means of a Bayesian optimisation algorithm, programmed in python [7]. The parameters of the algorithm were optimised in 30 steps using Bayesian with the values of the learning factors c1 and c2, the upper and lower bounds wmin and wmax for w in the adaptive weighting method, the particle size N and the maximum number of iterative steps max_iter. The parameters of the Bayesian optimisation iteration process are shown in Table 1

Table 1. Bayesian iterative process parameters

Steps	C1	C2	Max-iter	N	Wmax	Wmin	Target
1	1.67	2.88	50	40	0.21	0.13	8.47e-05
2	0.36	2.95	999	95	0.80	0.58	9.2le+01
3	0.75	1.38	426	63	0.59	0.96	3.73e-04
4	1.84	0.82	999	20	0.55	031	000
5	1.71	3.02	50	44	0.59	0.24	9.20e +01
6	3.72	2.86	673	98	1.35	0.42	7.19e +00
7	1.24	1.17	994	68	0.99	0.15	8.49e-04
8	1.54	1.72	295	33	0.73	0.23	1.14e +00
9	3.64	0.30	260	45	1.28	0.49	6.42e+00
10	0.99	0.76	125	41	0.82	0.18	9.2le +01
11	1.97	3.63	733	33	1.37	0.20	9.2le +01
12	1.44	3.75	801	32	1.32	0.20	9.2le +01
13	1.73	2.14	348	98	1.03	0.32	3.6le-01
14	1.58	2.14	445	61	1.08	0.22	3.6le-01
15	3.84	1.07	358	38	0.98	0.64	5.02e-04
16	0.43	1.76	505	45	0.82	1.11	9.2le+01
17	0.80	0.17	296	35	0.20	0.06	1.54e-01
18	0.29	2.92	860	86	0.50	0.08	2.02e-04
19	1.19	1.24	535	95	1.14	0.38	6.6le+01

20	1.90	3.47	375	85	1.10	0.57	3.30e +00
21	1.28	1.72	181	96	1.07	0.67	1.02e-01
22	1.40	2.74	207	50	0.77	0.06	9.2le +01
23	3.89	3.88	309	25	0.90	0.85	3.80e +01
24	3.01	2.39	741	82	0.07	0.02	7.50e-03
25	0.71	3.44	962	65	1.20	0.52	2.84e+01
26	3.48	3.44	536	84	1.38	0.24	6.78e-02
27	2.35	0.84	389	23	0.41	0.40	1.24e+00
28	0.57	3.39	628	52	1.22	0.84	9.2le+01
29	2.72	1.52	641	56	0.50	0.85	1.04e-02
30	0.68	3.24	916	45	0.33	0.18	0.00

As can be seen from Table 1, Bayesian optimization finds two minimum solutions over the course of thirty iterations [8]. The optimal solution is the case where the objective value of target is 0. Where minF1 is the objective function value, P_{CHP} and P_{SOC} represent the power generated by the cogeneration and the charging and discharging power of the battery, respectively. The iterative process of the standard particle swarm optimisation algorithm is shown in Figure 4(a) and the iterative process of the modified particle swarm algorithm (Bayesian-AWPSO) is shown in Figure 4 (b).

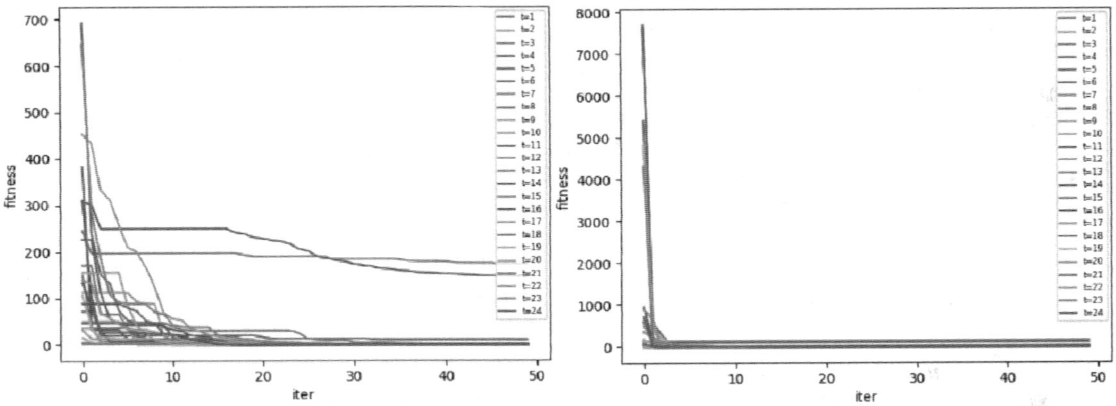

(a) Iterative process for each time step before improvement

(b) Iterative process for each time step after improvement

Figure 4. Iteration process of two algorithms before and after improvement

The curves in subplots (a) and (b) in Figure 4 show the convergence of the algorithm iterations at each time step over a 24-hour period, and we can see that the improved algorithm converges faster than before the improvement and is very stable [9]. (The iterative process at different time steps in (a) is more volatile and the algorithm is not stable. It can be found that the selection of parameters such as inertia weights and learning factors in the particle swarm optimization algorithm has a great influence on the performance of the algorithm. The improved algorithm shows certain advantages in terms of stability and convergence degree.

3. Conclusion

This chapter focuses on the reliability and economics of system operation and establishes an optimal configuration model containing wind turbines, cogeneration systems and energy storage devices. The minimum power interaction with the grid and the operating cost of the distributed energy system are taken as the objective functions for optimisation, with load balancing within the system and the output size of each module as constraints. The standard particle swarm algorithm is also described and the problems in the algorithm are analysed, pointing out several internal parameters that affect the performance of the algorithm. An improved particle swarm optimisation algorithm, Bayesian-AWPSO, is proposed to address the problems in the particle swarm algorithm, using adaptive weighting and Bayesian optimisation to optimise the internal parameters of the algorithm for different situations. Finally, the simulation is carried out with the optimisation objective function F1, i.e. minimising the interaction with the grid, as an example. The analysis of the results proves that the improved particle swarm algorithm has higher convergence and stability.

References

[1] Li P, Zhang YH, Li SONGRUI, Li HUIXUAN, Wang SQ, Zhang LH. (2022). Research on the structure of all categories of energy demand in rural China - an analysis based on rural energy consumption data in Henan Province [J/OL]. Price Theory and Practice.1-5.DOI:10.19851/j.cnki.CN11-1010/F.2021.10.359.

[2] Wang Li-Ying,Song Mei-Qin,Dong Hou-Qi,Fang Cheng,Zeng Ming. (2022). Research on the evolutionary development of energy big data ecosystem based on system dynamics [J/OL]. Journal of North China Electric Power University (Natural Science Edition).1-12. http://kns.cnki.net/kcms/ detail/ 13.1212.TM.20220311.1807.005.html

[3] Ma Linwei, Yuan Yuan, Li Zheng. (2022). Morphology, characterization and sensitivity analysis of low-carbon energy systems in China in 2050 [J/OL]. Journal of Tsinghua University (Natural Science Edition).1-8.DOI:10.16511/j.cnki.qhdxxb.2022.25.020.

[4] Zhang Hengshuo, Li Shaoping, Peng Min. (2022). Regional unevenness and dynamic identification of drivers of carbon emissions from rural energy consumption in China[J/OL]. China Rural Economy.(01):112-134 http://kns.cnki.net/kcms/detail/11.1262.F.20220208.1652.014.html

[5] Yang Y. (2022). The logic of interaction and pattern transformation between China and global energy networks[J]. Journal of Geography.77(02):295-314.

[6] Ma Xiaojun, Gu Yue, Song Yanqi. (2022). Energy satellite accounts: evolutionary background, development history and international experience[J]. Statistical Research.39(01):146-160.DOI: 10.19343/j.cnki.11-1302/c.2022.01.011.

[7] Shen Kli, Cheng Fan, Wu Qiong, Li Qifen, Chen Min. (2022). Reflections on the transformation of integrated energy services of electric power enterprises: the case of State Grid Shanghai Electric Power[J]. Shanghai Energy Conservation.(01):9-15.DOI:10.13770/j.cnki.issn2095-705x. 01.003.

[8] Li XQ, Yang XF. (2022). Challenges to China's energy security strategy and international experience to learn from [J]. Globalization.(01):107-115+136.DOI:10.16845/j.cnki.ccieeqqh.2022.01.011.

[9] Zhou Hongchun, Li Changzheng, Zhou Chun. (2022). Research and reflection on scientific low-carbon transition in China's energy sector[J]. China Coal, 2022,48(01):2-9. DOI:10.19880/j.cnki. ccm. 01.001.

Measurement for bubble size distribution in bubbly oil with digital imaging technology

Li Lu, Shuyun Jiang*

School of Mechanical Engineering, Southeast University, Nanjing, 210000, China

*Corresponding author's e-mail: jiangshy@seu.edu.cn

Abstract. The high viscosity of lubricating oil limits the working speed of sliding bearing, and the viscosity of bubbly oil is less than that of lubricating oil at high shear rate. The viscosity of bubbly oil is related to bubble size distribution. An experimental device for generating microbubbles in bubbly oil is designed, and a method for measuring bubble size distribution in bubbly oil with different aeration levels is proposed based on digital imaging technology. The experimental results show that the size distribution of bubbles accords with lognormal distribution basically, and the higher the air volume fraction is, the larger the average diameter of bubbles is.

1. Introduction

Compared with traditional machining technology, high-speed machining technology has the advantages of high production efficiency, good surface quality and smooth processing, which is suitable for automotive industry, aerospace industry and other fields with high machining precision requirement [1]. High-speed machine tools are the key equipment to realize high-speed machining, and bearings are important parts of high-speed machine tools, which play a role in supporting the spindle and reducing friction [2]. Therefore, high-speed machine tools have higher requirements for the performance of bearings, and it is of profound significance to carry out research on increasing the speed of bearings.

At present, exotherm invalidation of bearings is the biggest factor limiting the working speed of sliding bearings. The bearing temperature rise will increase with the increase of rotating speed of sliding bearings, and too high temperature rise will reduce the service life of bearings and even cause shaft holding [3]. Lubricating oil is the most widely used lubricating medium, and the high viscosity of lubricating oil is the main reason of bearing temperature rise sensitivity to rotating speed. Reducing the viscosity of lubricating oil can reduce the bearing temperature rise, which is an important way to increase the working speed of sliding bearings.

An Qi designed a new device for measuring viscosity of bubbly oil and studied the relationship between the viscosity of bubbly oil and the bubble content [4]. M J Goodwin et al. conduct the theoretical and experimental investigation of the effect of bubbly oil on the load-carrying capacity of a hydrodynamic journal bearing, and finally found that bubbles had little influence on the load-carrying capacity of the journal bearing [5,6]. It can be seen that bubbly oil is an excellent lubricating medium, which has low viscosity and barely affects the load-carrying capacity of bearings. However, there are some problems in the previous studies on bubbly oil: (1) The air volume fraction of bubbly oil is low. (2) Bubble size distribution in bubbly oil has not been studied.

The viscosity of bubbly oil is affected by air volume fraction and bubble surface tension, and bubble surface tension is related to bubble size distribution in bubbly oil. Therefore, it is of great significance to measure bubble size and study the size distribution. In this paper, a bubbly oil generation device with controllable air volume fraction is designed, and a measurement method of microbubble size distribution in bubbly oil is proposed. The overlapping bubbles are segmented by digital imaging technology, and the bubble size distribution law is obtained.

2. Generation of microbubbles

The microbubble generation device is shown in Figure 1, which generates bubbles in the lubricating oil by dispersing air. Firstly, the air from the air compressor is dispersed by the shear action of the impeller of the vortex pump to form large bubbles. Secondly, large bubbles are further refined through the micropores of the filter to generate microbubbles. Finally, the vortex pump circulates bubbly oil in the oil tank, so that the bubbles are cut repeatedly. After each cycle, the air volume fraction of bubbly oil will rise, and the air volume fraction can be controlled by adjusting the cycle time.

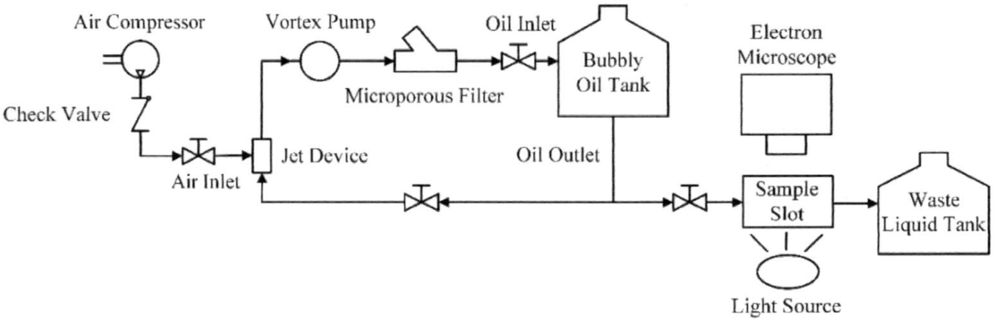

Figure 1. Microbubble generation device

The lubricating oil used in the experiment is spindle oil No. 22. Parameters of experimental instruments are shown in Table 1.

Table 1. Parameters of main experimental instruments

Name	Parameters
Vortex pump	Rated power: 1000W; rated flow: 3.0m^3/h
Microporous filter	Mesh number: 800; Micropore diameter: 19μm

3. Observation and measurement of microbubbles

3.1. Observation of microbubbles

In order to measure the air volume fraction and bubble size distribution of bubbly oil, this paper absorbs the advantages of sampling method which is convenient to observe bubbles and camera method which does not need to be in direct contact with bubbly oil, and the two observation methods are combined. Bubbly oil is introduced into the flat sample slot, and the bubbles are observed by an electron microscope in the sample slot. The sample slot is shown in Figure 2.

Figure 2. Sample slot

3.2. Measurement of microbubbles

After obtaining pictures of bubbles in bubbly oil, the geometric features of pictures can be extracted by digital imaging technology, so as to segment and identify the overlapping bubbles in pictures. The image processing flow includes image preprocessing, overlapping bubble segmentation and bubble fitting.

Preprocess the photos, including image graying, background removal, image binarization and hole filling, so as to reduce the amount of image data and facilitate subsequent processing. The background removal adopts the subtraction method, that is, the bubble image is subtracted from the background image to eliminate the influence of background noise. The calculation relation is as follows:

$$C(x,y)=A(x,y)-B(x,y) \tag{1}$$

Where C (x, y) is the gray value of image pixels after subtraction, A (x, y) is the gray value of bubble image pixels, and B (x, y) is the gray value of background image pixels.

Otsu algorithm is adopted for image binarization. The initial threshold i is set to divide the image into two parts. The part with gray value $0 \sim i$ is the foreground, the proportion of the whole image is a, and the average gray value of the foreground is recorded as b_1. The part with gray value $i \sim 255$ is the background, and the average gray value of the background is recorded as b_2. The average gray value of the whole image is recorded as b, and the variance of gray value between foreground and background is recorded as g:

$$b=ab_1+(1-a)b_2 \tag{2}$$

$$g=a(b_1-b)^2+(1-a)(b_2-b)^2 \tag{3}$$

Substituting Equation 2 into Equation 3 can be obtained:

$$g=a(1-a)(b_1-b_2)^2 \tag{4}$$

Take i corresponding to the maximum g as the threshold of binarization. The original image and the preprocessed image are shown in Figure 3.

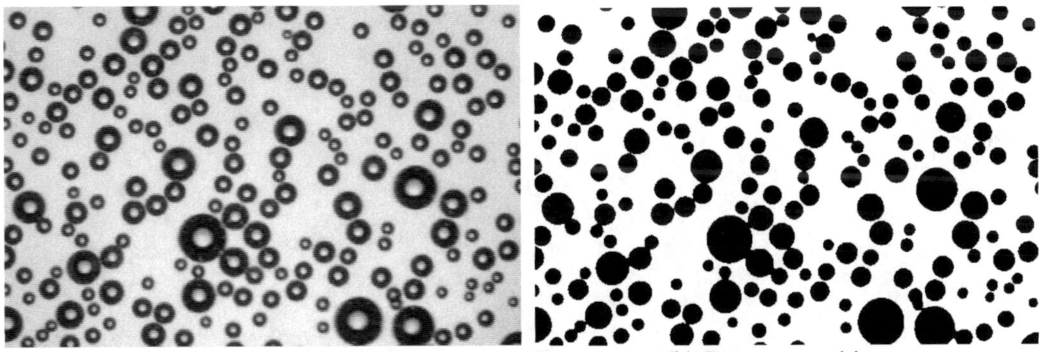

(a) Original image (b) Preprocessed image

Figure 3. Comparison between original image and preprocessed image

In the face of overlapping bubbles, watershed algorithm is used. The principle of watershed algorithm is to treat the gray image as a topographic map with ups and downs. The greater the gray value of a point, the higher the height of the point. Imagine that "water" starts to spread from the area with the lowest gray value, and gradually forms a ponding basin in the lower area. In this process, the ponding basins will be connected with each other. Build a dam at the confluence of two ponding basins, and the dam is the watershed. After the "water" submerges the point with the highest gray value, the watershed algorithm ends, and each ponding basin separated by the dam corresponds to an isolated area. In order to separate overlapping bubbles more accurately, the distance transformation needs to be carried out on the picture. The pixel value of each point in the target area is converted into the distance between the point and the background. The farther away from the background, the greater the pixel value. The image after segmentation by watershed algorithm is shown in Figure 4.

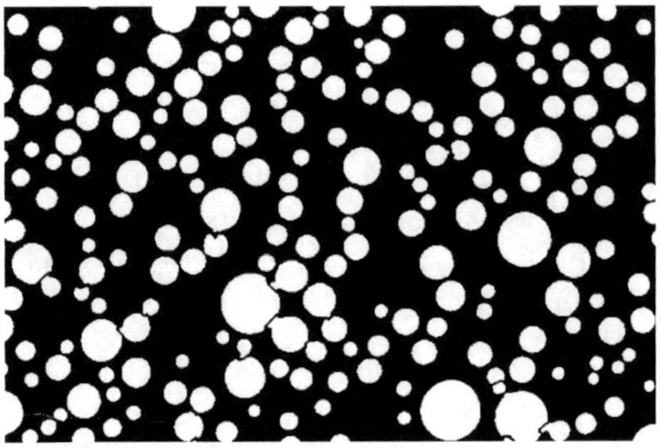
Figure 4. Image after segmentation by watershed algorithm

Use the least square method to fit the circular bubbles in the picture, that is, minimize the sum of squares of distances between a series of data points (x_i, y_i) and the edge of a circle:

$$f = \sum\left(\left(x_i - x_c\right)^2 + \left(y_i - y_c\right)^2 - R^2\right)^2 \tag{5}$$

Where (x_c, y_c) is the center coordinate and R is the radius. When partial derivatives of f to x_c, y_c and R are 0, f obtains the extreme value:

$$\begin{cases} \dfrac{\partial f}{\partial x_c} = -4\sum\left(\left(x_i - x_c\right)^2 + \left(y_i - y_c\right)^2 - R^2\right)\left(x_i - x_c\right) = 0 \\ \dfrac{\partial f}{\partial y_c} = -4\sum\left(\left(x_i - x_c\right)^2 + \left(y_i - y_c\right)^2 - R^2\right)\left(y_i - y_c\right) = 0 \\ \dfrac{\partial f}{\partial R} = -4R\sum\left(\left(x_i - x_c\right)^2 + \left(y_i - y_c\right)^2 - R^2\right) = 0 \end{cases} \tag{6}$$

By solving equation group 6, the center coordinates (x_c, y_c) and radius R can be obtained. The image after bubble recognition is shown in Figure 5.

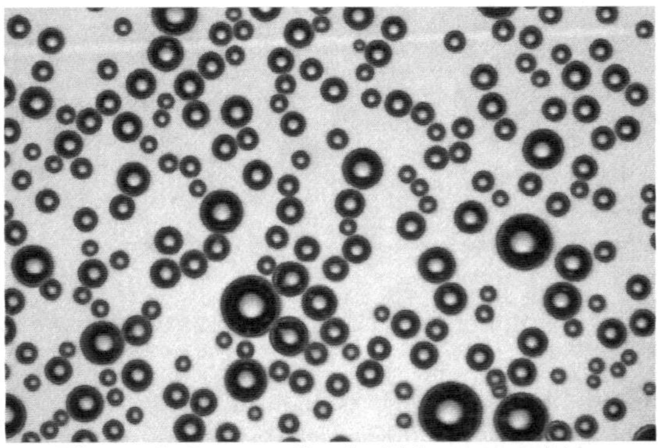
Figure 5. Image after bubble recognition

4. Results and Discussion

In order to obtain bubble size distribution, bubbles diameters are divided into several intervals. The histogram of bubble diameter in bubbly oil under different air volume fraction is shown in Figure 6.

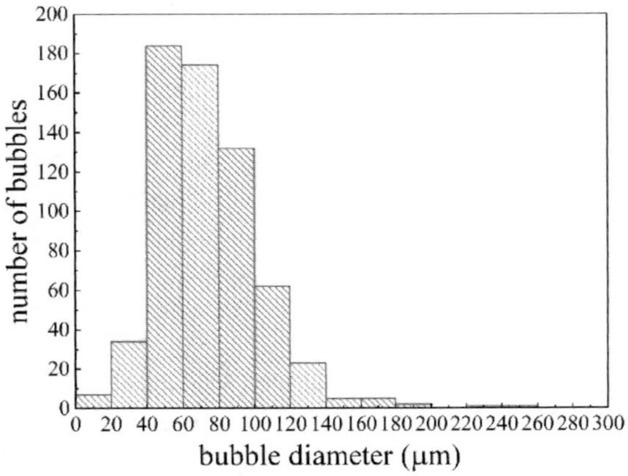

(a) 8% air volume fraction

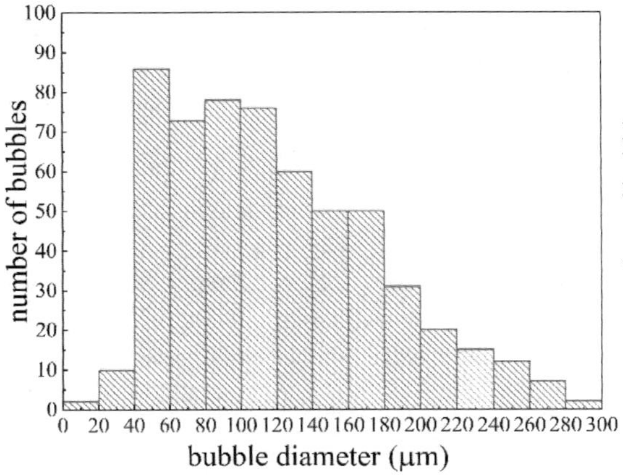

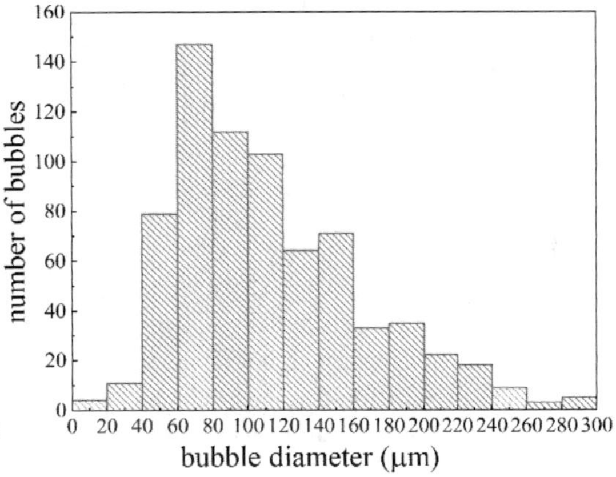

(b) 14% air volume fraction (c) 26% air volume fraction

Figure 6. Histogram of bubble diameter under different air volume fraction

According to the shape of the histogram, the diameter distribution of bubbles is very consistent with the lognormal distribution. The standard form of lognormal distribution is:

$$f(x)=\frac{1}{x\sigma\sqrt{2\pi}}e^{-\frac{(\ln x-\mu)^2}{2\sigma^2}} \tag{7}$$

Therefore, probability density functions of bubble diameter under different air volume fraction can be obtained by fitting histograms:

$$f(D)=\frac{1.077}{D}e^{-\frac{1}{2}\left(\frac{\ln D-4.25}{0.3788}\right)^2} \tag{8}$$

$$f(D)=\frac{0.7114}{D}e^{-\frac{1}{2}\left(\frac{\ln D-4.723}{0.5958}\right)^2} \tag{9}$$

$$f(D)=\frac{0.814}{D}e^{-\frac{1}{2}\left(\frac{\ln D-4.611}{0.4956}\right)^2} \tag{10}$$

Where D is the bubble diameter. The average diameter of bubbles $\overline{D}$ is:

$$\overline{D}=\int_0^{+\infty} f(D)D\,dD \tag{11}$$

Therefore, the average bubble diameter at 8% air volume fraction is 73.047 μm. The average bubble diameter at 14% air volume fraction is 108.312 μm. The average bubble radius at 26% air volume fraction is 121.793 μm.

5. Conclusion

Based on digital imaging technology, a method for measuring bubble size distribution in bubbly oil is proposed, which can effectively separate and identify overlapping bubbles. A bubbly oil generation device is designed, and the experimental study of microbubble size distribution under different air volume fraction is carried out. The main conclusions can be drawn as follows:

(1) Under different air volume fraction, bubble size distributions in bubbly oil all satisfy lognormal distribution.

(2) Bubbles with diameters of 40 μm to 80 μm are the most numerous.

(3) The higher the air volume fraction, the larger the average diameter of the bubble.

References

[1] Xuelin Chen, Jinyuan Tang, Han Ding, et al. (2021) A new geometric model of serrated chip formation in high-speed machining. Journal of Manufacturing Processes, 62: 632-645.

[2] S.B. Murgayya, H.N. Suresh, N. Madhusudhan, et al. (2021) Effective Rotordynamics Analysis of High Speed Machine Tool Spindle – Bearing System. Materials Today: Proceedings, 46: 8905-8909.

[3] Xianwen Zhou, Hao Zhang, Xu Hao, et al. (2019) Investigation on thermal behavior and temperature distribution of bearing inner and outer rings. Tribology International, 130: 289-298.

[4] An Qi, Zhou Yinsheng, Quan Yongxin. (1997) Study on the viscosity properties of bubbly oil and the static characteristics of journal bearing lubricated with bubbly oil. Wear, 213: 159-164.

[5] Goodwin M J, Dong D, Yu H, et al. (2007) Theoretical and experimental investigation of the effect of oil aeration on the load-carrying capacity of a hydrodynamic journal bearing. Proceedings of The Institution of Mechanical Engineers Part J-Journal of Engineering Tribology, 221: 779-786.

[6] Yang H, Wang X, Zhang Y, et al. (2014) Study on the bubbly lubrication of journal bearings at various shear rates and temperatures. Tribology International, 71: 132-139.

Optimal scheduling of wind energy storage microgrid considering demand response

Haibin Li[1a*], Li Sun[1b*], Bin Wu[1c]

[1]School of Business, Shanghai Dianji University Shanghai, China

[a*]1340293577@qq.com, [b] sunlitj13@126.com , [c] wub@sdju.edu.cn

Abstract—As an effective way to improve the utilization efficiency of renewable energy, microgrid has received more and more attention in ensuring energy security and sustainable supply. Microgrid generally consists of power supply, energy storage system and load. However, the uncertainty of renewable energy generation makes it difficult to solve the problem of peak regulation of microgrid. In addition, demand response strategies are playing an increasingly important role in renewable energy consumption because they can use load flexibility to maintain supply and demand balance, rather than just regulating generation on the power side. This paper mainly studies the scheduling problem and demand response strategy of photovoltaic energy storage in microgrid, and outputs the optimal scheduling scheme of microgrid by solving the model.

1. Introduction

With the rapid economic development of all countries in the world, the demand for energy is increasing day by day, resulting in more and more heavy consumption of traditional energy. The extensive use of fossil energy has also brought serious environmental pollution. The problem of energy shortage is becoming more and more serious. How to use clean new energy to replace traditional energy has become a common problem all over the world, looking for new green and renewable energy has become a consensus all over the world. The emergence of microgrid can effectively solve the two major problems of energy shortage and environmental pollution[1]. However, the current renewable energy has not been developed to the extreme.

Microgrid can carry out power loss nearby, reducing the transmission loss and high investment caused by long distance. Compared with conventional power generation systems, microgrid operation has the characteristics of low cost, low investment cost and low environmental pollution. As a whole, it can improve the flexibility and reliability of microgrid. It is a very important research topic in microgrid The important topic is the multi-objective optimal scheduling problem[2].Based on the power generation of each power supply this paper will analyze and study the economic dispatching problem of microgrid under the condition of grid connected operation.

2. Optimization model of microgrid

The optimal dispatching objective function of microgrid includes output power generation cost, system depreciation cost, system maintenance cost, environmental governance cost and power purchase cost from large power grid. Through the coordination of each processing unit and energy storage device, the operation index of microgrid is improved[3]. The simulation model of microgrid is designed as follows：

$$F_1 = \min \sum_{t=1}^{T} (C_H(t) + C_{WE}(t) + C_{FG}(t) + C_{DJ}(t) + C_c(t)) \tag{1}$$

Among them: T=24 , $C_H(t)$、$C_{WE}(t)$、$C_{FG}(t)$ is the fuel cost, maintenance cost, and abandoning the wind and wind within the time, and $C_{DJ}(t)$ is the electricity interaction cost. $C_c(t)$ is the carbon trading fee

2.1. *Fuel consumption cost*

$$C_H(t) = \sum_{i=1}^{N} C_{HJ} \frac{1}{LHV} \sum_{t=1}^{T} \frac{P_i(t)}{\eta_i(t)} \tag{2}$$

Among them : C_{HJ} is the price of natural gas fuel LHV is the low calorific value of natural gas. $P_i(t)$ 、 $\eta_i(t)$ are the output power and working efficiency of the first generating unit.

2.2. *Operation and maintenance costs*

$$C_{WE}(t) = \sum_{i=1}^{N} \mathrm{k}_i P_i(t) \tag{3}$$

Among them : k is the operating and maintenance cost coefficient of the i power supply. $P_i(t)$ is the output power of the i distributed power source at time.

2.3. *The cost of abandoning the scenery*

$$C_{FG}(t) = \sum_{t=1}^{T} \sum_{n=1}^{N} q_{\mathrm{n}} Q_{\mathrm{n},t} \tag{4}$$

Among them : N represents the nth wind and solar machinecause. q_{n} represents the cost per unit of abandoning the scenery. $Q_{\mathrm{n},t}$ represents the system abandoning wind and solar power in t period.

2.4. *Electricity interaction cost with the main grid*

$$C_{DJ}(t) = C_{BUY(t)} - C_{SELL(t)} \tag{5}$$

Among them : $C_{BUY(t)}$ is the electricity purchase cost from the main grid within t . $C_{SELL(t)}$ is the electricity purchase cost of the main grid from the microgrid at time t .

2.5. *Carbon trading fees*

The carbon trading fee in this article is mainly to constrain the carbon emissions of power sources to introduce a tiered carbon trading mechanism[4].

$$C_c(t) = \begin{cases} \lambda P_{CE,sys,t} & P_{CE,sys,t} \le l \\ \lambda l + \lambda(1+\alpha)(P_{CE,sys,t} - l) & l \le P_{CE,sys,t} \le 2l \end{cases} \tag{6}$$

Among them : $C_c(t)$ is the cost of carbon trading. λ is the basic price of carbon trading. l is the length of the carbon emission interval. α is the price increase.

3. RESTRICTIONS

Under the established objective function, some constraints of the system must be observed [5], as follows.

3.1. Power balance constraint

$$\sum_{i=1}^{N} P_i(t) - P_c(t) + P_j(t) + P_{DJ}(t) = P_z(t) \tag{7}$$

Among them : $\sum_{i=1}^{N} P_i(t)$ is the total output power of each distributed power generation in t period. $P_c(t)$ Is the power of the battery in t period. $P_{DJ}(t)$ is the power exchanged between the two grids during the t period. $P_z(t)$ is the load of the microgrid in the t time period.

3.2. Power limit constraints of each distributed power supply

$$P_{i,\min} \le P_i(t) \le P_{i,\max} \tag{8}$$

Among them : $P_{i,\min}$、$P_{i,\max}$ are the minimum and maximum values of the output power of each distributed power source, respectively.

3.3. Power constraints of microgrid and main grid

$$P_{l,\min} \le P_l(t) \le P_{l,\max} \tag{9}$$

Among them: $P_{l,\min}$、$P_{l,\max}$ are the minimum and maximum values of the interactive power between the microgrid and the main grid, respectively.

3.4. Energy storage restraint device

The following are the electric energy storage charge and discharge state constraints, electric energy storage charge and discharge power constraints, electric energy storage capacity constraints and electric energy storage beginning and end state constraints of the power supply system:

$$C_{t+1}^{ESS} = C_t^{ESS} + (\eta_{ch} P_{CH,t}^{ESS} - P_{DC,t}^{ESS} / \eta_{dc})\Delta t, \forall t \tag{10}$$

$$\begin{cases} 0 \le P_{DC,t}^{ESS} \le P_{DC,\max}^{ESS} \\ 0 \le P_{CH,t}^{ESS} \le P_{CH,\max}^{ESS} \end{cases} \forall t \tag{11}$$

$$C_{\min}^{ESS} \le C_t^{ESS} \le C_{\max}^{ESS}, \forall t \tag{12}$$

$$C_0^{ESS} = C_{\mathrm{T_{end}}}^{ESS} = C_*^{ESS} \tag{13}$$

Among them: $P_{CH,t}^{ESS}$ and $P_{DC,t}^{ESS}$ are the charging and discharging power of energy storage in t period respectively, and C_t^{ESS} is the energy storage capacity in t period.

3.5. Opportunity constraint

Backup is an important guarantee for balancing the power of the system and suppressing the uncertain power output of renewable energy generation[6]. This chapter considers the power grid and electric energy storage together to provide backup for the microgrid. Grid reserve constraints, electric energy storage reserve constraints, and gas turbine reserve constraints are as follows:

$$P_t^{grid} + R_t^{grid} \le P_{\max}^{grid}, \forall t \tag{14}$$

$$P_{Ress,t} \le \min\left\{\eta_{dc}(C_t^{ESS} - C_{\min}^{ESS}) / \Delta t, P_{DC,\max}^{ESS} - P_{DC,t}^{ESS}\right\} \tag{15}$$

$$P_{GT,t} + R_{GT,t} \le UP_{GT,\max}, \forall t \tag{16}$$

Among them: P_t^{grid} and R_t^{grid} are the purchased power and grid reserve power in t period respectively . $P_{GT,t}$ and $R_{GT,t}$ are the output power and standby power of the gas turbine in period t. U is the state variable of the gas turbineare the output power and standby power of the gas turbine.

Considering that renewable energy generation may not have power output during a certain period of time, under such a small probability, the microgrid must provide sufficient spinning reserve to maintain the continuity of system output, but this will incur additional costs[7]. Due to the small frequency of this situation, this paper has carried out a probabilistic modeling of the reserve constraints, and weighed the reliability and economy of microgrid operation. The opportunity constraint of scenery uncertainty can be described as:

$$P_{rob}\left\{\sum_{n=1}^{M_G} R_t^{grid} + P_{Ress,t} + R_t^{GT} \geq E(P_t^{DG}) - P_t^{WT} - P_t^{PV}\right\} \geq \alpha, \forall t \tag{17}$$

Among them: $P_{Ress,t}$ is the reserve capacity provided by energy storage during t period, and α is the system confidence level.

3.6. Load demand response constraint

As an important branch of controllable load, real-time scheduling can achieve system supply and demand balance. The time-shiftable load can be described by the following formula:

$$P_{t,\min}^{TSL} \leq P_t^{TSL} \leq P_{t,\max}^{TSL} \tag{18}$$

$$\sum_{i=1}^{T} P_t^{TSL} = 0 \tag{19}$$

Among them: P_t^{TSL} is the time-shifted electrical load power during t period.

During periods of insufficient power supply, the system can interrupt part of the load to relieve the power supply pressure during this period and improve the stability of system operation. The related constraints of interruptible load can be described as:

$$0 \leq P_t^{IL} \leq P_{t,\max}^{IL} \tag{20}$$

Among them: $P_{t,\max}^{IL}$ is the maximum interruptible load that can be interrupted during t period.

4. CPLEX algorithm

CPLEX is mainly used to solve discrete combinatorial optimization problems.The main steps used by CPLEX to minimize any objective function F are as follows:

step 1: Use heuristics to find the solution to the problem X. The stored value Y= F(X). X will be marked as the best solution found so far and used as the upper bound of the candidate solution.

step 2: Initialize a queue to hold partial solutions that do not contain problem variables.

step 3: Loop until the queue is empty.

5. Case analysis

5.1. Basic data

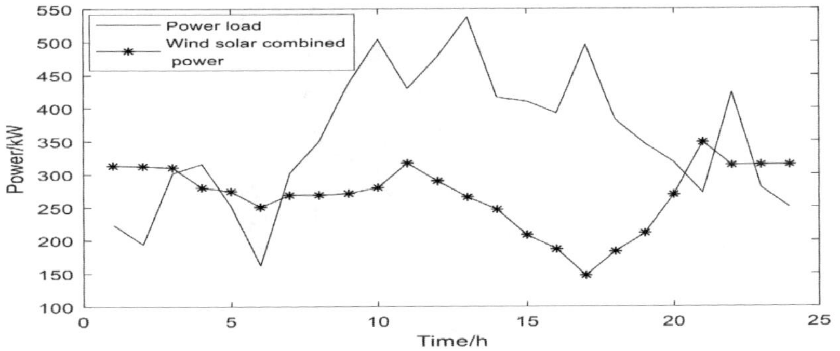

Fig 1 Basic data

5.2. Related parameters

Power grid parameters: The maximum purchase power is 300kW; the maximum sale power is 200kW.

Fan parameters: $v_{in} = 3\text{m/s}$, $v_* = 15\text{m/s}$, $v_{out} = 25\text{m/s}$, Rated power of fan $P_r = 300\text{kW}$.

PV module parameters: $\eta_{pv} = 0.093$, $A_{pv} = 3900\text{m}^2$, $P_{\max}^{PV} = 260\text{kW}$.

Electric energy storage parameters: The maximum capacity is, the charge and discharge coefficient of electric energy storage is 0.9, and the maximum charge and discharge power of electric energy storage is 60kW;.

Carbon trading mechanism: The transaction price is divided into two stages, the first stage price is 0.4￥/kg, the second stage price is 0.7￥/kg, The total carbon allowance is 80kg.

5.3. Optimizing results before and after carbon trading

The hourly dispatch output curve of the system after the implementation of carbon trading is shown in Figure 2

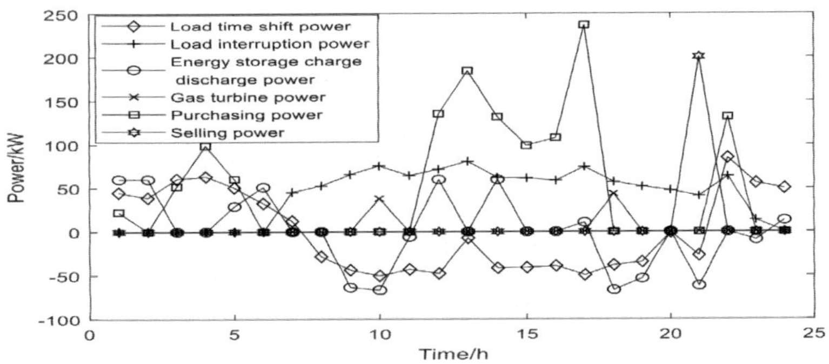

Fig 2 System hourly dispatch output curve

A comparison chart of carbon emissions and costs before and after the implementation of carbon trading is shown in Figure 3 and Figure 4.

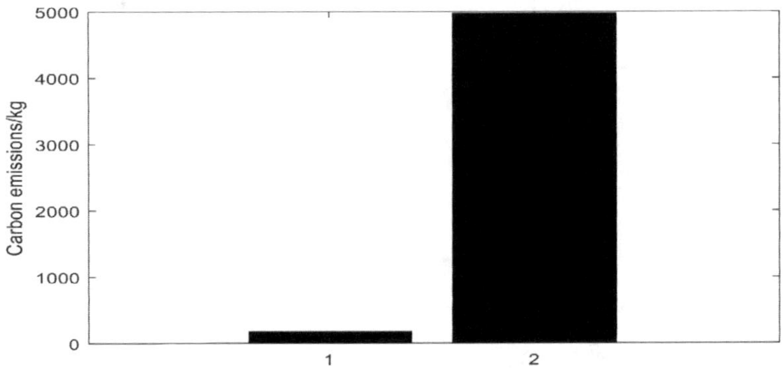

Fig 3 Comparison of carbon emissions with and without carbon trading

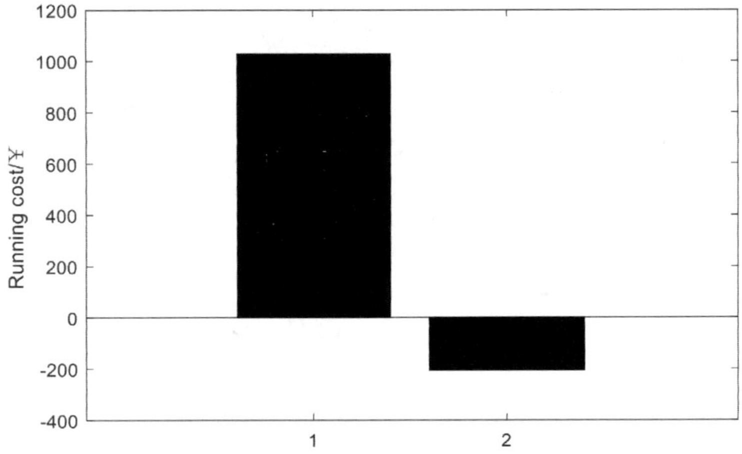

Fig 4 Comparison of operating costs with or without carbon trading

Table 1 Comparison of system daily comprehensive carbon emission cost

Operating status	carbon emission /kg	cost /￥
Before carbon trading	4997	1029.3
After carbon trading	179.7433	-206.3

It can be seen from Figure 3, Figure 4 and Table 1 that the changes in carbon emissions before and after the adoption of carbon trading are huge, and the feasibility of the algorithm and each micro power supply model in this paper is proved through simulation.

6. CONCLUSION

This paper establishes an optimal operation model with the lowest power generation cost and environmental starting cost of the microgrid. Combines various equality and inequality constraints based on carbon emissions during the operation of the microgrid, and uses the CPLEX algorithm to perform simulation solutions. Obtain the output power of the system hour and the carbon emissions before and after carbon trading and the total cost obtained a significant reduction in carbon emissions and a moderate increase in costs before and after carbon trading. The feasibility of the optimization model in this article is verified, and it has certain reference value.

References

[1] Chen X, Lv J, McElroy MB, et al. Power system capacity expansion under higher penetration of renewables considering flexibility constraints and low carbon policies[J]. IEEE Transactions on Power System, 2018, 33(6):6240-6253.

[2] Cecilia, M. VB, Biagio F, Feni A, et al. Cleaner production towards a sustainable transition[J]. Journal of Cleaner Production, 2017, 142:1-7.

[3] Vagropoulos S, Balaskas G, Bakirtzis A. An investigation of plug-in electric vehicle charging impact on power systems scheduling g and energy costs[J]. IEEE Transactions on Power Systems, 2017, 32(3):1902-1912.

[4] Jamali A, Aghaei J, Esmaili M, et al. Self-scheduling approach to coordinating wind power producers with energy storage and demand response[J]. IEEE Transactions on Sustainable Energy, 2020, 11:1210-1219.

[5] Li Y, Wang C, Li G. Optimal scheduling of integrated demand response-enabled integrated energy systems with uncertain renewable generations: a Stackelberg game approach[J]. Energy Conversion and Management, 2021, 235:113996.

[6] Rezvani A, Gandomkar M, Izadbakhsh M, et al. Environmental/economic scheduling of a micro-grid with renewable energy resources[J]. Journal of Cleaner Production, 2016, 87:216-226.

[7] Hu M, Wang Y, Xiao J, et al. Multi-energy management with hierarchical distributed multi-scale strategy for pelagic islanded microgrid clusters[J]. Energy, 2019, 185:910-921.

Multi-objective operation optimization method of integrated energy system considering energy saving

Xuan Yang*, Chunjie Zhu, Junhai Wang, Gang Wang, Hang Dong

State Grid Hangzhou Power Supply Company Hangzhou Zhejiang 310000, China

*Corresponding author's e-mail: 309365007yang_xuan@zj.sgcc.com.cn

Abstract. In order to improve the operating efficiency of the integrated energy system and reduce emissions, this paper proposes a multi-objective optimization method for the integrated energy system that considers energy saving. Set energy-saving goals, electric power balance, thermal power balance, operating cost constraints and other calculation equations. On the basis again, an optimization model of the integrated energy system is constructed. According to the characteristics of the constraint conditions, a penalty function is constructed, and the fuzzy satisfaction scheme is standardized to complete the operation optimization of the system. The analysis of calculation examples proves that the method in this paper can alleviate the environmental pollution problem, and can improve the operating efficiency and reduce the operating cost of the system.

1. Introduction

As an emerging energy supply mode, the regional integrated energy system can achieve a balance of energy supply and demand in a certain area. The system can integrate multiple renewable energy sources, increase the rate of renewable energy consumption, and achieve green energy supply [1-3]. In addition, the energy supply structure is diverse and the method is flexible, which can effectively realize the cascade utilization of energy and achieve the effect of energy saving and emission reduction. The integrated energy system can effectively improve the overall energy utilization efficiency and operating economy of the system through the complementary coupling between multiple energy sources [4]. The existing research on optimization design of integrated energy system is getting more and more attention, and these researches usually choose the design of CCHP system as the main research object. The energy supply structure of the integrated energy system is diverse, and it is necessary to study the composition of the structure. Compared with the innovation of capacity allocation and optimization algorithms, the existing research involves less consideration of energy-saving related issues.

Therefore, it is necessary to expand on the traditional electricity-gas combined network model, and consider adding other renewable energy parts, which is the content of this paper. For this reason, this paper proposes a multi-objective optimization method of integrated energy system considering energy saving.

2. Multi-objective operation optimization of integrated energy system

The integrated energy system is an integrated system that integrates energy production, conversion, storage and consumption. It can realize the coordination and optimization of different energy sources during the planning, construction and operation of the energy system [5]. The energy supply of the integrated energy system is shown in Figure 1.

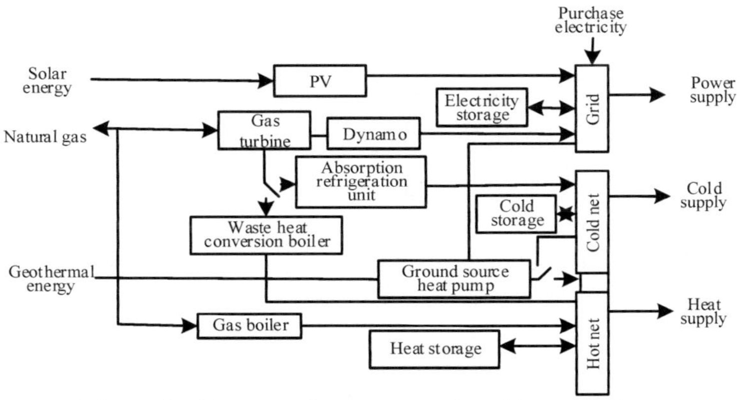

Figure 1 Schematic diagram of energy supply of integrated energy system

The main players participating in the regional integrated energy system include external energy supply companies and demand-side users. The external energy supply company is an important energy supplier for the regional integrated energy system. When the system energy supply is insufficient or the external energy price is low, it can provide the regional integrated energy system with a variety of energy required for operation. The external energy supply company involved in this article Mainly include external power companies, heating companies and gas companies.

2.1. *Energy-saving constraints*

In the capacity planning stage of the regional integrated energy system, system pollutant emissions and environmental costs are also one of the key considerations [6]. When the total amount of pollutant discharge exceeds the prescribed pollutant discharge standard, in addition to paying the basic discharge fee, additional tiered environmental transaction costs must be paid for the excess discharge. This article assumes that the electricity purchased from the grid is all thermal power, and the calculation equation for the energy-saving target is:

$$v(t) = v_k t \left(\frac{q}{q_k} \right)^2 \tag{1}$$

In equation (1), v_k is the power purchased from the grid, t is the power generation operation time, q is the output power of the emission source, and q_k is the system utilization efficiency.

The capacity planning stage of the regional integrated energy system needs to meet the rigid constraints of system energy demand, and the results of related equipment planning also need to meet the transmission constraints of the power grid, heating network, natural gas network and other energy grids in the system. Therefore, the main constraints considered in the capacity planning stage of this paper are power balance constraints and energy network transmission power constraints. Then the electric power balance calculation equation is:

$$A_w = \frac{Q\left[p(T_1 - T_w)\right]}{Q_w} \tag{2}$$

In equation (2), p is the electrical power loss, T_1 is the output power of the integrated energy system, T_w is the system DC output power, Q is the power loss, and Q_w is the exchange power between the system and the grid.

Thermal power balance mainly means that the heat generated by the system itself can meet its own needs. If it is not satisfied, the system needs to purchase a certain amount of heat from the heating network. Then the calculation equation of heat power balance is:

$$L = \delta \frac{U * \Delta t * \varpi}{H} \tag{3}$$

In equation (3), δ is the thermal load power, U is the output power of the gas boiler, ϖ is the output power of the electric boiler, and H is the heating power input to the system by the heating network.

Natural gas supply constraints mainly refer to ensuring that the real-time input system and the total amount of natural gas stored in the system can meet the real-time needs of gas turbines, gas boilers and residential users. Then the calculation equation for natural gas supply constraints is:

$$fitness\left(x_t\right) = \frac{\sum_{i=1}^{n} y\left(r'\right)_i}{q_t} \tag{4}$$

In equation (4), y is the power of the gas storage system, r' is the electricity load of residential users, and q_t is the power of the urban natural gas pipeline network inputting NG to the system.

In the regional integrated energy system, considering the safety of the energy grid, the exchange power between the system and the external energy supply system must be controlled within a certain range. Then the calculation equation of the energy network transmission power constraint is:

$$Y' = v * Y_j * \wp \tag{5}$$

In equation (5), v is the electric power delivered by the external grid to the system, Y_j is the heat power delivered by the external heating grid to the system, and $\wp$ is the natural gas power delivered by the external natural gas grid to the system.

2.2. Multi-objective operation optimization

In this research, a hierarchical design idea is adopted to design the energy supply structure of the system. The layered design framework consists of three parts: energy layer, equipment layer and demand layer. The energy layer contains different optional energy types, such as electrical energy from the grid, solar energy, wind energy, natural gas and other energy sources. The optimization model of the integrated energy system is a non-linear model that includes optimization goals, model constraints and optimization variables. The standard form of the model is as follows:

$$\Im = \frac{\sum_{i=1}^{365} (-P_{\Im(i)}) * \Delta x}{\sum_{i=1}^{365} S_{z(i)}} \tag{6}$$

In equation (7), $P_{\Im(i)}$ is the optimization variable, $S_{z(i)}$ is the target optimization coefficient, and Δx is the difference of the optimization variables.

This method takes the SP system as a reference, takes energy, economy and environment as the goals, and realizes the capacity configuration of the system. The linear weighted grouping method is used to transform the multi-objective optimization problem into a single-objective optimization problem. In order not to lose generality, the three weighting factors in this method are set to 1/3. Therefore, the objective function of IES can be obtained as follows:

$$\kappa = \mu\gamma \arg_{t\in T} \max \beta_j \alpha_j \tag{7}$$

In equation (8), t is the energy index weighting factor, β_j is the environmental index weighting factor, μ is the comprehensive optimization coefficient, and γ is the economic index weighting factor.

The NSGA-III used in this paper is a heuristic group search algorithm, and its composition mainly includes two parts: population initialization and population evolution. The NSGA-II algorithm is a fast non-dominated multi-objective optimization algorithm with an elite retention strategy. It is a multi-objective optimization algorithm based on Pareto optimal solution. The evolution of the population consists of crossover, mutation and individual selection. The flow of the NSGA-III algorithm is shown in Figure 2.

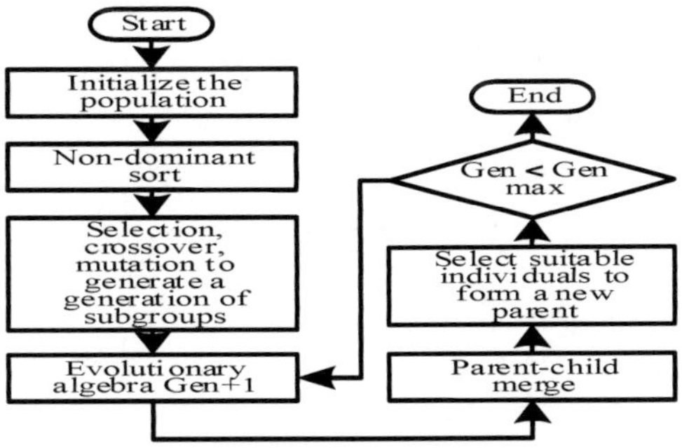

Fig. 2 The flow of the NSGA-III algorithm

Many constraints need to be met in the multi-objective power flow optimization of power systems. When any of the decision variables of power flow optimization including the active power output of the generator except the balance node, the voltage of the generator node, the adjustable transformer ratio and the parallel reactive power compensation, do not meet the constraint conditions. This article will use the boundary restriction strategy to restore the variable to the feasible space, and the repair equation is:

$$z = \begin{cases} \dfrac{\beta_j \alpha_j}{\sum\limits_{t \in T} \beta_j \Im_j}, t \in T \\ 0, t \in T \end{cases} \tag{8}$$

In equation (9), α_j is the upper limit of the decision variable, and $\Im_j$ is the lower limit of the decision variable. Then use the unconstrained optimization method to solve. Use equation (9) to standardize each non-dominated solution of the obtained Pareto front. At this time, the solution with the largest standardized fuzzy satisfaction degree is the optimal compromise solution.

$$\Im(z) = N\phi_i \frac{m(1+m)^2}{r_a(1+m)^2 - 1} \tag{9}$$

In equation (11), N represents the number of non-dominated solution sets of the Pareto frontier, m represents the number of objective functions of the optimization problem, and ϕ represents the weight coefficient of the i-th objective function. In this article, the weight coefficients are all set to 1. So far, the multi-objective operation optimization of the integrated energy system has been completed.

3. Case analysis

The integrated energy system used in this section is composed of IEEE37 calculation examples and a 14-node gas distribution system through 9 EC couplings. And from the multi-objective optimization indicators to analyze and verify the performance of the improved algorithm. The active load of the IEEE37-node standard simulation system is 283.4MW and 126.2Mvar. There are 30 system nodes, 6 generators, 24 load nodes, 9 reactive power compensation, and 4 adjustable transformers.

On the basis of the above experimental environment, calculate the corresponding integrated energy system operating cost (OC), ICES pollutant gas emissions (TE), power distribution system loss (PL), power distribution system voltage deviation sum (SVD) and distribution system Pneumatic system pressure deviation and (SPD) five attribute values. Compared with the traditional method (GSO algorithm), the results are shown in Table 1.

Tab. 1 Five attribute values corresponding to the final plan of integrated energy system optimization and dispatching

Attributes	Traditional method	Method of this paper
OC	254 $	221 $
TE	2.35 ton	1.89 ton
PL	53.12 kW	46.12 kW
SVD	1.97 p.u.	1.75 p.u.
SPD	0.063 p.u.	0.034 p.u.

It can be seen from Table 1 that the method in this paper is lower than the traditional method in terms of the five attributes. The operating cost of the integrated energy system optimized by the method in this paper is 221 $, and the pollutant gas emissions are significantly reduced. It can be seen that the energy-saving effect of the method in this paper is significant.

On this basis, under a variety of complex conditions, the energy storage system compensates the system according to different operating goals, and finally realizes economic and stable operation. Table 2 shows the situation of the microgrid purchasing and selling electricity from the distribution network under the fuzzy control strategy and the non-input control strategy of the energy storage system. The analysis is carried out from the monthly average data of the three typical weather types.

Tab. 2 System electricity purchase/sale statistics

Weather type	Control Strategy	Buy electricity/kWh	Sell electricity/kWh
Sunny day	Method of this paper	83	41.3
	Traditional method	102	86.5
Cloudy day	Method of this paper	98	4.3
	Traditional method	126	28.1
Rain day	Method of this paper	136	0
	Traditional method	179	2.3

It can be seen from Table 2 that adding the control strategy of the method in this paper can reduce the electricity sold to the distribution network during the peak power generation period of the integrated energy system. In the low power generation period, the demand for power purchase is reduced, and the loss of the power transmission process is reduced. Thereby reducing the impact on the power quality of the power grid, and improving the stability of the system operation.

The comparison of the power change of the energy storage system between the method in this paper and the traditional method is shown in Figure 3.

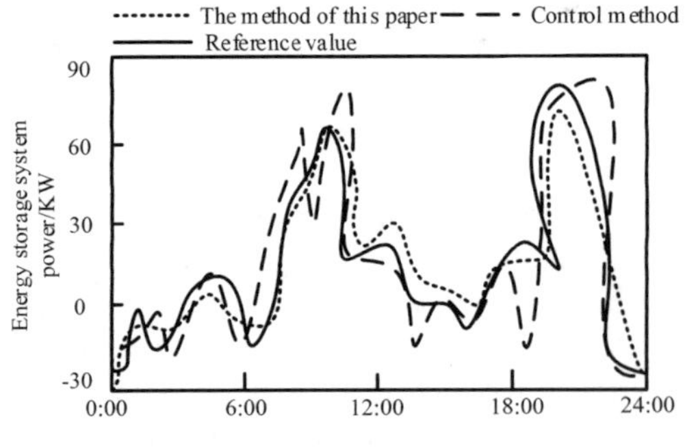

Fig. 3 Power change trend diagram of energy storage system

It can be seen from Figure 3 that the optimization results obtained by using the optimal control model are roughly consistent with the trend of the reference value. Compared with not using the optimal control model, the control method fluctuates greatly, and the power fluctuation of the method in this paper is significantly reduced, which is conducive to the stable operation of the system. In terms of energy storage characteristics, the cost of energy storage and energy conversion have an important impact on the cost.

4. Conclusion

With the integration and development of large-scale micro-grid, distributed renewable energy, etc. and natural gas power generation, as well as the continuous growth of regional cooling and heating load demand, the requirements for energy conservation are getting higher and higher. This paper studies the multi-objective optimization and coordinated operation of the system. Set energy-saving goals, electric power balance, thermal power balance, natural gas supply constraints, transmission power constraints and operating cost constraints equation functions, according to the constraints, construct an optimization model of the integrated energy system. According to the characteristics of the constraint conditions, a penalty function is constructed to standardize the fuzzy satisfaction plan. Through the analysis of the calculation examples, it can be seen that the method in this paper can alleviate the environmental pollution problem and reduce the operating cost of the system.

Acknowledgements

This work was supported by the science and technology project of the State Grid Zhejiang Electric Power Co. LTD. （B311HZ21000R）

References

[1] Wu T, Bu S, Wei X, et al, (2021) Multitasking multi-objective operation optimization of integrated energy system considering biogas-solar-wind renewables[J]. Energy Conversion and Management, 229: 113736.

[2] Zhou S, Pei R, Xie J, et al, (2021) Integrated energy system operation optimization with gas linepack and thermal inertia[J]. IET Renewable Power Generation, 15(15): 3743-3760.

[3] Wang Y, Wang Y, Huang Y, et al, (2019) Planning and operation method of the regional integrated energy system considering economy and environment[J]. Energy, 171: 731-750.

[4] Li J, Niu D, Wu M, et al, (2018) Research on battery energy storage as backup power in the operation optimization of a regional integrated energy system[J]. Energies, 11(11): 2990.

[5] Behzadi A, Gholamian E, Ahmadi P, et al, (2018) Energy, exergy and exergoeconomic (3E) analyses and multi-objective optimization of a solar and geothermal based integrated energy system[J]. Applied Thermal Engineering, 143: 1011-1022.

[6] Juanwei C, Tao Y, Yue X, et al, (2019) Fast analytical method for reliability evaluation of electricity-gas integrated energy system considering dispatch strategies[J]. Applied Energy, 242: 260-272.

Energy storage capacity planning method for urban integrated energy system considering uncertainty

Xibin Wu[1a*], Yi Lou[1], Xiaodong Wang[1]

[1]State Grid Wuxi Power Supply Company, Wuxi, Jiangsu, 214000, China

[a*]c154597918@126.com

Abstract—The conventional energy storage capacity planning method of urban integrated energy system has the problem of fuzzy coupling characteristics, which leads to the small energy storage capacity. A planning method of energy storage capacity of urban integrated energy system considering uncertainty is designed. Calculate the expected output value, to determine the optimal control targets, considering the uncertainty of power load forecasting, load combination forecast load density and space distribution, identification of urban comprehensive energy system coupling characteristics, optimize the distribution ratio of different energies, capture of power and steam energy, energy storage capacity planning design method. Experimental results: The average energy storage capacity planning method of the urban integrated energy system in this paper is 103.844MWh, 91.657MWh and 91.152MWh compared with the other two methods, indicating that the designed urban integrated energy system energy storage capacity planning method is more perfect after fully considering the uncertainty factors.

1. Introduction

Integrated energy system realizes the coordinated planning and construction of multiple types of energy, breaks the traditional single energy supply mode, and realizes the unified regulation and control operation mode of multiple energy sources, which is the development trend of the future energy system [1-3]. Increasing scarcity due to the overuse of fossil energy. Humans have to think about how to use more renewable energy, such as wind energy, solar energy, geothermal energy. The coordinated development of renewable energy and conventional energy has become the development path of my country's energy production. The core of the integrated energy system is electricity, which simultaneously integrates various energy forms such as electric energy, wind energy, photovoltaic energy and biomass energy, covering multiple energy supply links such as power generation, transmission, distribution, electricity consumption and energy storage [4]. However, most renewable energy belongs to distributed energy and should be utilized in the form of distributed system. During the operation of the integrated energy system, the addition of new energy has changed the traditional energy supply structure. It includes a variety of network structures such as electric power, natural gas, heat energy and transportation, as well as a variety of load types such as electric load, gas load, cooling load and heat load, as well as the coupling form between multiple energy sources, such as electricity to gas, cogeneration and gas boiler. Cooperate with computer, artificial intelligence and other different fields to develop in the direction of fine unity, combined with practical application and system planning and operation, become the current development trend of the energy industry[5]. With the speeding up of urbanization process, energy marketization, to alleviate the social contradiction of unbalanced development in China. We have to break through the development of new energy ideas.

The development of new energy technology innovation, set up energy comprehensive urban planning system, realize the energy development of fusion.

2. Energy storage capacity planning method for urban integrated energy system considering uncertainty

2.1. Calculating expected output value

Composite energy storage becomes an important development direction of energy storage technology in the future because of its complementary advantages to various energy storage media. In the process of transforming wind energy into electricity, wind turbine is the core equipment. The key to determine the desired suppression result lies in the selection of time window and expected output power. Wind turbine can be divided into two types according to principle, one is asynchronous generator, the other is synchronous generator. The selection of time window should take into account the actual situation of dispatching and wind power fluctuation. This paper takes the minimum variance of the target power offset as the objective to construct the desired suppression result, and the objective function is:

$$g = \min\left[\frac{1}{(\varepsilon - 1)}\right] \times e \tag{1}$$

In formula (1), ε represents the start time of the time window, and e represents the wind power of the selected time window. Wound asynchronous generator can also be called doubly-fed asynchronous motor, which is a very common motor in wind power generation system. Adjusting the power according to the actual situation, and undertake the main suppression task, which is defined as:

$$D_v = l\Delta\frac{1}{h}(1+v) \tag{2}$$

In formula (2), l represents charge and discharge power, h represents high-frequency component, and v represents reduction factor. On the basis of Formula (2), it can be concluded that the strategies to undertake the main overall suppression task are as follows:

$$H = l\Delta\left(1+\frac{1}{v}\right) \times \delta \tag{3}$$

In formula (3), δ represents the offset of wind power output, and the meanings of other quantities are the same as those in Formula (2). Thanks to power devices, it can constantly change its speed, which allows it to maximize wind power. In addition, the external power device can independently change the stator output reactive power and electromagnetic torque. The small capacity of the power device also results in the low cost, which only needs 30% of the output power. Under the condition of known operating cost and wind curtailment cost, the expression formula of insufficient punishment cost is obtained:

$$P = S + L + \frac{1}{\phi} \tag{4}$$

In formula (4), S represents the maintenance cost of a single group of different energy storage types, L represents the number of installed energy storage battery groups, and ϕ represents the battery capacity. The biggest disadvantage of his ion battery is poor safety and stability, but it has the advantages of high energy density, long cycle service life, and in recent years his ion battery's high current charge and discharge capacity has been greatly improved.

2.2. Power load forecasting considering uncertainty

Space load forecasting, also known as space load forecasting, refers to the prediction of the size and location of the future power load in the power supply area. The general definition of big climbing event in electrical industry is that the variation of photovoltaic power output exceeds a certain limit in unit time. If the time interval is short, the possibility of climbing events within the interval can be ignored, then the calculation formula of photovoltaic power generation climbing rate is as follows:

$$\varphi = \frac{\left|R(t+\Delta g) - R(t)\right|}{\Delta g} \tag{5}$$

In formula (5), R represents the output power of photovoltaic power generation, t represents the allowed climbing power per unit time, and g represents the output power of photovoltaic power generation. In the past, the energy transmission mode of the system was independent and the transmission relationship between different energy sources was not considered. However, the operation mode of the energy network of the current system is mutual influence and interaction. The definition of chance constrained programming is to obtain a series of distribution curves to describe the uncertainty by fitting and analyzing the existing experience on the basis of historical data statistics. If the output power of photovoltaic power generation suddenly decreases and the climbing rate is relatively large, it is necessary to immediately discharge to compensate for the reduction of system power, and the specific expression is as follows:

$$T = \frac{1}{\gamma}(\Delta g - \eta) \tag{6}$$

In formula (6), γ represents rated charge and discharge power, and η represents continuous discharge time. Correlation refers to the relationship between the existence of objective phenomena, that is, each value of independent variable is affected by random factors, and its corresponding value is non-deterministic. Recombination sequencing method is used to improve the uncertainty of distributed power output. The fluctuation of distributed wind power generation is analyzed. Based on the above description, complete the steps of power load forecasting.

2.3. Identifying coupling characteristics of urban integrated energy system

Heterogeneous energy flow, coupled with electricity, heat, cold and gas, is a prominent feature and research focus of urban integrated energy system. The analysis of heterogeneous energy flow has reference value for monitoring and control, optimal scheduling and safety evaluation of the system. The device works by using an array of cells to convert solar energy into electricity that can be stored in a battery or used directly by the user. Photovoltaic power generation is not only related to the energy conversion efficiency of photovoltaic panels, but also related to illumination and external temperature. The mathematical model of photovoltaic is as follows:

$$G = f_c z \times \frac{1}{\left[1 + (z_c - \mu)\right]} \tag{7}$$

In formula (7), μ represents the energy conversion efficiency of photovoltaic power output, z represents actual radiation intensity, c represents rated light intensity, and f represents temperature power coefficient. According to the basic principle of steady-state heat transfer, the heat loss per unit length of pipeline can be expressed as follows:

$$\Delta q = 2\pi \frac{Q - Q_\sigma}{K} \tag{8}$$

In formula (8), Q represents the temperature of the heat medium in the pipeline, K represents the total thermal resistance of the pipeline per kilometer, and σ represents the average temperature of the surrounding medium. In order to adapt to the continuous development of urban order, the energy system must constantly adapt to the new forms. In the integrated energy system, the core technology of energy conversion is energy conversion technology. That is, according to the characteristics of energy itself, some preferential price or excess energy can be converted into the energy required by other users, which can be used to meet the load supply and reduce the system expenditure.

2.4. Designing storage capacity planning methods

Large-scale application of energy storage capacity in integrated energy system is a necessary prerequisite for renewable energy to access power system. With the continuous request for clean and efficient energy utilization in China, the application of renewable energy in the integrated energy system is more and more extensive, but in practice, the intermittent and fluctuating characteristics of renewable energy have a great impact on the internal power grid of the energy system. Thermal efficiency is usually used in power generation systems to express the overall efficiency, that is, the fuel consumed per kilowatt-hour of electricity. However, CHP systems also produce effective heat in the process of generating electricity. The gas boiler uses natural gas as fuel to output heat energy to meet the demand of heat load. Because the gas boiler heating has the advantages of high combustion efficiency, almost no smoke and dioxide emissions, high economy and environmental protection, and gradually replace the traditional coal-fired boiler. The simple structure diagram of the gas boiler is shown in Figure 1:

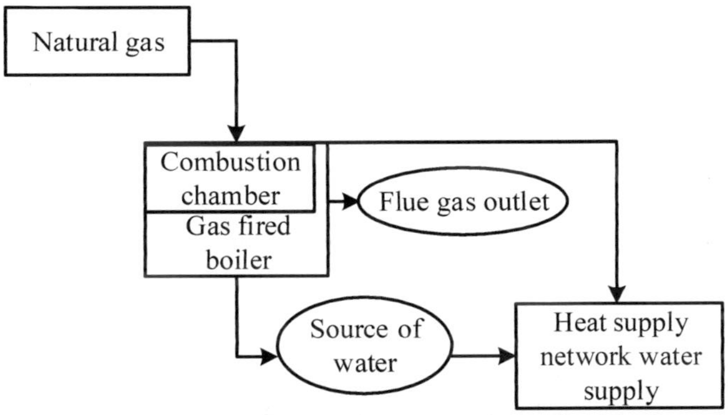

Fig. 1 Working principle diagram of gas-fired boiler

As can be seen from Figure 1, the external natural gas is fed into the combustion chamber and mixed with air to generate combustion heat. The heat generated by combustion heats the hot water in the boiler, while the waste heat steam is discharged from the airport, and the heated water in the boiler is sent to the heat network for use after passing through the water distributor. Cogeneration in small and medium-sized regions. Such cogeneration units typically have a capacity of 1MW to 10MW, and are usually heated and supplied to a small area of the local area by such small and medium capacity cogeneration units with a heating network of no more than 1000 customers. In the process of scene aggregation, the total number of scenes at a certain moment is obtained by permutation and combination of scenes among all uncertain variables. According to the principle of permutation and combination, its value is equal to the product of the total number of uncertain variables and a group of related uncertain variables. Load areas where users are relatively concentrated, such as some residential quarters, schools, commercial buildings, etc. This form of small and medium central heating is usually composed of multiple units. Some of them carry the base load, and some are used for peak shaving. The total number of scenes at one point in time would be very large. Some thermal power stations also have the ability to convert heat into cold volume through absorption chiller, so as to provide users with cold volume to make full use of the heat energy produced. If a set of correlated uncertain variables is considered, the number of scenarios for this set of variables alone can reach thousands at a time. And this set of uncertain variables still needs to be permuted and combined with other uncertain variables. In addition to generating electricity to meet local loads, such small regional cogeneration systems can also generate surplus electricity to supplement the larger grid. Based on the above description, complete the steps of designing the storage capacity planning method.

3. Experimental analysis

3.1. Experimental preparation

A specific comprehensive energy system park is selected as the research object. In this park, the load demand includes electricity load, heat load and cooling load, and there are power grid and natural gas on the external supply side to provide the required energy for the park. The installed capacity is 90MW and 450MW respectively. Power generation equipment is photovoltaic, fan, gas turbine. Heating equipment includes waste heat boiler, gas boiler. The cooling equipment includes the absorption chiller and electric chiller. The pv power generation permeability is 16.7%, and the specific installed capacity of the generator set and the regulating rate of the AGG regulating set range from 12 to 100MW. MATLAB is used to simulate the control process. The computer is configured with i7CPU, 4G memory, and energy storage equipment including battery. The operation strategy of the park is to ensure that users' demand load (cold, heat and electricity) is satisfied first. With Windows10 operating system, the average running time of each control cycle in the running process is less than the control time 1 min. On this basis, the model of "surplus power online, power shortage purchase" is adopted, the service life of the device is set as 20 years, with the accuracy of one hour, and the capacity expansion planning model of the comprehensive energy system proposed above is used for research.

3.2. Experimental result

In the experiment, the energy storage capacity planning method of urban comprehensive energy system based on cluster analysis and the energy storage capacity planning method of urban comprehensive energy system based on genetic algorithm are selected for experimental comparison with the energy storage capacity planning method of urban comprehensive energy system in this paper, The larger the energy storage capacity of the three urban integrated energy system energy storage capacity planning methods, the better the energy storage performance. The experimental results are shown in table 1-3:

Table 1 Tracking 90% energy storage capacity (MWh)

Number of experiments	Energy storage capacity planning method of urban comprehensive energy system based on cluster analysis	Energy storage capacity planning method of urban integrated energy system based on genetic algorithm	The planed method of energy storage capacity of urban comprehensive energy system in this paper
1	116.593	123.114	136.794
2	121.506	116.589	139.554
3	118.478	121.057	141.002
4	122.331	118.669	134.978
5	119.054	122.313	137.466

Table 2 Tracking 70% energy storage capacity (MWh)

Number of experiments	Energy storage capacity planning method of urban comprehensive energy system based on cluster analysis	Energy storage capacity planning method of urban integrated energy system based on genetic algorithm	The planned method of energy storage capacity of urban comprehensive energy system in this paperc
1	88.947	85.201	96.331
2	87.516	86.339	95.847
3	85.337	87.115	93.616
4	86.414	85.914	94.588
5	87.218	86.318	96.373

Table 3 Tracking 50% energy storage capacity (MWh)

Number of experiments	Energy storage capacity planning method of urban comprehensive energy system based on cluster analysis	Energy storage capacity planning method of urban integrated energy system based on genetic algorithm	The planned method of energy storage capacity of urban comprehensive energy system in this paper
1	63.641	69.485	77.818
2	67.845	67.511	79.623
3	68.194	68.320	75.488
4	71.231	64.119	76.948
5	70.548	65.214	81.236

According to Table 1, the average energy storage capacity of the planned method for urban integrated energy system in this paper is 137.959 MWh, 119.592 MWh and 120.348 MWh compared with the other two methods. According to Table 2, the average energy storage capacity of the planned method of urban integrated energy system in this paper is 95.351MWh, 87.086MWh and 86.177mWh compared with the other two methods. According to Table 3, the average energy storage capacity of the planned method of urban integrated energy system in the paper is 78.223mWh, 68.292MWh and 66.930MWh compared with the other two methods. The storage capacity of the planned method of urban integrated energy system in the text is larger.

4. Conclusion

The design of the urban integrated energy system energy storage capacity planning method, considering the photovoltaic, wind power output, electricity, heat, natural gas load uncertainty, and not considering the correlation of uncertain variables, through energy management optimization of the comprehensive energy system energy storage plan. The influence of scenario reduction on the planning result of energy storage capacity of urban integrated energy system is studied. When considering the correlation between electrical and thermal loads, the joint probability density function may also obey other distributions. The influence of uncertain variables on energy storage capacity planning of integrated energy system can be further studied in the future.

Acknowledgments

The study was supported by "The Science and Technology Project of Jiangsu Electric Power Company (J2021135)".

References

[1] Kotowicz J, Uchman W. (2021) Analysis of the integrated energy system in residential scale: photovoltaics, micro-cogeneration and electrical energy storage[J]. Energy.

[2] Wei S, Gao X, Zhang Y, et al. (2021) An improved stochastic model predictive control operation strategy of integrated energy system based on a single-layer multi-timescale framework[J]. Energy, 2021, 235.

[3] Yan C A, Had B, Nf C, et al. (2021) Multi-objective Bat Optimization for a Biomass Gasifier Integrated Energy System Based on 4E Analyses[J]. Applied Thermal Engineering.

[4] Liu H, Tian S, X W ang, et al. (2021) Optimal Planning Design of a District-level Integrated Energy System Considering the Impacts of Multi-dimensional Uncertainties: A Multi-objective Interval Optimization Method[J]. IEEE Access, PP(99):1-1.

[5] Zhong J, Li Y, Cao Y, et al. (2020) Stochastic optimization of integrated energy system considering network dynamic characteristics and psychological preference[J]. Journal of Cleaner Production, 275:122992.

[6] Xza B, Yza B. (2020) Environment-friendly and economical scheduling optimization for integrated energy system considering power-to-gas technology and carbon capture power plant[J]. Journal of Cleaner Production, 276.

Ultra-short-term Forecast of Multi-energy load for Integrated Energy System based on Attention Mechanism and BiLSTM

Zhijie Zheng[1a], Liang Feng[1b], Rui Liu[1c*], Xian Wang[1d] and Yi Sun[1e]

[1]Economic & Technology Research Institute of State Grid Shandong Electric Power Company, Shandong 250021, China.

[a]85182664@163.com, [b]243268141@qq.com, [c*]liu_rui@zju.edu.cn, [d]stillalice10@163.com, [e]442028471@qq.com

Abstract—Under the background of the continuous promotion of the construction of the integrated energy system (IES), accurate and rapid multi-type energy load prediction is of great significance to the safe and economic operation of IES. Traditional forecasting models are difficult to model complex fluctuations of multi-energy loads, cannot consider the correlation among different types of energy loads. For this reason, this paper presents an ultra-short-term prediction method for multi-energy load of integrated energy system based on attention mechanism and bi-directional long short-term memory (BiLSTM) network. It uses deep learning model to learn the feature information of time series, and uses attention mechanism and BiLSTM to improve the effectiveness of features. The validity of the algorithm is validated in the example section. The results show that the proposed method reduces the prediction error effectively compared with traditional methods.

1. Introduction

The integrated energy system (IES) often integrates many forms of energy, such as cooling, heating and electric energy. The optimal operation of IES needs to coordinate the use of complementary characteristics among different energy sources to meet the energy requirement of users. Therefore, accurate and efficient prediction of multi-energy load of users is of great significance to the safe, stable and economic operation of IES.

For the prediction of multi-energy load such as heating, cooling, electric energy, it can be divided into two categories: the traditional machine-learning-based method and the deep-learning-based method. For example, ref [1] proposed a NARX-based method to realize the forecast of heating, cooling, and electrical load. In [2], ANN, SVMN, fuzzy network were applied to prediction of energetic performance, and the applicability of different methods was analyzed. Ref [3] proposed a medium-term forecasting of cold, electric and gas load in multi-energy system based on VAR model.

For the deep-learning-based method, ref [4] proposed a multi-energy load prediction method of IES based on LSTM network and multi-task learning. Ref [5] proposed a short-term forecast method of multi-energy load of electrical heating and cooling based on deep LSTM and RNN. Both of them make full use of the modeling ability of LSTM for time series data. A GBRT-based integration method is proposed in [6] to improve prediction accuracy by combining the output of multiple DNNs. Ref [7] first analyzed the correlation among multi-energy loads in combined cooling heating and power (CCHP) system, and then predicted them through LSTM-RNN.

Previous researches have shown that attention mechanism and BiLSTM can improve the feature extraction ability of deep neural network [7]. Therefore, this paper proposes a multi-energy load forecasting method of IES based on attention mechanism and BiLSTM. Firstly, the attention mechanism is used to improve BiLSTM, and a deep neural network model is constructed. Then, the external characteristics such as historical energy consumption and weather data are used as inputs to train DNN. The trained model will be able to realize the accurate prediction of ultra-short-term multi-energy load.

The organization of this paper is as follows: in the second section, the multi-energy load ultra-short-term forecasting based on BiLSTM and attention mechanism is proposed. In the third section, the proposed method is verified on the actual data set. In the fourth section, the characteristics of the proposed method are summarized.

2. Ultra-short-term prediction method of multi-energy load for IES

The flow of the ultra-short-term prediction method of multi-energy load for IES proposed in this paper is shown in Figure 1. First of all, the historical data of four energy consumption types in the IES are extracted, cleaned and sorted out. Then the weather data are extracted together as the input.

Through the BiLSTM layer, the change trend and correlation of historical data of multi-energy load are modeled, and the influence of meteorological factors is considered. The attention level is used to further identify the key components in the high-level features, so as to improve the effectiveness of the features. Finally, the ultra-short-term forecast results of 4 types of load are output simultaneously through the fully connected layer.

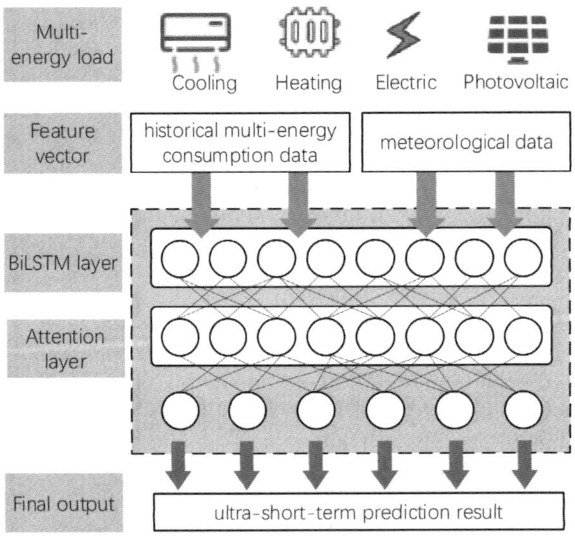

Fig.1 The prediction process

2.1. *Feature input*

The multi-energy load dataset used in this paper is provided by Campus Metabolism Project in Arizona State University [8]. We select the energy consumption data of the building named Barrett in the Honors College, its location is as shown in Figure 2. Specifically including the cooling, heating, electric and photovoltaic data in 2019. The historical data is shown in Figure 3. The sampling rate is 15 minutes. Other meteorological factors used include apparent temperature, dew point, cloud cover, pressure, uv index and visibility.

The goal of ultra-short-term prediction is to output the multi-energy load value at the next sampling time. According to existing research, we select the historical data of 60 minutes before the target time and the corresponding meteorological prediction characteristics of the target time as input features. So the dimension of input features is 4×4 (historical data) + 16 (meteorological factors) = 32.

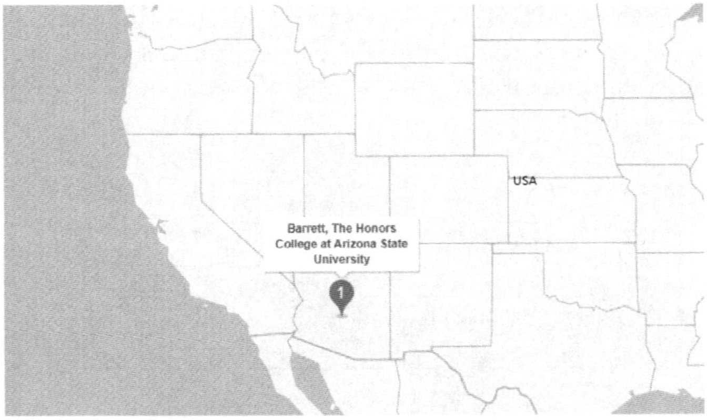

Fig.2 Location of the target building

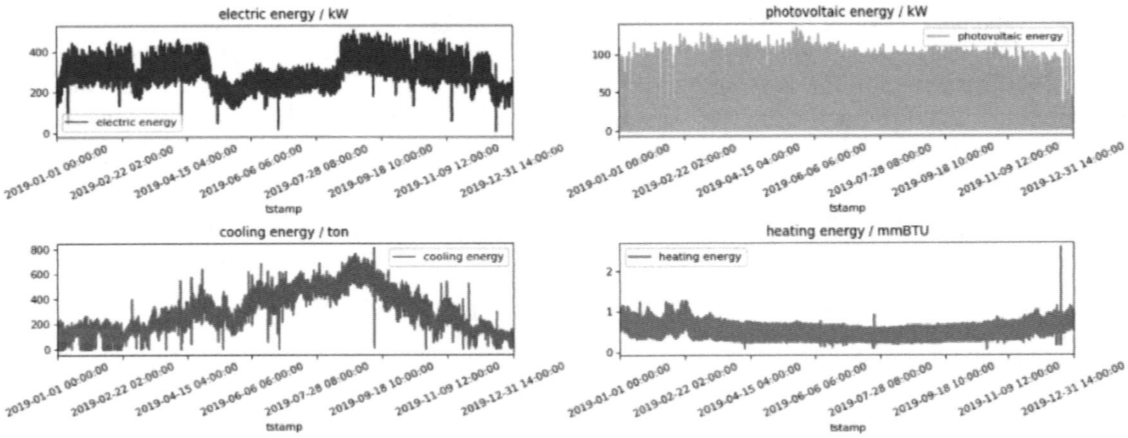

Fig.3 Multi-energy consumption in 2019

2.2. *BiLSTM model and attention layer*

First we introduce the LSTM model, as shown in Figure 4, LSTM has forget gate, input gate and output gate [9]. The first forget gate f_t is responsible for forgetting part of the unit state information through sigmoid function. The second input gate i_t controls the information that can be added to the LSTM unit. The updated candidate state is obtained by combining the past memory information with the present state information. The third output gate o_t controls the current output. The formulas involved are listed as follows.

$$f_t = \sigma(W_f \cdot [h_{t-1}, x_t] + b_f) \tag{1}$$

$$i_t = \sigma(W_i \cdot [h_{t-1}, x_t] + b_i) \tag{2}$$

$$\tilde{C}_t = \tanh(W_C \cdot [h_{t-1}, x_t] + b_C) \tag{3}$$

$$C_t = f_t * C_{t-1} + i_t * \tilde{C}_t \tag{4}$$

$$O_t = \sigma(W_O \cdot [h_{t-1}, x_t] + b_O) \tag{5}$$

$$h_t = O_t * \tanh(C_t) \tag{6}$$

Where W_f, W_i, W_o and W_C are weight matrices for the corresponding inputs of the network activation functions, b_f, b_i, b_o and b_C are bias for the corresponding inputs of the network activation functions, σ(·) is a sigmoid function, while [·] represents the connection of two vectors, symbol * represents element-wise multiplication.

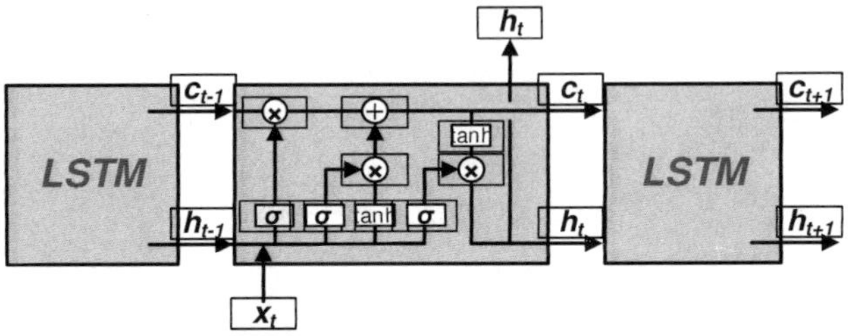

Fig. 4 Location of the target power station

Bi-directional LSTM means that the results are calculated once from front to back, and once from back to front, respectively, and stitched together. This is because one-way LSTM can only infer the information behind from the previous information, and the closer to the prediction point, the more important the time information is. Attention mechanism for processing sequential data that considers the context for each timestamp. Attention mechanisms can be used to assign significant weight to the representation of input features, which can determine the factors most relevant to the output while ignoring noise and redundancy in the input. In this paper, it is implemented by package *keras-self-attention*, and the detailed principle will not be described here.

2.3. Deep neural network structure based on BiLSTM and attention layer

Based on the BiLSTM model and attention layer, we proposes a deep neural network structure as shown in Table 1.

Table 1 Deep neural network structure

Name	Parameter
Input layer	dimension: 32
LSTM layer	LSTM unit number: 128
Dropout 1	0.2
Bi-directional LSTM layer	BiLSTM unit number: 64
Dropout 2	0.1
SeqSelfAttention layer	Attention unit number: 16
Flatten layer	----
Dense 1	Dense unit number: 4
Output layer	dimension: 4
Learning rate	0.01
Optimizer	Adam
Loss function	MAE
Epoch	50
Parammeters	185,893

It can be seen from Table 1 that the input of the proposed network includes the historical power of 4 types of energy consumption in the past 60 minutes (dimension 4×4) and the meteorological information of forecast time point (dimension 16), so the input has 32 dimensions. The proposed network structure can consider the historical data of different data, and can simultaneously predict the ultra-short-term load of cooling, electric, heating, photovoltaic energy, so the output dimension is 4. Except for BiLSTM layer and attention layer, we also use dropout layer between layers to prevent over fitting [10]. For the training process, we use the mean absolute error (MAE) index as loss value, and use Adam as the optimizer for training.

3. Case study

3.1. Experimental settings

We randomly divided the training set and the test set according to the ratio of 9:1, and prepared 4 groups of experiments to avoid the interference of random factors.

The comparison methods include LSTM, BiLSTM and support vector machine regression (SVR) with RBF kernel. The purpose of the first 2 groups of experiments is to verify that BiLSTM and attention mechanism can work together to improve the accuracy of prediction. SVR is a representative of traditional machine learning methods. Our experimental environment is AMD Ryzen 3700X (8 cores, 16 threads), 32GB DDR4 RAM. The proposed deep learning method is realized by *Keras*, and the traditional method SVR is based on *Sklearn* package [11].

3.2. Experiment result

As mentioned above, we repeated 4 experiments and calculated the forecast errors for different types of loads. Including 2 error measurement indicators, MAE and root mean squared error (RMSE), as shown in Table 2.

Table 2 Formatting sections, subsections and subsections

Substation		BiLSTM+Attention		BiLSTM		LSTM		SVR	
		MAE	RMSE	MAE	RMSE	MAE	RMSE	MAE	RMSE
1	Electric	38.051	60.236	49.258	66.980	36.349	62.846	50.495	64.378
	Cooling	25.059	31.137	23.630	28.912	19.197	26.152	27.845	47.011
	Heating	22.783	34.264	26.584	38.929	29.501	41.249	0.171	0.198
	Photovoltic	44.544	67.524	47.220	71.973	47.239	70.274	137.494	163.873
2	Electric	7.067	11.966	10.106	17.049	12.626	18.803	50.128	63.444
	Cooling	5.367	7.526	6.069	8.550	11.863	20.274	28.185	47.413
	Heating	5.553	7.753	3.657	5.956	6.037	10.858	0.172	0.203
	Photovoltic	4.392	6.662	35.297	45.633	42.855	53.912	137.979	163.923
3	Electric	20.307	28.805	22.247	28.531	20.357	27.540	50.389	64.376
	Cooling	10.639	14.906	26.497	38.641	28.781	37.615	29.010	48.127
	Heating	37.392	54.630	39.332	54.164	33.319	50.475	0.170	0.201
	Photovoltic	20.217	25.423	15.804	23.421	17.314	23.967	137.483	163.873
4	Electric	12.362	16.852	13.184	18.050	29.423	41.869	50.316	63.560
	Cooling	39.964	58.217	28.215	48.983	43.124	67.090	28.786	48.265
	Heating	15.320	23.941	0.917	1.273	19.039	27.094	0.170	0.200
	Photovoltic	26.564	38.621	16.564	28.202	5.356	6.409	135.766	161.580
	Average	20.974	30.529	22.786	32.828	25.149	36.652	54.035	68.789

It can be seen from Table 2 that the proposed BiLSTM-attention-based method has best performance in both error indexes on average. Compared with LSTM and BiLSTM, the proposed method has smaller prediction error. On the one hand, the proposed method can further explore the correlation between different types of historical data of multi-energy loads; on the other hand, through the attention mechanism, the features learned by DNN are more effective.

For different types of energy consumption data, it can be seen from Table 2 that the error of the proposed method is lower than the other 3 comparison methods in most cases. However, for heating load, the error of SVR is always the lowest, which may be due to the large difference between the range of heating load and the other 3 types of data. However, this paper does not carry out normalization treatment, resulting in the poor attention of DNN model on heating load. As far as the photovoltaic prediction results are concerned, the LSTM model is more effective in some of the prediction results, but the average error of the proposed method is the smallest in general.

4. Conclusion

In this paper, an ultra-short-term forecast method of multi-energy load for IES based on attention mechanism and BiLSTM is proposed. The input eigenvector is constructed by historical data of multi-energy loads and external meteorological factors. The temporal autocorrelation and the correlation between different types of energy are modeled by BiLSTM network, and the attention mechanism is further used to improve the effectiveness of the obtained features. The effectiveness of the proposed algorithm is demonstrated by the comparison of several methods in the case study section.

Acknowledgments

This work was financially supported by the Science and Technology Project of SGCC [grant numbers SGSDJY00GPJS2000248].

References

[1] Powell K.M., Sriprasad A., Cole W.J., Edgar T.F. (2014) Heating, cooling, and electrical load forecasting for a large-scale district energy system. Energy, 74: 877–885.

[2] Shahsavar A., Bagherzadeh S.A., Afrand M. (2021) Application of artificial intelligence techniques in prediction of energetic performance of a hybrid system consisting of an earth-air heat exchanger and a building-integrated photovoltaic/thermal system. Journal of Solar Energy Engineering. Transactions of the ASME, 143(5).

[3] Yujie L., Xiaoling Y., Jieyan X., Zheng C., Fei M., Haoming L. (2018) Medium-term forecasting of cold, electric and gas load in multi-energy system based on VAR model. In: 13th IEEE Conference on Industrial Electronics and Applications (ICIEA). pp. 1676–1680.

[4] Sun Q., Wang X., Zhang Y., Zhang F., Zhang P., Gao W. (2021) Multiple Load Prediction of Integrated Energy System Based on Long Short-term Memory and Multi-task Learning. Automation of Electric Power Systems, 45(5): 63–70.

[5] Liu E., Wang Y., Huang Y. (2020) Short-term forecast of multi-load of electrical heating and cooling in regional integrated energy system based on deep LSTM RNN. In: IEEE 4th Conference on Energy Internet and Energy System Integration (EI2), Wuhan. pp. 2994–2998.

[6] Xuan W., Shouxiang W., Qianyu Z., Shaomin W., Liwei F. (2021) A multi-energy load prediction model based on deep multi-task learning and ensemble approach for regional integrated energy systems. International Journal of Electrical Power and Energy Systems, 126.

[7] Zhou P., Shi W., Tian, J., Q Z., Xu, B. (2016) Attention-based bidirectional long short-term memory networks for relation classification. In: Proceedings of the 54th Annual Meeting of the Association for Computational Linguistics. pp. 207–212.

[8] Arizona State University. (2021) Campus Metabolism. http://cm.asu.edu/.

[9] Hochreiter S., Schmidhuber J. (1997) Long Short-Term Memory. Neural Computation, 9(8): 1735–1780.

[10] Srivastava N., Hinton G., Krizhevsky A., Sutskever I., Salakhutdinov R. (2014) Dropout: A simple way to prevent neural networks from overfitting. The Journal of Machine Learning Research, 15(1): 1929–1958.

[11] Kingma D., Ba J. (2015) Adam: A method for stochastic optimization", International Conference on Learning Representations (ICLR).

Daily streamflow forecasting using hybrid long short-term memory model

Mengfei Xie[1], Bangcan Wang[1], Shuang Zhu[2*], Gaoquan Ma[1], Zhelin Yang[1], Bin Liu[1], Yugong Jia[1]

[1] Kunming Power Exchange Center, Kunming 650011

[2]China University of Geosciences, Wuhan 430074

[*]Corresponding author's e-mail: Shuang Zhu: zhushuang@cug.edu.cn

Abstract. Characterized by heterogeneity, complexity and non-stationary, streamflow forecasting has always been a challenge in hydrological sciences. In this study, a multiscale wavelet decomposition method with long short-term memory model (WLSTM) is developed to handle the daily streamflow forecasting. Discrete wavelet transform (DWT) is employed to extract multiscale features, which are then simulated by long short-term memory models (LSTMs), respectively. The outputs of the different scales LSTMs are finally reconstructed toward the forecasting results. Seven years daily streamflow sequences of three tributaries and one mainstream in the upper reaches of the Yangtze River are investigated by the WLSTM models. For comparison, standard LSTM, and traditional ANN are chosen for the same streamflow forecasting task. Experimental results show that the proposed hybrid model is better than other comparison models in terms of evaluation indicators. Considering that large floods often occur in the Yangtze River Basin, the performance of flood forecasting is also analyzed and the result shows that forecasting errors of WLSTM are more concentrated, which means WLSTM outperforms traditional ANN and LSTM for the extreme flood forecasting.

1. Introduction

Streamflow forecasting is essential for effective water resources planning, development, utilization, and allocation management, especially in flood- and drought-prone areas[1]. The formation of streamflow is affected by many related factors such as weather, surface vegetation type, land type, slope, aspect, watershed size, shape, etc. Therefore, the runoff at the outlet of the watershed usually presents very complex nonlinear and non-stationary characteristics, which gives the runoff simulation extremely difficult[2].

In the past few decades, with the enhancement of water resources development, streamflow simulation of river basins of different scales has become more and more frequent and has received great attention. According to different mechanisms, streamflow prediction models can be divided into physical models, conceptual models and data-driven models[3]. Physical models and conceptual models need to have insights into the process of runoff formation, and use complex or simplified mathematical formulas to describe the runoff generation, confluence, and river channel evolution. Physical and conceptual models need meteorological, topographic, landform, and soil texture data support, which results in the model being too complex and restricted to a certain extent[4]. Data-driven models have the advantages of relying only on historical hydro-meteorological data, not directly considering potential physical processes, and requiring much less input and parameter data. Therefore, data-driven models have become suitable substitutes for physical models and conceptual models in streamflow forecasting[5]. So far, various data-driven models have been successfully applied to the hydrological field to obtain accurate streamflow predictions[6, 7]. The simplest forms of data-driven models include univariate autoregressive models, autoregressive moving average models, multivariate linear regression models. These models are based on the assumption that current streamflow is linearly related to the past values. Obviously, this is not very suitable for non-linear and non-stationary runoff processes[8]. The machine learning model is a relatively advanced type of data-driven model. By introducing mechanisms such as neuron structure, activation function, and kernel function, the machine learning model has a powerful ability to solve high-dimensional nonlinear prediction problems[9-12]. Applications of machine learning models, such as artificial neural network (ANN), support vector machine (SVM), random forest (RF), extreme learning machine[13] on streamflow forecasts can be found in lots of researches[14-17].

As we all know, the reason for the successful promotion and application of the data-driven models is the model's powerful mining of the internal connections of historical data[18]. Continuously improving the data mining capabilities of the data-driven based streamflow forecasting models is the key to ensuring the accuracy of predictions. Generally, there are two ways of improvement measures. The first way is to improve the data-driven model structure itself. Deep learning techniques developed in recent few years are confirmed to outperform traditional machine learning methods in numerous streamflow forecasting applications owing to its ‘‘deeper’’ representations[18, 19]. When solving time-correlated input signals, the traditional ANNs’ straight-forward and straight-out structural characteristics make it process only the current input information, cannot use the previous information. The structure of recurrent neural network (RNN)[13] involves feedback connection, therefore, RNN has a "memory" function to capture additional information from the previous steps[20]. This feature is very effective in simulating complex and continuous hydrological time series. LSTM is a more advanced type of RNN. It adds a "processor" to the RNN structure to determine whether the information in the previous steps is useful. The structure optimization effectively avoids the vanishing/explosive gradient problem resulting from learning long-term dependence of the sequence[21]. Inspired by the success of LSTM in the field of natural language processing, some studies have explored the capabilities of LSTM in streamflow series forecasting and obtained exciting results[22-24].

The second way to improve the data mining capabilities of the data-driven based streamflow forecasting models is the pre-analysis of cycle and trend changes of complex hydrological data. Wavelet transform (WT) can reveal the in-depth characteristics of various frequency signals of time series, which makes it possible to analyze runoff series in different time scales, and reveal more detailed localized information in time and frequency domains[25]. Lots of studies have applied wavelet transform technology to remove the nonlinear dynamic noise of the original data, analyzed the multi-time scale and nonlinear characteristics of the time series, and then combined the wavelet transform and machine learning models to predict the runoff[15, 26-28]. The performance and potential of the mixed models have been studied in their works. However, it was pointed out by[29] that the common mistake in the large amount of hydrology papers using the wavelet transform is the usage of WT as preprocessing methods. If WT is applied to the entire observed original time series, and a set of reconstructed or decomposed time series is obtained and selected as the input of the ANN or SVM model, then the model will use the future information to be predicted, resulting in an abnormally high prediction accuracy of the model.

In this article, we combined the two strategies mentioned above and proposed a wavelet coupled long short-term memory network (WLSTM), and studied its use in predicting daily streamflow in the upper reaches of the Yangtze River and its three main tributaries. To this end, wavelet transform is used to extract the multi-scale features of the incoming sequence, and then they are represented by LSTMs, respectively. Continuous segmentation decomposition mechanism is used to avoid the widespread problem of incorrect use of wavelet transform. The output of the LSTM is reconstructed to generate predictions. The proposed WLSTM is compared with generalized regression neural network (GRNN) and conventional LSTM models. The rest of this article is structured as follows. The study area and runoff data are described in Section 2. Forecasting methods and predictive performance indicators are described in Section 3. Section 4 illustrates the results and discussion. The conclusion is given in Section 5.

2. Study area and runoff data

In order to study the performance of the proposed WLSTM method, we considered the daily streamflow forecast in the upper reaches of the Yangtze River, which refers to the section from the source of the Yangtze River to Yichang city. The main tributaries in this section are the Yalong River, Minjiang River, Jialing River on the left bank, and Wujiang River on the right bank. The schematic diagram of the main tributary rivers and measuring stations in the area is shown in the Figure. 1. The basic statistics of rivers, basin area, length are shown in Table 1.

Figure 1. A sketch map of study area, rivers, and observation stations

Table 1. Basic features of the rivers and the available data

Rivers	Catchment area (km2)	Length (km)	Annual streamflow (m^3/s)
Yalong River	144200	1571	1860
Jinsha River	485099	3481	4750
Min River	135400	711	2830
Jialing River	157900	1571	1860
Wu River	87920	1037	1650
Upper Yangtze River	1005501	4504	13600

3. Methodology

3.1. Wavelet transform

Wavelet transform (WT) is a localized analysis of time (space) frequency. It gradually refines the signal in multiple scales through expansion and translation operations, and finally achieves time subdivision at high frequencies and frequency subdivision at low frequencies. WT can reveal the local time and frequency information of non-stationary time series, focus on any details of the time series. Therefore, the wavelet transform is suitable for capturing the detailed information of the streamflow process with a high degree of nonlinearity and a large number of random factors.

For streamflow time series f(t) with a fixed sampling interval, the discrete wavelet transform is defined as:

$$W_{\psi} f(j,k) = 2^{-j/2} \sum_{t=0}^{N-1} f(t) \bar{\psi}(2^{-j} t - k) \tag{1}$$

where the mother wavelet $\psi(t)$ satisfies $\int_{-\infty}^{+\infty}\psi(t)dt=0$, $\bar{\psi}(t)$ is its complex conjugate function. N is the length of the streamflow time series. $W_{\psi}f(j,k)$ represents the wavelet coefficient with scale a = 2^j and location b = 2^jk.

Continuous segmentation decomposition mechanism is used to avoid the incorrect use of wavelet transform in widespread hydrology forecasting researches. If the streamflow value at time T will be predicted, only the streamflow series from time 0 to T-1 is to be decomposed, N=T. If the runoff value at time T+1 will be predicted, only the streamflow series from time 0 to T is re-decomposed, N=T+1. The decomposed sub-sequence updated in real time is used as the model input to make predictions.

3.2. LSTM

LSTM is an evolved RNN[13], which can effectively use the state information of hidden neurons in the previous time step to have a memory function[20]. The LSTM network adds a "processor" to the RNN structure to determine whether the information in the previous steps is useful. Useful information will be retained and useless information will be removed. This process is controlled by structures called gates consisting of sigmoid layers and dot multiplication operations. In this way, LSTM can avoid the explosion gradient problem so as to learn long-term time series dependence[30].

The hidden layer in the LSTM consists a memory cell and three gates. The state of the hidden layer is updated as shown in the following formula.

$$f_t=\sigma(W_f\cdot[h_{t-1},x_t]+b_f) \tag{2}$$

$$i_t=\sigma(W_i\cdot[h_{t-1},x_t]+b_i) \tag{3}$$

$$C_t'=\tanh(W_C\cdot[h_{t-1},x_t]+b_C) \tag{4}$$

$$C_t=f_t*C_{t-1}+i_t*C_t' \tag{5}$$

$$o_t=\sigma(W_o\cdot[h_{t-1},x_t]+b_o) \tag{6}$$

$$h_t=o_t*\tanh(C_t) \tag{7}$$

Where f_t is the forget gate, i_t is the input gate, C_{t-1} is the old state of the network, C_t is the updated state, o_t is the output gate.

3.3. The proposed WLSTM

The key of developing data-driven based streamflow forecasting models is to improve the data mining capabilities. The first way is to improve the data-driven model structure itself. The second way is the pre-analysis of cycle and trend changes of complex hydrological data. In this paper, combining the two strategies mentioned above, WLSTM is proposed to study the streamflow forecasts in the upper Yangtze River. In the hybrid model, Long short-term memory network (LSTM) is used as a driven model. Wavelet decomposition is used to analyze time series in different scales independently, that makes it possible to analyze runoff series in different time scales, and reveal more detailed localized information. The architecture of WLSTM model is described in Figure 2. We also develop standard LSTM and GRNN model to make a comparison. Detail description of each step of WLSTM model is given below.

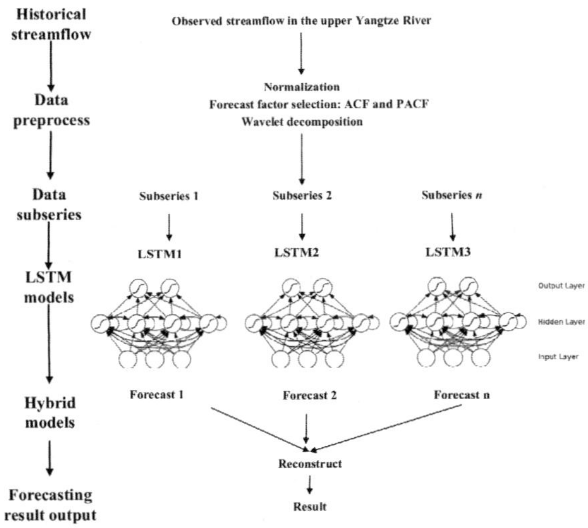

Figure 2. The architectures of WLSTM streamflow forecast model

3.4. Performance evaluation

In this article, predictive performance evaluation system is studied to measure the predictive ability of the WLSTM model in terms of the goodness of fit for total runoff, correlation, high and low streamflow. Among them, root mean squared error (RMSE) is a commonly used metric to measure the difference between the predicted value and the actual observed value. R^2 is the coefficient of determination, which reflects how much of the fluctuation of the predicted value can be described by the fluctuation of the observed value. Mean relative error (MRE) measures the averaged percentage error between the observed value and the predicted value. Mean squared logarithmic error (MSLE) and mean fourth-power error (M4E) are designed to emphasize on low flows and high flows, respectively, due to the logarithm transformation algorithm in MLSE and power algorithm in M4E. The definitions of RMSE, R^2, MRE, MSLE, M4E can be found in the reference [15].

4. Results and discussions

We adopted wavelet transform to analyze the multi-scale time-frequency information of the time series, obtained the details of the highly non-linear streamflow process. Introduced the LSTM model that has proven to be powerful in data mining, and combine the detailed decomposition components as the input of the LSTM model to further construct the daily streamflow forecast model. This section also built streamflow forecast models of GRNN and standard LSTM, and compare the forecasting results for a comprehensive and detailed analysis of the application of the proposed WLSTM. The selected time series data are daily streamflow at Beibei, Gaochang, Pingshan and Yichang from 1/1/2004 to 31/12/2010. The datasets that used for training models are the first 5 years streamflow data, and the datasets that used for testing models are the remaining 2 years data. In order to ensure that the constructed models are robust, the L2 weight penalty regularization mechanism is used in the training mode to prevent overfitting.

4.1. Models and candidate predictors

Determining the potential input variables is an important step for developing a data-driven forecasting model[31]. In our study, the previous streamflow at the Beibei, Gaochang, Pingshan and Yichang stations with adequate and reasonable lags were considered as model predictors. The actually selected predictors were determined based on ACF and PACF statistics. The statistics results show that previous streamflow with 1, 2, 3 lags were selected as predictors for the streamflow forecasting models of Beibei and Pingshan, streamflow with 1, 2, 3, 4, 5 lags were predictors for the streamflow forecasting models of Gaochang and Yichang. Then, the original observation data needs to be decomposed by DWT. The Mallat algorithm, which has been proved to be suitable for hydrological time series decomposition, is used to obtain sub-time series at three decomposition levels of four control station runoff process Nourani, Komasi[32]. Maheswaran and Khosa[33] pointed out that wavelets with wider support and higher vanishing moments, such as the Daubechies wavelet with five vanishing moments (db5), are more suitable for nonlinear and stochastically oscillating hydrological sequences. Therefore, the Mallat algorithm and the db5 mother wavelet are used to decompose the streamflow time series of the main and tributary streams of the Yangtze River.

4.2. Performance of ANNs

4.2.1. Daily streamflow forecasting

The results of the WLSTM, GRNN, LSTM daily streamflow forecasting in the testing period (1/1/2009-31/12/2010) are provided in Figure 3, 4, 5. Figure 3 presents the WLSTM streamflow forecasting for Beibei (Figure 3(a)), Gaochang (Figure 3(b)), Pingshan (Figure 3(c)) and Yichang (Figure 3(d)). Black line represents the observations and red line represents the forecasts. Forecasted values follow the observed values well. In the WLSTM model, wavelet transformation is applied to analyze time-frequency localization, makes it possible to reveal more detailed localized information in time and frequency domains on the basis of LSTM. Figure 4 shows the GRNN streamflow forecasting for Beibei (Figure 4(a)), Gaochang (Figure 4(b)), Pingshan (Figure 4(c)) and Yichang (Figure 4(d)). In Figure 4(a), we can observed that, GRNN has a better performance at low streamflow. However, GRNN overestimates the high streamflow at the first year and then underestimates the high streamflow at the following year. This over and under estimation phenomenon also can be found in Fig. 4(b), (c) and (d). When streamflow fluctuates significantly, GRNN seems unable to catch the streamflow change. This may be due to the limited learning ability of GRNN. Figure 5 is LSTM streamflow forecasting. We can observe that relatively better forecasts are achieved at both high streamflow and low streamflow compared with GRNN owing to its “deeper” representations ability and the gate structure updating the memory cell state, forgetting irrelevant content.

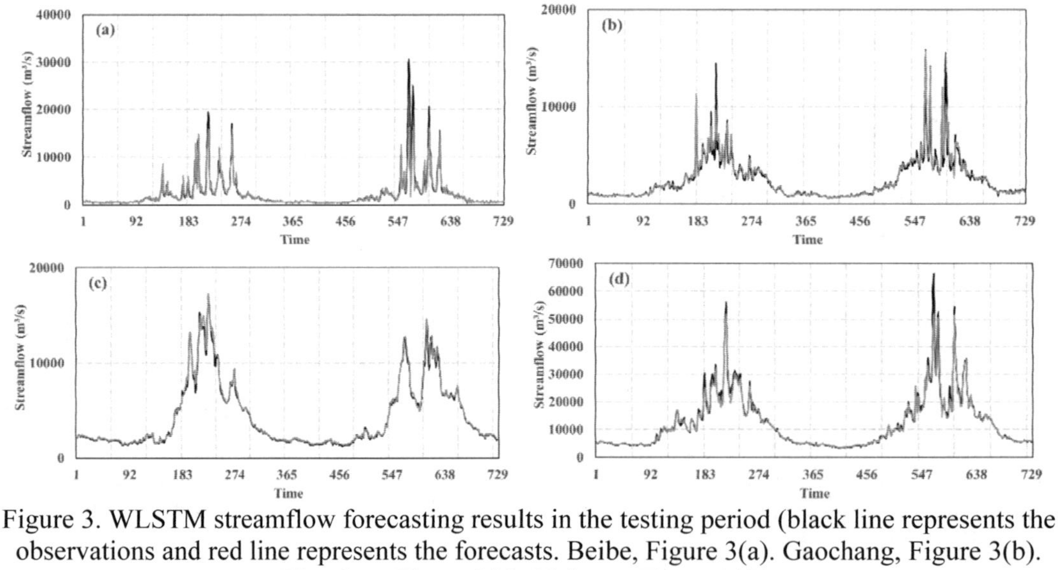

Figure 3. WLSTM streamflow forecasting results in the testing period (black line represents the observations and red line represents the forecasts. Beibe, Figure 3(a). Gaochang, Figure 3(b). Pingshan, Figure 3(c). Yichang, Figure 3(d))

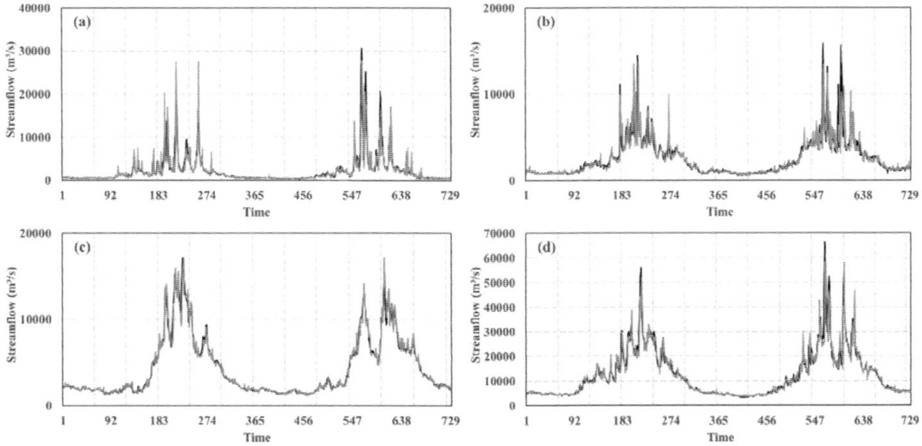

Figure 4. GRNN streamflow forecasting results in the testing period (black line represents the observations and red line represents the forecasts. Beibe, Figure 4(a). Gaochang, Figure 4(b). Pingshan, Figure 4(c). Yichang, Figure 4(d))

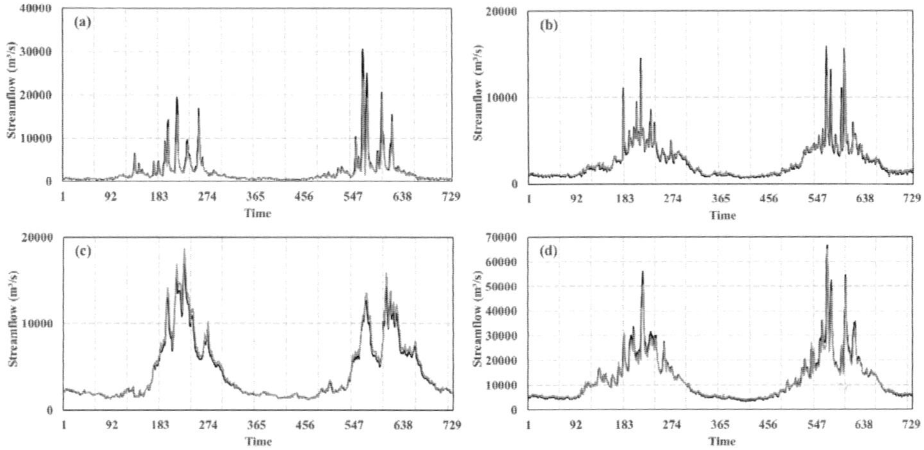

Figure 5. LSTM streamflow forecasting results in the testing period (black line represents the observations and red line represents the forecasts. Beibe, Figure 5(a). Gaochang, Figure 5(b). Pingshan, Figure 5(c). Yichang, Figure 5(d))

Table 2 presents the detail streamflow forecasting error statistics of the selected three models and four stations. The statistical indicators are MSLE, M4E, RMSE, R^2 and MRE, the five error statistical indicators are explained in the previous section. Models deemed to have better predictive performance have lower RMSE, MRE, M4E and MLSE values, and higher R^2 value. When comparing the forecast performance of the three models, a stepwise comparison mode will be used, that is, the forecast performance of GRNN-LSTM, and GRNN-LSTM-WLSTM will be compared successively, and finally the streamflow forecasting model that is most suitable for the streamflow characteristics of the main and tributary streams of the upper reaches of the Yangtze River is summarized.

Select Yichang streamflow forecasting for analysis, RMSE, MRE, R^2, M4E, MLSE of GRNN are 2918, 0.09, 0.91, 2.56E+15 and 0.017. Compared with the GRNN prediction, the RMSE value of LSTM prediction decreased from 2918 to 2206, the MRE value remained unchanged, the R^2 value increased from 0.91 to 0.95, the M4E value reduced from 2.56E+15 to 8.53E+14, the MLSE value reduced from 0.017 to 0.013. As the MSLE and M4E put more emphasis on goodness-of-fit to low flows and high flows, respectively. It reveals that in the low streamflow parts and high flow parts, LSTM both performs better than the GRNN model and on the whole, LSTM model performs better.

Importantly, RMSE, MRE, R^2, M4E, MLSE of the proposed WLSTM is 1957, 0.07, 0.96, 4.24E+14 and 0.010 for Yichang daily streamflow forecasting. Due to the lowest RMSE, MRE, M4E and MLSE value, and the highest R^2 value, WLSTM is more suitable for streamflow forecasting in the upper reaches of Yangtze River than GRNN and LSTM. For Beibei, Gaochang, and Pingshan stations, the same results can be found, that is, WLSTM streamflow forecasting model is more suitable than GRNN and LSTM streamflow forecasting models.

Analyzing Table 2, it can be found that the streamflow forecasting errors of Pingshan is smaller than that of the other three selected stations. The streamflow forecasting errors of Gaochang and Beibei are relatively higher. According to analysis, the Jialing River and Minjiang River where Beibei and Gaochang are located often suffer heavy rains. Regional floods are usually torrential rain-type floods, with steep fluctuations and weaker autocorrelation characteristics than that of Pingshan, so forecasting is more difficult. For Jialing River and Minjiang River, more meteorological and hydrological data are needed to develop better streamflow forecast models.

Table 2. Streamflow forecast error of selected stations using GRNN, LSTM and WLSTM

Model	Station	RMSE	MRE	R^2	M4E	MLSE
GRNN	Beibei	1721	0.27	0.76	2.75E+14	0.117
	Gaochang	1060	0.16	0.73	3.27E+13	0.047
	Pingshan	562	0.07	0.97	1.19E+12	0.008
	Yichang	2918	0.09	0.91	2.56E+15	0.017
LSTM	Beibei	1322	0.17	0.86	1.19E+14	0.059
	Gaochang	942	0.14	0.78	3.24E+13	0.032
	Pingshan	547	0.07	0.97	6.52E+11	0.007
	Yichang	2206	0.09	0.95	8.53E+14	0.013
WLSTM	Beibei	1067	0.15	0.90	1.15E+13	0.048
	Gaochang	851	0.11	0.83	2.75E+13	0.024
	Pingshan	365	0.06	0.99	2.04E+11	0.006
	Yichang	1957	0.07	0.96	4.24E+14	0.010

4.2.2. Extreme flood forecasting

As we know, affected by heavy rainfall, floods often occur in the upper Yangtze River, threatening the safety of downstream life and property. Relying on accurate floods forecasts and formulating optimal reservoir operation strategy can effectively reduce downstream flood risks. Therefore, a further evaluation of the streamflow forecasting at flood season is needed to be discussed. Here, to further reveal the ability of LSTM-based models to learn long-term time series with continuous high flow and great volatility, RNN flood forecasting model is added, and forecast errors are analyzed. Forecast errors frequency histograms of GRNN, RNN, LSTM and WLSTM are illustrated in Figure 6. Forecast errors are calculated by forecasted value minus observed values. The sharper the error distribution and the more concentrated it is near 0, the better the model prediction performance. The Figure 6(a) shows the flood forecast errors frequency histogram of the station Beibei. Forecast errors of WLSTM are more concentrated on 0, followed by RNN and LSTM, the forecast errors of GRNN is the most decentralized, WLSTM, RNN and LSTM models tend to underestimate the floods because the errors are skewed distributed on negative values, while the flood forecast error of GRNN is relatively evenly distributed on both positive and negative regions. For Beibei, performance of flood forecasts is WLSTM>RNN>LSTM>GRNN. Figure 6(b) shows similar results with Figure 6(a), WLSTM>LSTM≈RNN>GRNN but the errors of all models are distributed on both positive and negative numbers. For Pingshan, Figure 6(c) shows that performance of flood forecasts is WLSTM>LSTM≈RNN>GRNN. For Yichang, WLSTM>LSTM>GRNN>RNN. The results show that WLSTM is superior to the other models in flood forecasts in the upper Yangtze River. RNN model prediction is not robust compared to the LSTM-based model.

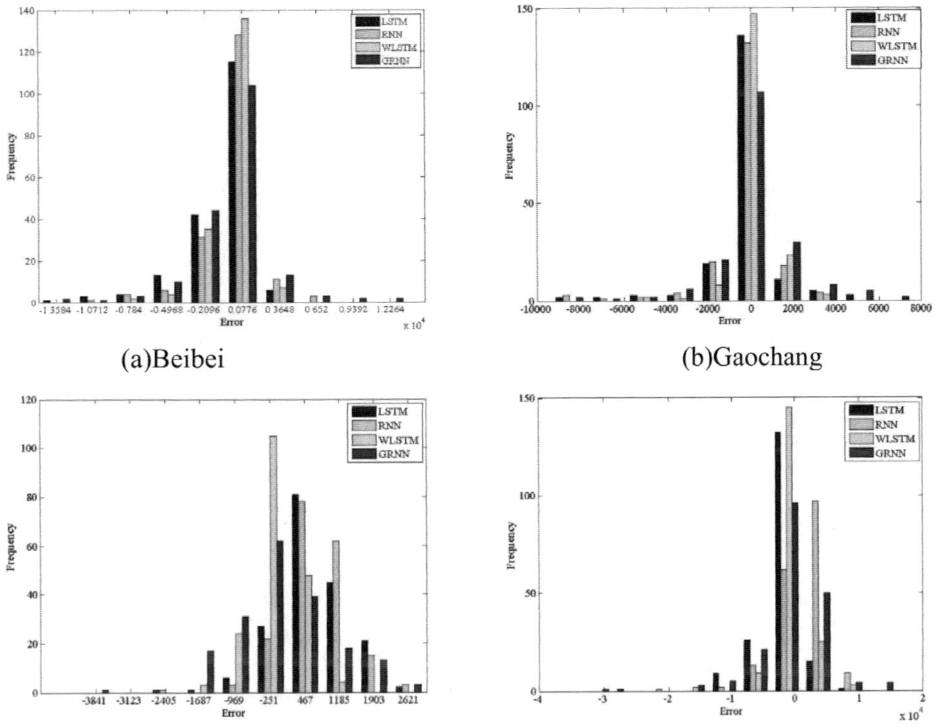

(a)Beibei (b)Gaochang

(c)Pingshan (d)Yichang

Figure 6. Forecast error frequency histograms of GRNN, RNN, LSTM and WLSTM

5. Conclusion

In this paper, the WLSTM model was proposed to study the daily streamflow forecasts in the upper Yangtze River and its main tributaries. GRNN, LSTM and RNN models were also developed to make a comparison. A predictive performance evaluation system was studied to measure the predictive ability in terms of MSLE, M4E, RMSE, R^2, and MRE. The preliminary conclusions summarized below.

With the lowest RMSE, MRE, M4E and MLSE value, and the highest R^2 value, WLSTM streamflow forecasting model is more suitable than GRNN and LSTM streamflow forecasting models. A further evaluation of the flood forecasts shows the errors distribution of WLSTM is more concentrated on 0, which means WLSTM also outperforms GRNN, RNN and LSTM for flood forecast. RNN model prediction is not robust compared to the LSTM-based model. Streamflow forecasting errors of Pingshan is smaller than that of the other three selected stations. The streamflow forecasting errors of Gaochang and Beibei are relatively higher. Torrential rain-type floods with steep fluctuations and weaker autocorrelation characteristics made forecasting of Jialing River and Minjiang River more difficult. GRNN has a good performance at low streamflow, as the streamflow increased, GRNN seems unable to catch the streamflow change. This is probably due to the limited learning capability of GRNN. The structure of LSTM-based model improves the study of long-term series dependence, and WT can reveal the local time and frequency information of non-stationary time series. Therefore, the combined WLSTM model has the capacity in modelling complex, nonlinear and non-stationary streamflow series.

Data availability
The datasets used and/or analyzed during the current study available from the corresponding author on reasonable request.

Conflicts of Interest
There are no relevant financial or non-financial competing interests to report.

Reference
[1] Kisi, O. and M. Cimen, A wavelet-support vector machine conjunction model for monthly streamflow forecasting. Journal of Hydrology, 2011. 399(1): p. 132-140.
[2] Tikhamarine, Y., et al., Improving artificial intelligence models accuracy for monthly streamflow forecasting using grey Wolf optimization (GWO) algorithm. Journal of Hydrology, 2020. 582: p. 124435.
[3] Devia, G.K., B.P. Ganasri, and G.S. Dwarakish, A Review on Hydrological Models. Aquatic Procedia, 2015. 4: p. 1001-1007.
[4] Cheng, M., et al., Long lead-time daily and monthly streamflow forecasting using machine learning methods. Journal of Hydrology, 2020. 590: p. 125376.
[5] Papacharalampous, G., H. Tyralis, and D. Koutsoyiannis, Predictability of monthly temperature and precipitation using automatic time series forecasting methods. Acta Geophysica, 2018.
[6] Wagena, M.B., et al., Comparison of short-term streamflow forecasting using stochastic time series, neural networks, process-based, and Bayesian models. Environmental Modelling & Software, 2020. 126: p. 104669.
[7] Yaseen, Z.M., et al., Application of the hybrid artificial neural network coupled with rolling mechanism and grey model algorithms for streamflow forecasting over multiple time horizons. Water Resources Management, 2018. 32(5): p. 1883-1899.
[8] Chen, L., et al., Copula-based method for multisite monthly and daily streamflow simulation. Journal of Hydrology, 2015. 528: p. 369-384.
[9] Zhou, Y., et al., Inverse Sparse Group Lasso Model for Robust Object Tracking. IEEE Transactions on Multimedia, 2017. 19(8): p. 1798-1810.
[10] Yuan, X., L. Xie, and M. Abouelenien, A regularized ensemble framework of deep learning for cancer detection from multi-class, imbalanced training data. Pattern Recognition, 2018. 77: p. 160-172.
[11] Karimi, S., et al., Forecasting daily streamflow values: assessing heuristic models. Hydrology Research, 2017.
[12] Lu, X., et al., Improving forecasting accuracy of river flow using gene expression programming based on wavelet decomposition and de-noising. Hydrology Research, 2017.
[13] Elman, J.L., Finding structure in time. Cognitive Science, 1990. 14(2): p. 179-211.
[14] Moeeni, H. and H. Bonakdari, Forecasting monthly inflow with extreme seasonal variation using the hybrid SARIMA-ANN model. Stochastic Environmental Research and Risk Assessment, 2016.
[15] Zhu, S., et al., Streamflow estimation by support vector machine coupled with different methods of time series decomposition in the upper reaches of Yangtze River, China. Environmental Earth Sciences, 2016. 75(6): p. 531.
[16] Zhang, Z., Q. Zhang, and V.P. Singh, Univariate streamflow forecasting using commonly used data-driven models: literature review and case study. Hydrological Sciences Journal, 2018. 63(7): p. 1091-1111.

[17] Rahmani-Rezaeieh, A., M. Mohammadi, and A.D. Mehr, Ensemble gene expression programming: a new approach for evolution of parsimonious streamflow forecasting model. Theoretical and Applied Climatology, 2020. 139(1): p. 549-564.
[18] Zuo, G., et al., Two-stage variational mode decomposition and support vector regression for streamflow forecasting. Hydrology and Earth System Sciences, 2020. 24(11): p. 5491-5518.
[19] Xiang, Z. and I. Demir, Distributed long-term hourly streamflow predictions using deep learning–A case study for State of Iowa. Environmental Modelling & Software, 2020. 131: p. 104761.
[20] Ishak, S., et al., Optimization of Dynamic Neural Network Performance for Short-Term Traffic Prediction. Transportation Research Record Journal of the Transportation Research Board, 2003. 1836(1): p. 27-31.
[21] Sak, H., A. Senior, and F. Beaufays, Long short-term memory recurrent neural network architectures for large scale acoustic modeling. Computer Science, 2014: p. 338-342.
[22] Zhu, S., et al., An improved long short-term memory network for streamflow forecasting in the upper Yangtze River. Stochastic Environmental Research and Risk Assessment, 2020. 34(9): p. 1313-1329.
[23] Yuan, X., et al., Monthly runoff forecasting based on LSTM–ALO model. Stochastic environmental research and risk assessment, 2018. 32(8): p. 2199-2212.
[24] Alizadeh, B., et al., A novel attention-based LSTM cell post-processor coupled with bayesian optimization for streamflow prediction. Journal of Hydrology, 2021. 601: p. 126526.
[25] Zhang, D., Wavelet transform, in Fundamentals of Image Data Mining. 2019, Springer. p. 35-44.
[26] Freire, P.K.d.M.M., C.A.G. Santos, and G.B.L. da Silva, Analysis of the use of discrete wavelet transforms coupled with ANN for short-term streamflow forecasting. Applied Soft Computing, 2019. 80: p. 494-505.
[27] Hadi, S.J. and M. Tombul, Streamflow forecasting using four wavelet transformation combinations approaches with data-driven models: a comparative study. Water Resources Management, 2018. 32(14): p. 4661-4679.
[28] Kambalimath, S. and P.C. Deka, Performance enhancement of SVM model using discrete wavelet transform for daily streamflow forecasting. Environmental Earth Sciences, 2021. 80(3): p. 1-16.
[29] Du, K., Z. Ying, and J. Lei, The incorrect usage of singular spectral analysis and discrete wavelet transform in hybrid models to predict hydrological time series. Journal of Hydrology, 2017. 552: p. 44-51.
[30] Hochreiter, S. and J. Schmidhuber, Long Short-Term Memory. Neural Computation, 1997. 9(8): p. 1735-1780.
[31] Adamowski, J., et al., Comparison of multiple linear and nonlinear regression, autoregressive integrated moving average, artificial neural network, and wavelet artificial neural network methods for urban water demand forecasting in Montreal, Canada. Water Resources Research, 2012. 48(1): p. 273-279.
[32] Nourani, V., M. Komasi, and A. Mano, A Multivariate ANN-Wavelet Approach for Rainfall–Runoff Modeling. Water Resources Management, 2009. 23(14): p. 2877-2894.
[33] Maheswaran, R. and R. Khosa, Comparative study of different wavelets for hydrologic forecasting. Computers & Geosciences, 2012. 46: p. 284-295.

Numerical Simulation of Wake Flow Field in Pitch Motion of Underwater Vehicle

Bo Huang[1a], Zhongyan Liu [1b], Yujing Xu[1c], Qiaochu Ding[1d], Guanjie You[1e], Mengchun Pan[1f], Jiafei Hu[1g], Qi Zhang[1h*]

[1]College of Intelligence Science and Technology, National University of Defense Technology, Changsha, Hunan, China

[a]1061086179@qq.com,[b]lansedexin26@126.com,[c]xyj@nudt.edu.cn, [d]596055844@qq.com,[e]402084699@qq.com,[f]13141317443@163.com, [g]garfield_nudt@163.com,[h*]zhangqi@nudt.edu.cn

Abstract—The frequent pitching motion of underwater vehicle during the mission will produce a more complex hydrodynamic wake, which is important for the detection and tracking of underwater vehicles. In this paper, a numerical method is presented to simulate the wake field during the pitching motion of an underwater vehicle by combining the dynamic overset grid method. The evolution law of the wake field is analyzed based on the DARPA SUBOFF simplified model. The effects of underwater vehicle speed on the velocity and turbulent kinetic energy of the wake field and the effects of the longitudinal inclination of the underwater vehicle on the extent of the wake field are discussed.It is found that the wake velocity is positively correlated with the sailing speed of the underwater vehicle, the turbulent kinetic energy of the wake field is directly proportional to the 1.75 power of the sailing speed of the underwater vehicle, and the range of wake field is positively correlated with the pitch angle of underwater vehicle. The larger the pitch angle is, the larger the intensity and size of vortex formed in the flow field are, and the farther the vortex can propagate to the rear of underwater vehicle wake.

1. Introduction

Submarines are stealthily maneuverable and powerful underwater vehicles used for maritime strategy and national security. In its operation, especially in the execution of the mission, the submarine will frequently carry out maneuvering motion such as rising, diving and so on. In the course of the submarine's pitching maneuvering, complex flow separation occurs in the submarine's hull, which leads to strong eddy currents being discharged into the wake and many flow phenomena with rich characteristics.

The hydrodynamic wake left in seawater during submarine pitching motion can be maintained for a long time, and the position and movement of submarine can be determined according to the shape characteristics and velocity changes of the wake, so as to track and detect the submarine. At present, hydrodynamic wake has become a important research target of submarine indirect detection [1-2]. In 1999, a study by the University of Arizona found that for a submarine with a diameter of 10m and a length of 100m, when it accelerates by 10% or turns 5°, its tail eddy current attenuation period can reach 6 days, the horizontal scale can reach 2km and the vertical scale can reach 80m[3]. In 2000, a study by Arizona State University showed [4] that when a submarine in a stratified fluid accelerates, decelerates or changes course, it will transfer enormous energy to the surrounding fluid, resulting in the creation of unusual vortices. In 2015, Ashok[5-6] carried out wind tunnel experiments on a 1:120 DARPA SUBOFF naked boat with no appendage installed in a low-speed wind tunnel with a flow Reynolds number of 2.4×106 by using PIV and hot line technology to study and analyze the wake at different pitch angles, revealing that when the submarine is in pitch motion, a pair of asymmetrical vortices will appear in the wake. These vortexes have become research targets based on eddy current detection in non-acoustic detection [7]. The early information of submarine pitching flow field mainly comes from experiment, but the experiment is not only long period but also very high cost. With the upgrading of computer technology and the rapid development of computational fluid dynamics, the numerical simulation of submarine flow field has been rapidly developed.

In this paper, a numerical simulation of the wake and large eddy generated during the pitching motion of the submarine is carried out by using the overset grid technology and parallel computing.The influence of underwater vehicle speed on flow field velocity and turbulent kinetic energy and the influence of underwater vehicle trim angle on flow field range were explored.It provides a theoretical basis for the future underwater vehicle detection based on vortex wake.

2. Pitching motion and governing equations

2.1. Underwater vehicle pitching motion coordinate system

Underwater vehicle pitching motion is a maneuvering motion occurring in two degrees of freedom in vertical plane. Different translational velocity and angular velocity may be obtained by observing submarine pitching maneuvering motion from different reference frames. To ensure that the result has practical physical significance, the reference frame is given first,as shown in Fig.1. The barycentric coordinate of the initial position of the submarine, O-MN, is taken as inertial reference coordinates, which are referred to as inertial system. Inertial system is stationary relative to the earth, and point O represents the origin of inertial system. The submarine's connected coordinate system is G-XY, referred to the moving system, where point G is the origin of the moving system and the submarine's centroid. At the initial moment, inertial and dynamic systems overlap. The submarine's pitching motion can be divided into the translation of the submarine's centroid point G relative to point O and the rotation of any point S on the submarine around point G.

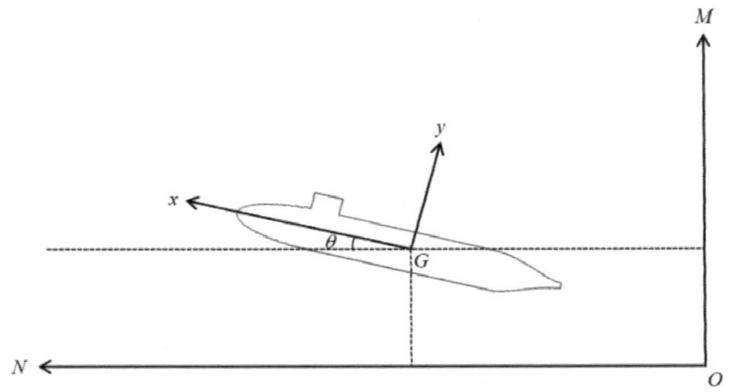

Fig.1 Underwater vehicle pitching motion coordinate system

When elevator is operated to a fixed rudder angle during a fixed depth voyage, the submarine will enter a steady straight pitch motion. If elevator is moved back to its original position, the submarine will make a steady depth voyage at a new depth. The pitch maneuver of the submarine, in the case of elevator, can be operated by a single rudder, and in the case of fast diving and buoyancy, can be operated by a relative rudder. In addition, the minimum depth change ΔH_{min} suitable for a given command trim angle θ should match the ship speed.

2.2. Basic theory of turbulence
The solution idea of computational fluid dynamics technology is to discret the governing equations of flow field in space and time, approximate them by means of gradual iteration, replace the original continuous physical field with physical quantities on discrete nodes, and connect different nodes by interpolation. According to the different simplified forms of governing equations, the numerical simulation methods of viscous flow mainly include direct numerical simulation , large eddy simulation and Reynolds average method.

The flow control equation for incompressible viscous flow using tensor index symbol can be expressed as follows:

$$\frac{\partial u_i}{\partial x_i} = 0 \tag{1}$$

$$\rho\frac{\partial u_i}{\partial t} + \rho u_j\frac{\partial u_i}{\partial x_j} = -\frac{\partial p}{\partial x_i} + \mu\frac{\partial}{\partial x_j}\left(\frac{\partial u_i}{\partial x_j}\right) + S_i \tag{2}$$

Where, u_i , u_j are the components of velocity in i and j direction respectively; x_i , x_j are the components of displacement in i and j direction respectively; ρ is fluid density; μ is the dynamic viscosity coefficient; p is the static pressure; S_i is the component of the volume force on.

The above equations are transient governing equations of turbulence, and the calculation of them is very powerful. So far, they are only limited to some simple flows with low Reynolds number. Reynolds average method and Standard $k-\omega$ turbulence model are used in the numerical simulation of submarine pitch wake.

3. Meshing and numerical simulation
The numerical simulation of submarine pitching motion flow field model is carried out by using computational fluid dynamics technology. The calculation process can be roughly divided into geometric modeling, meshing, material property setting, boundary condition and initial condition setting, solver algorithm selection and iterative calculation, display and output calculation results and so on.

3.1. Grid Division
The DARPA SUBOFF model is simplified and used as a simulation object to establish the appropriate fluid domain and construct the geometric model.

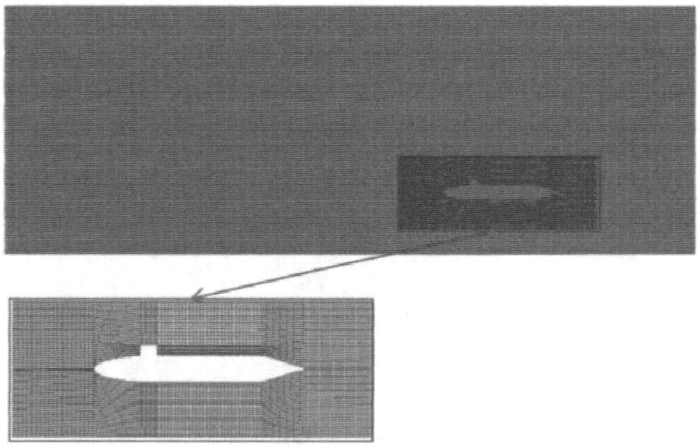

Fig.2 The overset grid of submarine pitching motion

The overset grid method is used in submarine pitch motion simulation,which is presented in Fig.2. The overset grid of submarine pitch motion is composed of the background grid and the surrounding grid of submarine. The two grid regions overlap in space, but do not have a connected relationship and exist independently of each other. Before calculation, holes are dug in the grid and interpolation points are matched to establish the connection relation. In the numerical discretization process, the overset grid is processed into cavity elements, discrete elements and interpolation elements. The fluid control equations are solved on the background grid and the submarine surrounding grid respectively. The interpolation elements constitute the internal boundary condition and are used to transfer data, and finally the flow field information is obtained in the whole computational domain.

3.2. Simulation Results

The evolution of wake field in submarine pitching motion is shown in Fig.3. The simulation results are described in detail below, and the formation reasons are briefly analyzed.

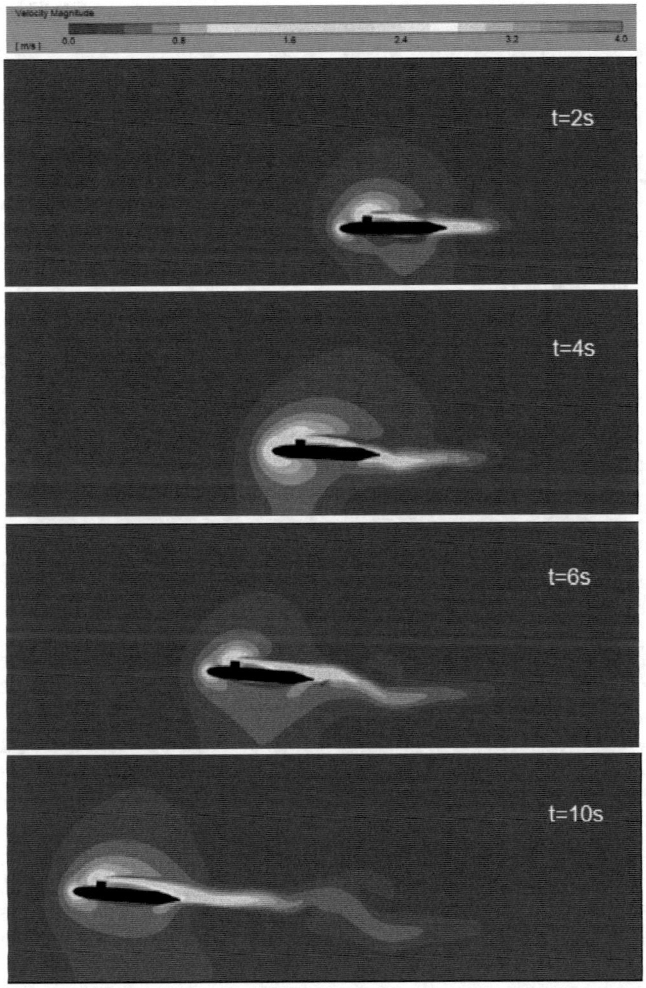

Fig.3 Evolution of wake field in submarine pitching motion

When the submarine is moving in a fixed depth straight line(t=2s in Fig.3), a large range of flow field is formed above the observation platform and behind the stern, and the maximum velocity is located behind the observation platform. When the submarine begins to turn the rudder to carry out pitching motion(t=4s in Fig.3), a wide range of flow field is generated in the front of the submarine. This is because the submarine's pitching motion transfers huge energy to the forward flow, and the wake flow field begins to deform and bend. In the steady linear pitching process(t=6s and t=10s in

Fig.3), the forward flow field begins to transfer backward, resulting in the formation of large range of flow field in the tail, and the deformation and bending part of the wake flow field becomes more obvious and the shedding vortex appear.

4. Comparative analysis of flow field

According to relevant data, the speed of the submarine in the cruising state is about 3m/s, and the speed of the submarine in high speed is generally about 10-15m/s. For the pitch maneuver of the submarine, the single rudder can be operated, and the relative rudder can be operated in the time of fast diving and floating. The general depth maneuver command has a longitudinal angle of 3°-5°, and a fast maneuver of 5°-7°, or greater than 7°. In order to explore the influence of submarine sailing speed and pitch angle on wake field, three different calculation conditions,as shown in Tab.1, are set for numerical simulation.

Tab.1 Three different calculation conditions

condition	I	II	III
speed（m/s）	3	6	3
pitch angle（°）	5	5	10

4.1. Influence of speed on flow rate

The figure on the left in Fig.4 shows the vertical plane velocity cloud diagram of working conditions I and II when the pitching motion reaches T =10s. The figure on the right in Fig.4 shows the velocity change curve of the wake survey line in the far flow field.

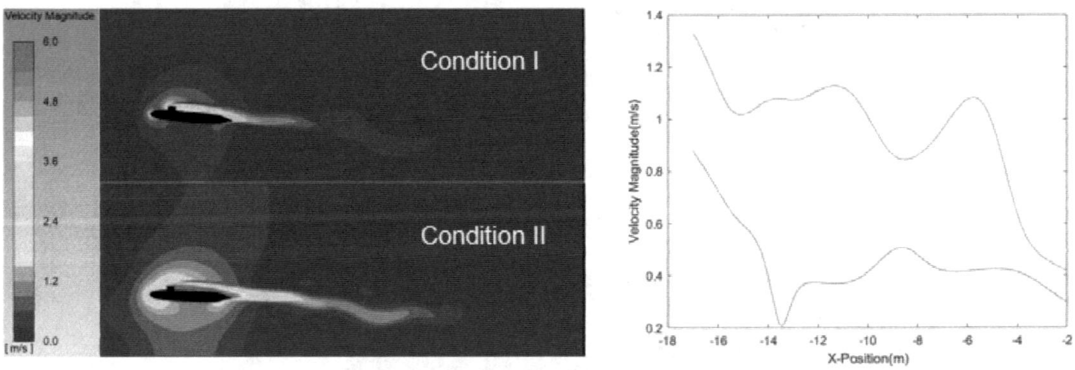

Fig.4 Influence of speed on flow rate

A representative survey line in the wake region is selected and its velocity change with position was plotted. The average velocity of the survey line in condition I and condition II was 0.45m/s and 0.95m/s respectively. Thus, when the ship speed is twice, the velocity of the surrounding flow field is also about twice; The range of near flow field around the submarine increases, but the range of far flow field does not change obviously.

4.2. Influence of speed on turbulent kinetic energy

The figure on the left in Fig.5 shows the vertical plane turbulent kinetic energy cloud diagram when the pitching motion of condition I and condition II reaches T =10s. The figure on the right in Fig.5 shows the change curve of turbulent kinetic energy on the wake survey line in the far flow field.

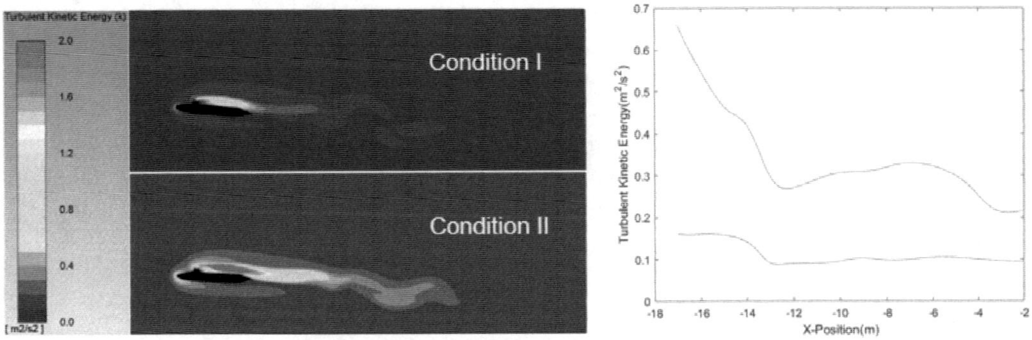

Fig.5 Influence of speed on turbulent kinetic energy

A typical survey line in the wake region was selected, and the variation of its turbulence kinetic energy with position was plotted. The average turbulence kinetic energy of the survey line in condition I was $0.11 m^2/s^2$, and that in condition II was $0.34 m^2/s^2$. Thus, when the ship speed is doubled, the turbulent kinetic energy of the surrounding flow field is about 3.09 times. The range of turbulence kinetic energy in the near flow field around the submarine does not change obviously, but the range of turbulence kinetic energy in the far flow field increases obviously.

In engineering practice, turbulence intensity I is commonly used to estimate turbulence kinetic energy k , and the formula is as follows:

$$I = 0.16 \times \mathrm{Re}^{-\frac{1}{8}} \tag{3}$$

$$k = \frac{3}{2}(v \cdot I)^2 \tag{4}$$

$$\mathrm{Re} = \frac{\rho v d}{\eta} \tag{5}$$

Where, Re is Reynolds number, ρ is the density of the medium, v is the average velocity, d is the characteristic length and η is the viscosity coefficient of the medium.

Therefore, $k \propto v^{1.75}$,when the flow velocity increases to twice, the turbulent kinetic energy should increase to 3.36 times theoretically. Therefore, considering the effect of discrete error and truncation error in numerical simulation, the simulation results are basically consistent with the theory.

4.3. *Influence of pitch angle on flow field range*

The figure on the left in Fig.6 shows the vertical plane velocity cloud diagram when the pitching motion of condition I and condition III reaches T =10s. The figure on the right in Fig.6 shows the velocity change curve on the wake survey line in the far flow field.

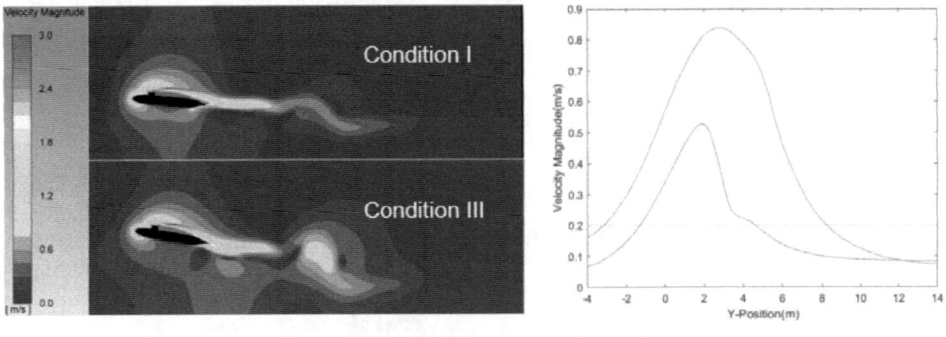

Fig.6 Influence of pitch angle on flow field range

The measuring line parallel to the vertical axis in the center of the shedding vortex of the wake flow is selected to draw the velocity change curve with position on the measuring line, and the range of the flow field with the velocity of 0.2m/s was compared. The longitudinal range of working condition I is 5.8m, and that of condition III is 11.3m. It can be seen that when the pitch angle is doubled, shedding vortex is formed obviously in the far flow field, and the range of the far flow field is significantly increased, which is about twice as large as before.

5. Conclusion

The underwater vehicle pitch motion wake field is simulated and analyzed, and its evolution law is introduced. The effect of underwater vehicle speed and pitch angle on wake field is investigated by setting three different calculation conditions.

(1)When the underwater vehicle changes from uniform linear motion to pitch motion, a wide range of eddy current field will be generated in the tail of the underwater vehicle. The intensity and range of turbulent kinetic energy generated in the pitch motion of the underwater vehicle will be larger than that in the straight motion.

(2)The velocity of wake field is positively correlated with the underwater vehicle's sailing speed, and the turbulent kinetic energy of wake field is proportional to the 1.75 power of underwater vehicle's sailing speed.

(3)The range of wake field is positively correlated with the pitch angle of underwater vehicle. The larger the pitch angle is, the larger the intensity and size of vortex formed in the flow field are, and the farther the vortex can propagate to the rear of underwater vehicle wake.

Therefore, this vortex can become the research target based on eddy current detection in non-acoustic detection. Compared with acoustic guidance, eddy current guided exploration and submersible technology has the advantages of long detection distance, strong anti-interference ability, not easy to deceive and induce.

References

[1] Yujie Shi, Haigang Ren. (2015)Trends of Foreign Non-Acoustics Exploration Potential and Stealth Technology.Ship Electronic Engineering,35(1):5-9.

[2] Jun Zhang, Xiaoci Zhang, Feng Zhao. (2003)Progress of Investigation on Non-acoustic Detection for Submarine Wake Originated from Naval Hydrodynamics.Journal of Ship Mechanics,7(2):121-128.

[3] Voropayev S I, Smirnov S A. (2003)Vortex Streets Generated by A Moving Momentum Source in a Stratified Fluid.Physics of Fluids,15(3):618-624.

[4] Lillberg E, Alin N, Fureby C. (2000)A computational Study of Wakes behind Submarines and Surface Ships.Defense research establishment weapons and protection division. pp. 1-31.

[5] Ashok A, Smits A J.(2013)The turbulent wake of a submarine model in pitch and yaw.TSFP Digital Library Online.

[6] Ashok A, Van Buren T, Smits A. (2015)Asymmetries in the wake of a submarine model in pitch. Journal of Fluid Mechanics,774:416-442.

[7] Ning Wu, Lianyun Mu, Wei Li. (2017)Hitting Efficiency Analysis of Submarine-launched Torpedo Based on Vortex Guidance.Ship Electronic Engineering,37(2):102-104.

Numerical study on wave attenuation performance of arc layout floating breakwater

Shuai Zhao[1], Yong Cheng[1], Chunyan Ji[1], Jianting Guo[1*]

[1] School of Naval Architecture and Ocean Engineering, Jiangsu University of Science and Technology, Zhenjiang, 212100, China

* Corresponding author's e-mail: guojianting11@163.com

Abstract: The floating breakwater has the functions of resisting wave propagation and maintaining stability of water surface, and its layout form will have an impact on the shielding effect and the shielding area. This paper uses the CFD method for numerical calculation based on the self-compiled program. The hydrodynamic performance and wave attenuation performance of floating breakwater with arc layout are mainly studied, and the calculation results of linear layout are compared and analyzed. The results show that the rolling response of linear layout under the action of the cross wave is intense, and there is transverse inclination under the action of the oblique wave. The motion of arc layout is relatively stable under the action of cross wave and oblique wave, and the hydrodynamic performance is better. The wave attenuation performance of arc layout is better than that of linear layout under transverse wave condition, while the wave attenuation performance of arc layout is further improved under the oblique wave, and the wave attenuation performance of linear layout is decreased under the oblique wave. It can be seen that the arc layout has better shielding performance than the linear layout under the oblique wave.

1. Introduction

As a new type of breakwater structure, floating breakwater has strong foundation adaptability, and the cost is less affected by water depth, which can be used as wave protection floating body in waters.

As an important basis for designing the floating breakwater, wave attenuation performance has attracted the attention of scholars at home and abroad. Lei Lin takes square box floating breakwater as a single module to carry out numerical analysis of multi floating breakwater [1].

For the hydrodynamic performance of floating breakwater with irregular layout, Ding Ning et al. studied the determination of wave force in the structural design of V-shaped floating breakwater. By changing the height and width of the rectangular section, the relationship between the change of relevant geometric parameters and the wave dissipation effect of the floating breakwater was studied[2]. Based on the potential flow theory, Zheng Zhi et al. established a hydrodynamic model of the interaction between waves and double-body parabolic floating breakwater and studied the influence of double-body spacing on the wave height distribution on the front and back waves of the system[3]. Ge Jiangtao calculated the hydrodynamic performance of floating breakwater with arc layout based on three-dimensional potential flow theory and calculated the motion response and mooring tension under arc layout[4]. The US Army Engineers Research and Development Center explores the wave attenuation performance of V-shaped floating breakwaters that can be quickly installed[5]. Luca Martinelli et al. examined two layouts characterized by different complexity (I-shaped and J-shaped) and three inclination angles (0°,30°,60°). Study on the Influence of Wave Inclination on Wave

Propagation, Mooring, and Connector Force between Modules [6].

The research of scholars in China and abroad is mainly carried out on the linear layout. However, considering the application of the future engineering, it is imperative to explore the design of irregular layout breakwaters to ensure the safe berthing of ships and as a cover for wave and energy dissipation of marine structures such as ports and offshore platforms. Therefore, in this paper, two arc layout schemes are taken as an example. Based on the CFD method, STAR CCM+ software and self-compiled program are used to explore the hydrodynamic performance of the irregular layout. According to the comparison of the wave attenuation performance between the arc layout and the linear layout.

2. Materials and methods

2.1. Theoretical basis of numerical calculation

2.1.1. Governing equation

Computational fluid dynamics has many forms, but they are based on continuity equation, energy equation, and momentum equation. The above three equations also meet Newton's second law, mass conservation law, and energy conservation law. The governing equation is simplified in this paper, assuming that the fluid is viscous and incompressible.

Based on the principle of mass conservation, the increase of mass in the volume element per unit time is equal to the net mass flowing into the micro-mass at the same time interval, and the continuity equation is:

$$\frac{dm}{dt}=\frac{d}{dt}\iiint_{V(t)}\rho dV=0 \tag{1}$$

Where: $V(t)$ is the volume of fluid elements; m is the quality of fluid in the control body; ρ is the fluid density.

The continuity equation can be derived from the arbitrariness of fluid element volume $V(t)$, and its equation is:

$$\frac{d\rho}{dt}+\rho\nabla\bullet\vec{v}=\frac{\partial\rho}{\partial t}+\nabla\bullet(\rho\vec{v})=0 \tag{2}$$

The change rate of fluid momentum relative to the time in the fluid element is equal to the force acting on the fluid element, and its expression is:

$$\frac{d}{dt}\iiint_V\rho\vec{v}\,dV=\iiint_V\rho\vec{F}\,dV+\iint_S\tau_n dS \tag{3}$$

Where: S is boundary condition; $\vec{F}$ is body force; τ_n is surface pressure.

The formula of the random derivative is:

$$\frac{d}{dt}\iiint_V\vec{a}\,dV=\iiint_V[\frac{d\vec{a}}{dt}+\vec{a}\nabla\bullet\vec{v}]dV=\iiint_V\frac{d\vec{a}}{dt}dV+\iint_S v_n\vec{a}dS \tag{4}$$

Where: $\vec{a}$ is any vector.

Referring to formulas (3) and (4), the momentum equation in integral form can be obtained:

$$\iiint_V\frac{\partial(\rho\vec{v})}{\partial t}dV+\iint_s\rho v_n\vec{v}\,dS=\iiint_V\rho\vec{F}\,dV+\iint_s\tau_n dS \tag{5}$$

2.1.2. Turbulence model

The K-Epsilon turbulence model is a two-equation model that solves transport equations for the turbulent kinetic energy k and the turbulent dissipation rate ε to determine the turbulent eddy viscosity. In actual sea conditions, considering the uncertainty of the external environment and wave load, the realizable K-Epsilon turbulence model is widely used to solve such problems, so this paper chooses this model for numerical calculation.

The transport equations of kinetic energy and turbulent dissipation rate are as follows:

$$\frac{\partial}{\partial t}(\rho k)+\nabla\cdot(\rho k\bar{v})=\nabla\cdot\left[\left(\mu+\frac{\mu_t}{\sigma_k}\right)\nabla k\right]+P_k-\rho(\varepsilon-\varepsilon_0)+S_k \tag{6}$$

$$\frac{\partial}{\partial t}(\rho\varepsilon)+\nabla\cdot(\rho\varepsilon\bar{v})=\nabla\cdot\left[\left(\mu+\frac{\mu_t}{\sigma_\varepsilon}\right)\nabla\varepsilon\right]+\frac{1}{T_e}C_{\varepsilon1}P_\varepsilon-C_{\varepsilon2}f_2\rho\left(\frac{\varepsilon}{T_e}-\frac{\varepsilon_0}{T_0}\right)+S_\varepsilon \tag{7}$$

Where: $\bar{v}$ is the mean velocity; μ is the dynamic viscosity; σ_ε , σ_k , $C_{\varepsilon1}$ and $C_{\varepsilon2}$ are Model Coefficients; P_ε and P_k are Production Terms; f_2 is a Damping Function; S_k and S_ε are the user-specified source terms.

2.1.3. calculation of wave attenuation performance of floating breakwater

The transmission coefficient K_t directly reflects the ability to float breakwater to resist waves and is also one of the important indexes to test the design of floating breakwater. The expression is:

$$K_t=\frac{H_t}{H_i} \tag{8}$$

Where: H_t represents the height of the transmitted wave behind the embankment, H_i represents the incident wave height.

2.2. Verification of the numerical method

2.2.1. Accuracy validation of numerical wave-making

In this paper, a numerical pool of 60m long, 100m wide and 1m deep is established based on CFD code, the regular wave with wave height H=0.075m and period T=1.1s is taken as an example for wave generation verification. The calculation results are shown in Fig.1.

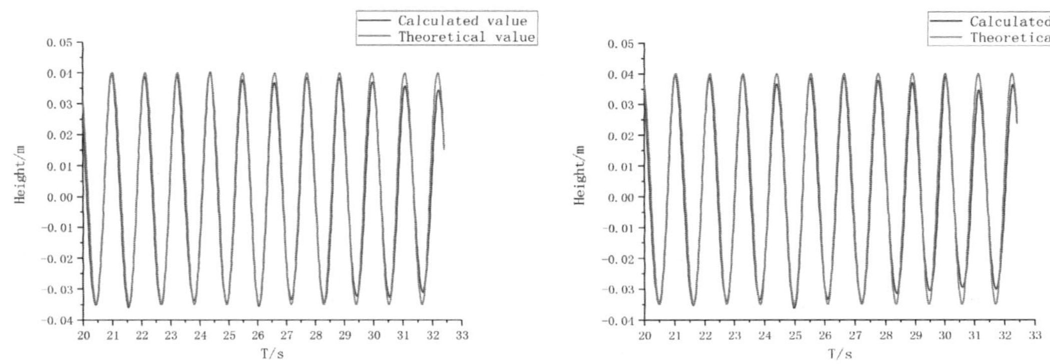

X=10m X=20m

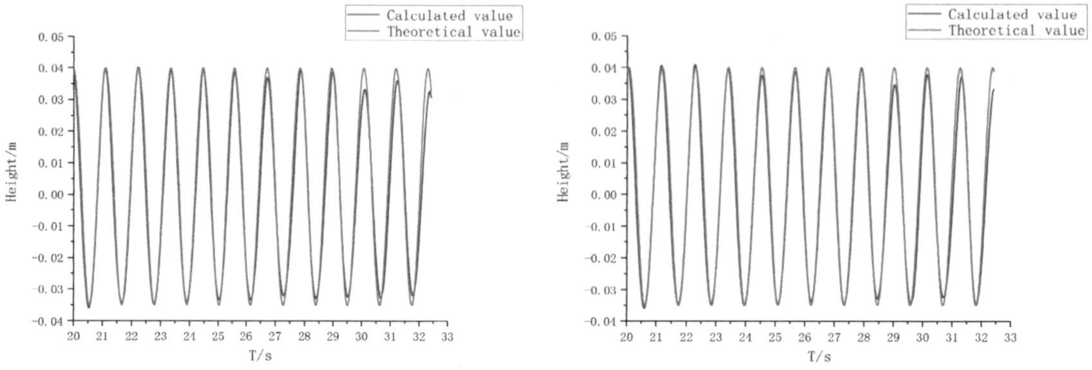

X=30m X=40m

Fig. 1 Comparison between the calculated value and theoretical value

The wave height time history curves of 10m, 20m, 30m, and 40m from the wave-making region are given in the diagram. It can be seen that the calculation results of each monitoring point are in good agreement with the theoretical value, and the attenuation is small, which can provide support for the following research.

2.2.2. Accuracy verification of wave attenuation performance

The STAR-CCM + software developed by the CD-cdapco company is widely used in many fields and has high precision and high reliability. Therefore, this paper uses self-compiled program and STAR-CCM + software for numerical calculation of linear layout of the floating breakwater, selects typical working conditions (wave height H = 0.25m, wave period T = 1.79s), and compares and analyzes the transmission coefficient according to the arrangement of measuring points.

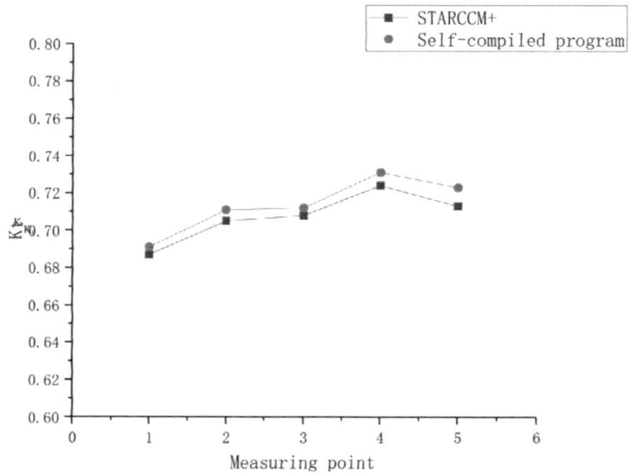

Fig. 2 Comparison of the transmission coefficient of different measuring points

Fig. 2 is the comparison of the numerical results between the self-compiled program and the STAR-CCM + software, and the average error is about 6.7 %. Therefore, it can be obtained that the numerical calculation results of both are close to each other and have good agreement. Therefore, the self-compiled program calculation method can accurately simulate the transmission coefficient of floating breakwater under different layouts.

2.3. numerical calculation

2.3.1. Arc layout design

According to the operational requirements and layout idea of the floating breakwater, this paper puts forward the plane layout scheme of arc layout and linear layout. To facilitate discussion, the projection length of arc type 1 layout is as same as linear layout in x-axis, and the total length of arc type 2 layout is as same as linear layout. This paper only studies the regularity characteristics because the numerical calculation time is too long. According to the scale ratio of λ=20, the numerical model is established, and the wave period is changed. The hydrodynamic performance of the three layouts is analyzed and studied. The relevant main scale of the layout is shown in Table 1, and the specific layout scheme is shown in Fig. 3.

Table 1 Main dimensions related to arc layout

Layout	Parameter	Prototype	Model
Arc 1	Arc length /m	1047.19	52.36
	Distance of vertical projection /m	1000	50
	Moulded beam /m	20	0.5
	Molded depth /m	8	0.2
	Draft /m	4	0.1
	Angle between arc tangent and x axis /°	30	30
Arc 2	Arc length /m	1000	50
	Distance of vertical projection /m	954.91	47.74
	Moulded beam /m	20	0.5
	Molded depth /m	8	0.2
	Draft /m	4	0.1
	Angle between arc tangent and x axis /°	30	30

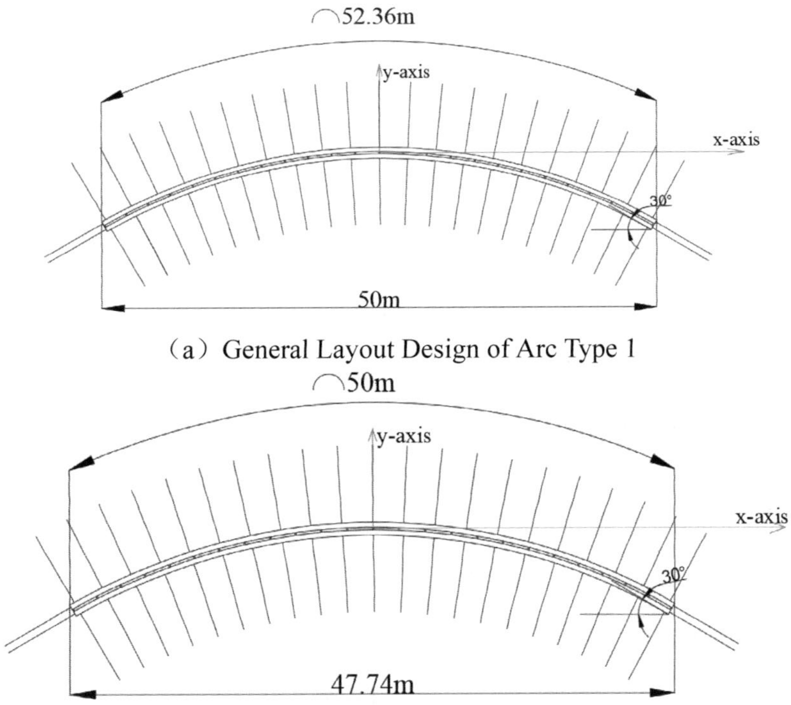

(b) General Layout Design of Arc Type 2

Fig.3 General Layout Design of Arc Layout

2.3.2. General Layout Design of Linear Layout

In order to explore the hydrodynamic performance of floating breakwater under arc layout, the linear layout adopts the same number of mooring anchor chains and parameter settings. The main dimensions of linear layout are shown in Table 2, and the layout scheme is shown in Fig.4.

Table 2 Main dimensions related to linear layout

Layout	Parameter	Prototype	Model
	Distance of vertical projection /m	1000	50
linear	Moulded beam /m	20	0.5
layout	Molded depth /m	8	0.2
	Draft /m	4	0.1

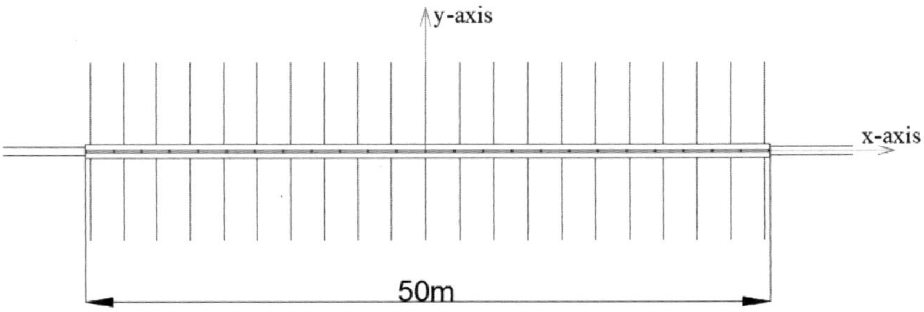

Fig.4 Linear general layout design

2.3.3. Design and Parameters of Mooring System

The mooring system is symmetrically arranged by 46 anchor chains behind the front embankment. The R3 level anchor chain is selected. The specific mooring parameters are shown in Table 3. And the left view of the overall mooring system is shown in Fig.5.

Table 3 Mooring parameters

material quality	Length (m)	Mass in air per unit length (g/m)	Diameter (mm)	Axial stiffness (N/m)	Break load (N)	Mooring radius (m)
R3	7.75	630	5.35	693.17	10055	7.54

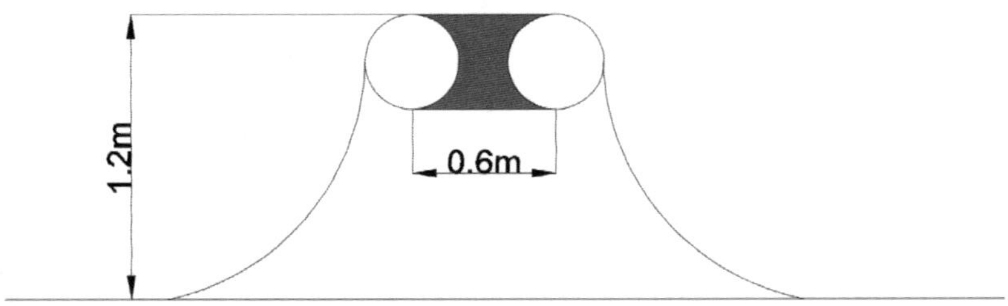

Fig.5 Left view of the mooring system

2.3.4. Distribution setting of monitoring points

According to the calculation requirements of the transmission coefficient, the distance between monitoring points ΔL is between 0.05 L and 0.45 L, and the distance between wave altimeter and floating breakwater x_i >0.2L (L is wavelength). In this paper, the spacing ΔL between two adjacent monitoring points in the y-direction is 0.8 m, and the distribution details are shown in Fig. 6.

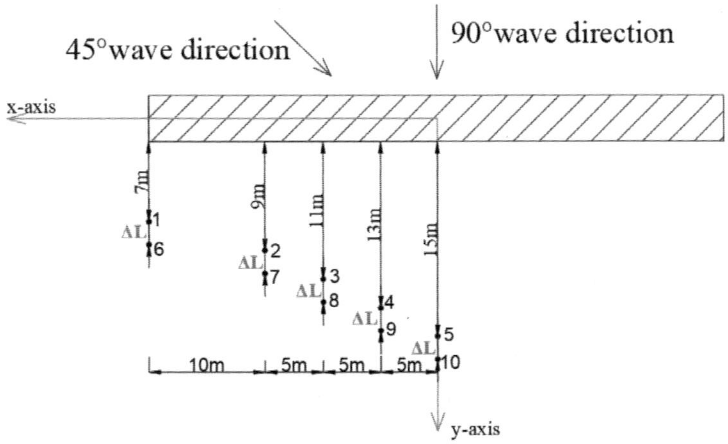

Fig.6 Distribution of monitoring points

2.3.5. Numerical experiment water depth and working condition

Experimental water depth is 1m, using regular waves, specific parameters as shown in Table 4.

Table 4 Numerical conditions

Condition number	A1	A2	A3	A4	A5	A6
Wave height/m	0.25	0.25	0.25	0.25	0.25	0.25
Period/s	1.34	1.57	1.79	2.01	2.24	2.45
Wave direction/°			45，90			

3. Results & Discussion

3.1. Analysis of Calculation Results of Transmission Coefficient

3.1.1. Analysis of transmission coefficient calculation results under transverse wave condition

According to the numerical results, the typical measuring points are selected as measuring points 1-3, and the wave attenuation performances of the three layouts are compared. The change of transmission coefficient is shown in Fig. 7.

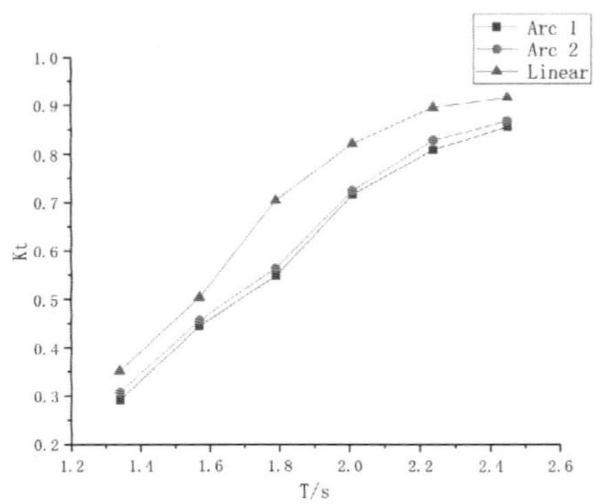

（a）Measuring point 1

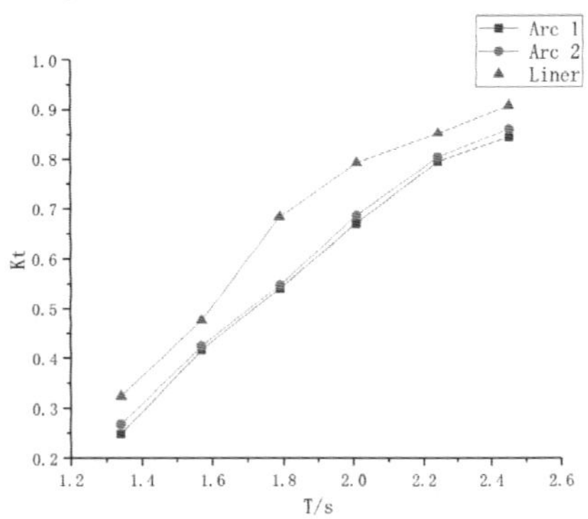

（b）Measuring point 2

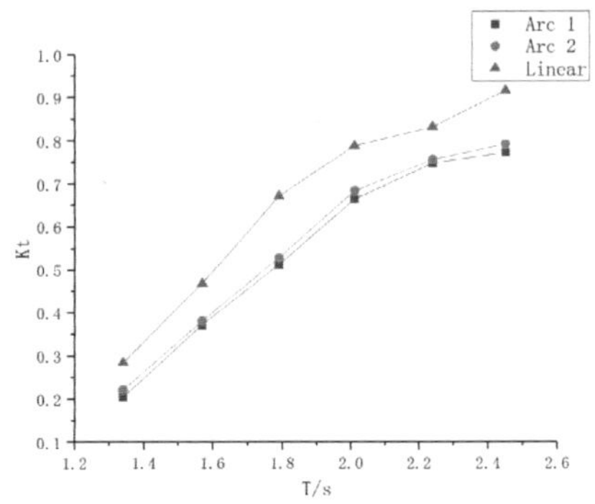

（c）Measuring point 3

Fig.7 Transmission coefficient comparison

From Fig. 7, it can be seen that the transmission coefficient of three kinds of floating breakwaters increases with the increase of wave period T under transverse wave conditions. Comparing the transmission coefficient of three kinds of floating breakwater layout, the linear layout is the largest, and the arc type 1 is the smallest. Since the variation trend of the corresponding measuring points in each working condition of the calculation results is similar, taking the NO.3 measuring point and A3 working condition as an example, when the total length of the arc type and the linear type is the same, the transmission coefficient of the arc type layout is 19.88 % lower than that of the linear type layout, and the wave elimination effect is improved. When the projection distance of arc type and linear type is the same, the transmission coefficient of arc type layout is 21.05 % lower than that of linear type layout, and the wave elimination effect is improved. The difference of projection distance between arc type 1 and arc type 2 is 2.26 m, and the difference of wave attenuation performance is 1.48 %. Compared with the linear layout, the arc layout has better wave attenuation performance no matter under the condition of equal total length with the linear layout or equal projection distance with the linear layout.

3.1.2. Comparative analysis of transmission coefficient under oblique wave condition

According to the numerical results, the A3 condition is selected, and the transmission coefficient is compared and analyzed according to the arranged measuring points. The change of transmission coefficient is shown in Fig.8.

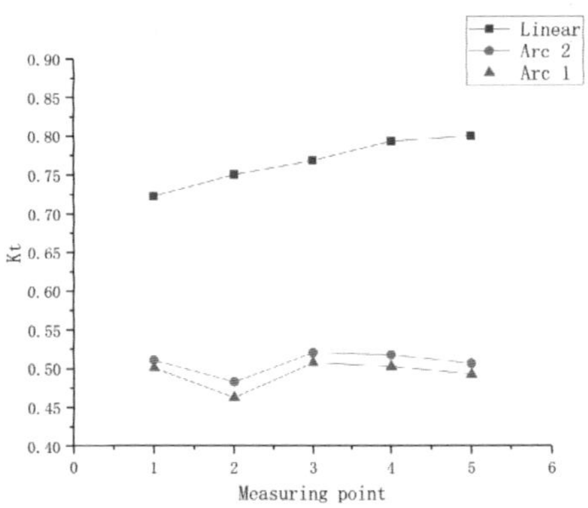

（a）Oblique wave condition

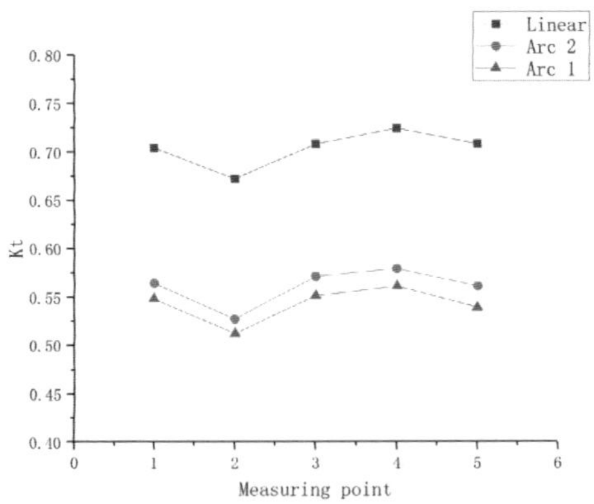

（b）Cross wave condition

Fig.8 Comparison of the transmission coefficient of different measuring points

Under the same regular wave, the change trends of transmission coefficients of different measuring points are roughly the same for the three layouts. It can be seen from the figure that the wave attenuation performance of the linear layout is significantly reduced under the oblique wave condition, while the wave attenuation performance of the arc layout is improved. Taking measuring point 3 as an example, the transmission coefficient of the arc layout 1 under the oblique wave condition is reduced by 8.75 % than that under the cross wave condition, the arc layout 2 is reduced by 7.8 %, and the linear layout increases by 8.61 %. Under oblique wave conditions, the transmission coefficient of arc 1 layout is 33.94 % lower than that of linear layout, and the transmission coefficient of arc 2 layout is 32.49 % lower than that of the linear layout. Therefore, under oblique wave conditions, arc layout has a better wave attenuation performance and shielding performance compared with linear layout.

3.2. Analysis of Calculation Results of Motion Response

3.2.1. Analysis of Calculation Results of Motion Response under Cross Wave Condition

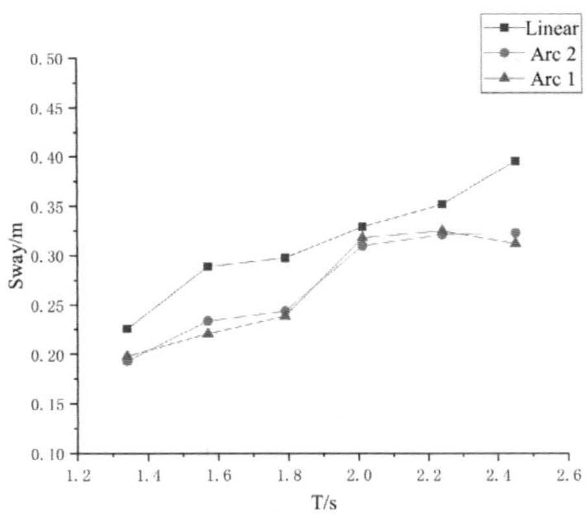

(a) Sway amplitude

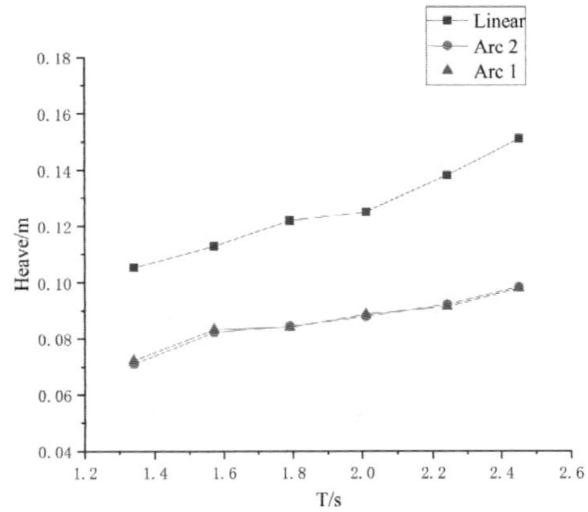

(b) Heave amplitude

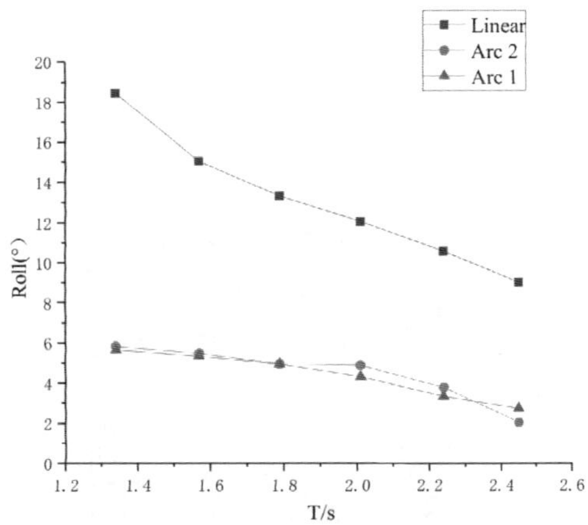

(c) Roll amplitude

Fig. 9 Comparison of Three Layout Motion Responses under Cross Wave Condition

The motion responses of three layouts under cross wave conditions are shown in Fig. 9. Comparing the sway amplitude of three layouts, the sway amplitude of the linear layout is greater than that of arc type 1 and arc type 2 layouts. With the increase of wave period T, the sway amplitude of linear layout increases gradually, from 0.226 m to 0.395 m, with an increase of 74.78 %. The changing trend of sway amplitude of arc type 1 and arc type 2 layout is roughly similar. The sway amplitude of arc type

1 layout is about 71.3% of linear layout, and the sway amplitude of arc type 2 layout is about 73.21 % of that of the linear layout.

By comparing the heave amplitudes of floating breakwaters with three layouts, the heave amplitude of linear layout is significantly greater than that of arc layout 1 and arc layout 2. With the increase of wave period T, the heave amplitude of linear layout increases gradually, from 0.105 m to 0.151 m, with an increase of 43.81 %. The heave amplitude of arc 1 and arc 2 layouts is opposite to that of sway. In general, the heave amplitude of two arc layouts is about 66.73 % of that of the linear layout.

The figure shows that the roll amplitudes of arc 1 and arc 2 decrease first and then increase with the increase of wave period T, and the motion is relatively mild. The roll amplitudes of linear layout decrease with the rise of wave period T, from 18.47° to 9.02°, and the decrease is 51.16 %. The roll amplitudes of two arc layouts are about 23.11 % of that of linear layouts, and the motion is relatively mild.

3.2.2. Comparative analysis of motion response under oblique wave condition

According to the numerical calculation, the A3 working condition is selected for the comparative analysis of the motion response under oblique waves.

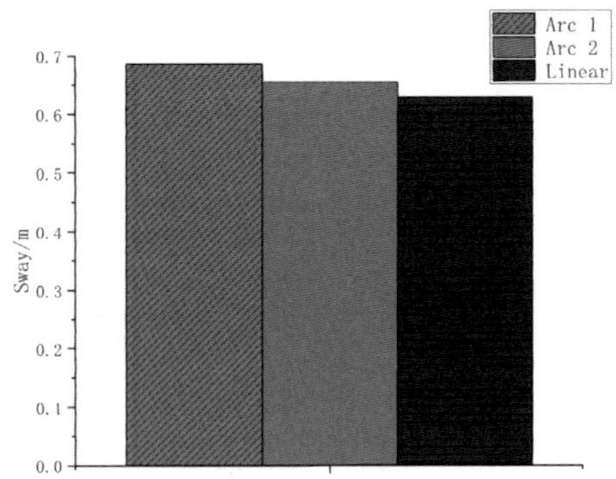

(a) Sway amplitude

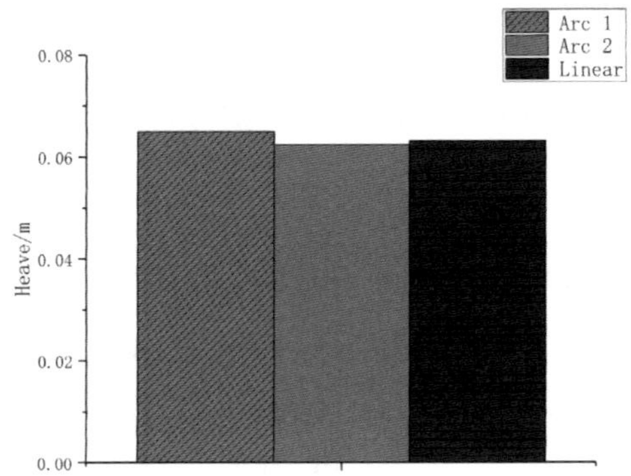

(b) Heave amplitude

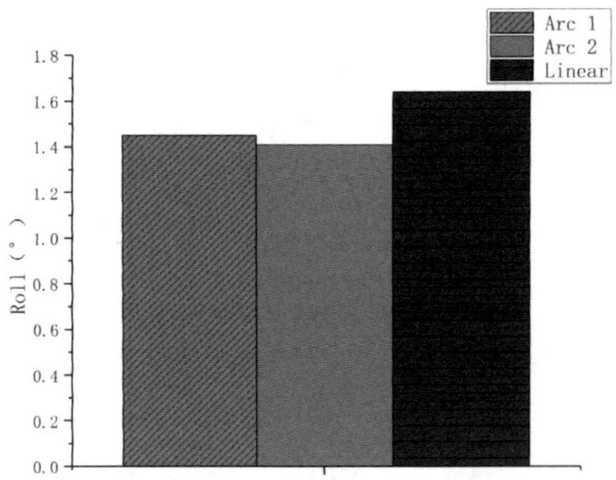

(c) Roll amplitude

Fig.10 Comparison of motion response of three layouts under oblique wave condition

It can be seen from Fig.10 that the sway motion response of floating breakwaters with three layouts under oblique waves is smaller than that under transverse waves. The sway amplitude of arc type 1 is

0.686 m, 9.06% larger than that of linear type, and the sway amplitude of arc type 2 is 0.655 m, 4.13 % larger than that of linear type. In heave motion, the response of the three layouts of the floating breakwater is small, and the difference between the two arc layouts and the linear layout is less than 8 %. The rolling amplitude of arc type 1 is 1.45°, which is 11.58 % smaller than that of linear type, and the rolling amplitude of arc type 2 is 1.41°, which is 14.02 % smaller than that of linear type.

4. Conclusions

In this paper, the arc layout scheme is designed, and the linear layout is taken as the comparison object. The calculation results of two wave directions are obtained in the numerical simulation, and the transmission coefficient, motion response, and mooring cable tension are obtained. Comparative analysis of the hydrodynamic performance of the three layouts, the following conclusions :

(1) Compared with the linear layout, the roll motion response of the arc layout is gentle in the cross-wave condition, which is about 26% of the linear layout in the small period condition and about 30 % of the linear layout in the long period condition.

(2) Under oblique wave conditions, when the projection distance of arc layout and linear layout is the same in the x-axis, the roll amplitude of arc layout is 12% smaller than that of the linear layout. When the total length of arc layout and linear layout is the same, the rolling amplitude of the arc layout is 14% smaller than that of the linear layout. There is a certain roll angle in the roll motion of linear floating breakwater.

(3) Under the cross-wave condition, according to the typical numerical conditions, when the total length of arc and linear layouts is the same, the transmission coefficient of arc layout is 20% lower than that of linear layout, and the wave elimination effect is improved. When the projection distance of arc type is as same as that of linear type, the transmission coefficient of arc type layout is 21 % lower than that of linear type layout, and the wave elimination effect is improved.

(4) Under oblique wave condition, according to the typical measuring points, the transmission coefficient of arc type layout is reduced by 34 % than that of the linear type under the condition of the same projection distance between arc type and linear type, and the transmission coefficient of arc type layout is reduced by 32 % than that of the linear type under the condition of the same total length between arc type and linear type.

(5) By comparing the calculation results of oblique wave and transverse wave, it can be seen that under the condition of a transverse wave, the rolling response of linear layout is significantly stronger than that of arc layout hydrodynamic performance is not good. The mooring tension of the two arc layouts is generally equivalent to that of the linear layout, the rolling response in the two wave directions is relatively mild, and the overall hydrodynamic performance is better. In the cross wave condition, the transmission coefficient of the arc layout is slightly smaller than that of the linear layout, but in the oblique wave condition, the arc layout has a better wave attenuation performance and better shielding performance.

Acknowledgments
National Natural Science Foundation Project 52071161

References:

[1] Lei L, Zhou C, Yuan P Y, Liu J L(2018) Hydrodynamic performance analysis of multi-floating breakwater. Water transportation in China (second half month), 18(09):160-161.
[2] Ding L, Yu J X, Hou J J(2014). Dynamic pressure characteristics of V-shaped floating breakwater. Journal of Drainage and Irrigation Machinery Engineering, 32(01):29-32.
[3] Zheng Z, Jin P, Zhou B Z, Zhu L S(2021) Study on the wave-proof and wave-gathering performance of double parabolic floating breakwater.In: Sixteenth National Hydrodynamics Conference (Thirty-second National Hydrodynamics Seminar). Wu Xi in Jiangsu province: 653-659.
[4] Ge J T, Ji C Y, Guo J T, Bian X Q(2021) Hydrodynamic response analysis of floating

breakwater with arc arrangement. Marine Engineering, 43(09):152-157.
[5] BRIGGS M, YE W, DEMIRBILEK Z, et al(2002) Field and numerical comparisons of the RIBS floating breakwater. Journal of Hydraulic Research, 40(3) : 1-13.
[6] Luca Martinelli, Piero Ruol, Barbara Zanuttigh (2008)Wave basin experiments on floating breakwaters with different layouts. Applied Ocean Research, 30(3).

Study on array fish schools based on lattice Boltzmann method

Haotian Xue[1,a], Jian Xia[1,*]

[1]Nanjing University of Aeronautics and Astronautics, Nanjing, Jiangsu, 210016, China

[a]haotian@nuaa.edu.cn;[*]jxia@nuaa.edu.cn

Abstract—In order to reveal the energy-saving mechanism in the process of fish swimming, an IB-LB model based on implicit boundary force correction is used to study the same-frequency swimming of the basic unit of the fish schools composed of 6 fish. Through research, it is found that lagging fish can absorb energy from the tail vortex of the fish in front, achieving the effect of energy saving, and the fish directly in front provide more energy than the fish in front of the side, but the overall energy loss of the fish schools is basically the fish alone swim the same.

1. Introduction

The group migration movement of fish is a very common phenomenon in nature, and in the course of movement, they always maintain a certain arrangement law. This phenomenon has attracted the attention of scientists as early as the middle of the last century[1,2]. Why fish schools need to maintain a regular formation? After years of research by predecessors, people now have a certain understanding of this problem. Through observation and calculation, it is generally believed that maintaining a certain regular formation can effectively utilize the flow induced by the vortex street next to it and reduce energy consumption during exercise. Magnuson[3] believes that the swimming of a school of fish can save an average of 10% to 20% of energy.

In order to better explain the energy-saving mechanism in the regular arrangement of fish schools, this paper will adopt a direct force format IB-LB method. Different from the basic unit of fish school composed of 3 or 4 fish in previous studies, this article studied the combination of 6 fishes containing the two basic units of fish school (triangle and diamond) at the same time, and compared the lift resistance in different arrangements.

2. Numerical method

The LB model uses the temporal and spatial evolution of the mesoscopic density distribution function to describe the movement of the fluid. The lattice Boltzmann method is associated with the lattice velocity model. In the present work, the 9-bit lattice velocity model (D2Q9) is applied, which consists of the following lattice velocity directions:

$$c_i = \begin{cases} (0,0) & ,(i=0) \\ c\left(\cos\left[(i-1)\frac{\pi}{2}\right], \sin\left[(i-1)\frac{\pi}{2}\right]\right) & ,(i=1\sim4) \\ \sqrt{2}c\left(\cos\left[(2i-1)\frac{\pi}{2}\right], \sin\left[(2i-1)\frac{\pi}{2}\right]\right) & ,(i=5\sim8) \end{cases} \tag{1}$$

where $c = \delta_x / \delta_t$ is set to be 1 in this work; and δ_x is the lattice spacing, δ_t is time step. The lattice Boltzmann equation[4] can be expressed as:

$$f_i(x + c_i\delta_t, t + \delta_t) - f_i(x,t) = -\frac{1}{\tau}\left[f_i(x,t) - f_i^{eq}(x,t)\right] + \left(1 - \frac{1}{2\tau}\right)\delta_t F_i \tag{2}$$

where $\boldsymbol{\tau}$ is the relaxation time, the forcing term $\boldsymbol{F_i}$ can be written:

$$\boldsymbol{F_i(f,u)} = \boldsymbol{\omega_i}\left(\frac{\boldsymbol{c_i - u}}{\boldsymbol{c_s^2}} + \frac{\boldsymbol{c_i^2 \cdot u}}{\boldsymbol{c_s^4}}\right)\cdot \boldsymbol{f} \tag{3}$$

The equilibrium distribution functions are given by

$$\boldsymbol{f_i^{eq}} = \boldsymbol{\omega_i \rho}\left[\boldsymbol{1} + \frac{\boldsymbol{c_i u}}{\boldsymbol{c_s^2}} + \frac{(\boldsymbol{c_i u})^2}{\boldsymbol{2c_s^4}} - \frac{\boldsymbol{u^2}}{\boldsymbol{2c_s^2}}\right] \tag{4}$$

the sound speed $\boldsymbol{c_s} = \frac{\boldsymbol{1}}{\sqrt{3}}\boldsymbol{c}$. The weighting coefficient $\boldsymbol{\omega_i}$ are

$$\boldsymbol{\omega_i} = \begin{cases} \boldsymbol{4/9}, & \boldsymbol{i = 0} \\ \boldsymbol{1/9}, & \boldsymbol{i = 1 \sim 4} \\ \boldsymbol{1/36}, & \boldsymbol{i = 5 \sim 8} \end{cases} \tag{5}$$

The constant relaxation time $\boldsymbol{\tau}$ and the fluid viscosity $\boldsymbol{v}$ are linked as follows:

$$\boldsymbol{v} = \boldsymbol{c_s^2}\left(\boldsymbol{\tau} - \frac{\boldsymbol{1}}{\boldsymbol{2}}\right)\boldsymbol{\delta_t} \tag{6}$$

the basic hydrodynamic quantities, such as the density and the velocity components, are obtained through moment summations in the velocity space:

$$\boldsymbol{\rho} = \textstyle\sum_0^8 \boldsymbol{f_i},\quad \boldsymbol{\rho u} = \sum_0^8 \boldsymbol{c_i f_i} + \frac{1}{2}\boldsymbol{f\delta_t},\quad \boldsymbol{p} = \boldsymbol{c_s^2 \rho}$$

In order to better simulate the constantly changing boundary position of the fish in the swimming process, the immersion boundary method [5-8] can be used to calculate the moving boundary. Compared with the traditional body-fitted grid method, the IB method has more advantages in adaptability and computational efficiency.

In this paper, an IB method based on the direct force format of Richardson iteration is adopted. On the basis of ensuring the accuracy of the numerical solution, the non-physical vibration near the boundary is suppressed.

First, given the initial conditions of the flow field $(\boldsymbol{\rho_0}, \boldsymbol{u_0})$ and the body force density $\tilde{\boldsymbol{f}}$, the initial distribution function is constructed through Eq.(3) and Eq.(4). Among them, the distribution function, density and velocity related to $\tilde{\boldsymbol{f}}$,are expressed as $\widetilde{\boldsymbol{f_\alpha}}$, $\tilde{\boldsymbol{\rho}}$, $\tilde{\boldsymbol{u}}$. Since $\tilde{\boldsymbol{f}}$ is an estimated value, the speed cannot accurately meet the no-slip condition, so the volume force $\boldsymbol{f}$ and the speed $\boldsymbol{u}$ need to be corrected. This process is completed by the IB model. According to the real velocity at the Lagrangian node (the boundary of the object immersed in the fluid) and the interpolation velocity, $\boldsymbol{U'}$ that satisfies the no-slip constraint is calculated and the boundary force $\boldsymbol{F'}$ is constructed, and then it is diffused to the Euler node. The corrected body force density and velocity are:

$$\boldsymbol{f} = \tilde{\boldsymbol{f}} + \boldsymbol{f'} \tag{7}$$

$$\boldsymbol{u} = \tilde{\boldsymbol{u}} + \boldsymbol{u'} \tag{8}$$

Use f on the left side of Eq.(7) as the new $\tilde{f}$ and repeat the above process until convergence.

3. Physical model

This paper is based on a two-dimensional simulation. The computational domain is shown in Fig.1.Construct a fish model based on NACA0012 airfoil. The fish body length is L, the calculation domain area is $8L \times 15L$, the distance between two adjacent fish in the x direction is $0.75L$, and the distance in the y direction is $0.5L$. The position parameters are shown in Fig.1.In order to understand the impact of wake vortices on fish, the oscillation frequency of the fish is the same $f = 0.00054$,the Strouhal number $S_t = Lf\ /\ U_\infty = 0.54$.

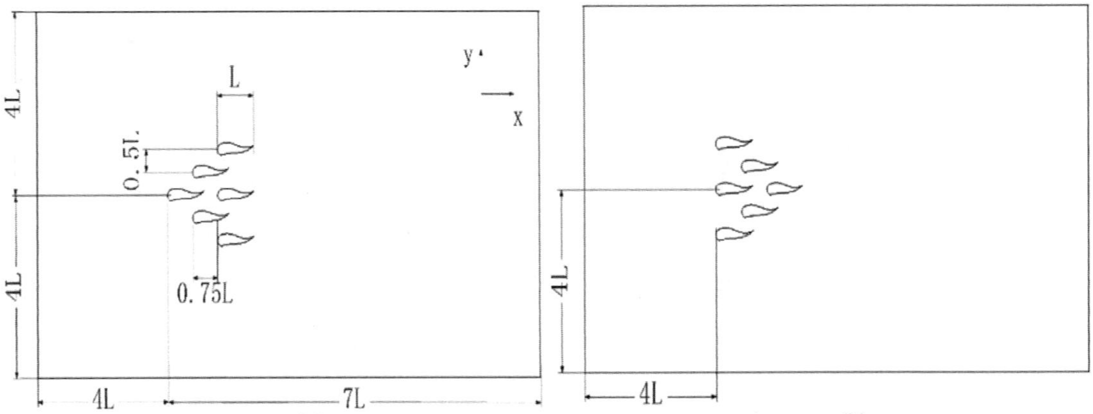

(a) (b)

Fig.1 Schematics of computational model. (a) The first type of formation. (b) the second type of formation.

The motion of the chamber line of the fish can be described by the following function:

$$\begin{cases} translation: \quad y_{(t)} = H_0 \cdot sin(2\pi f t) \\ undulation: \; y_{(t)} = A_{(x)} \cdot sin(2\pi f t + \phi) \end{cases}$$

and $A_{(x)} = 0.01Lf(x-0.4) + 0.51Lf(x-0.4)^2$ represents the amplitude of the chamber line, x represents the horizontal distance from the chamber line to the fish head, ϕ is undulation phase angle, $H_0 = 0.2L$.

4. Results and discussion

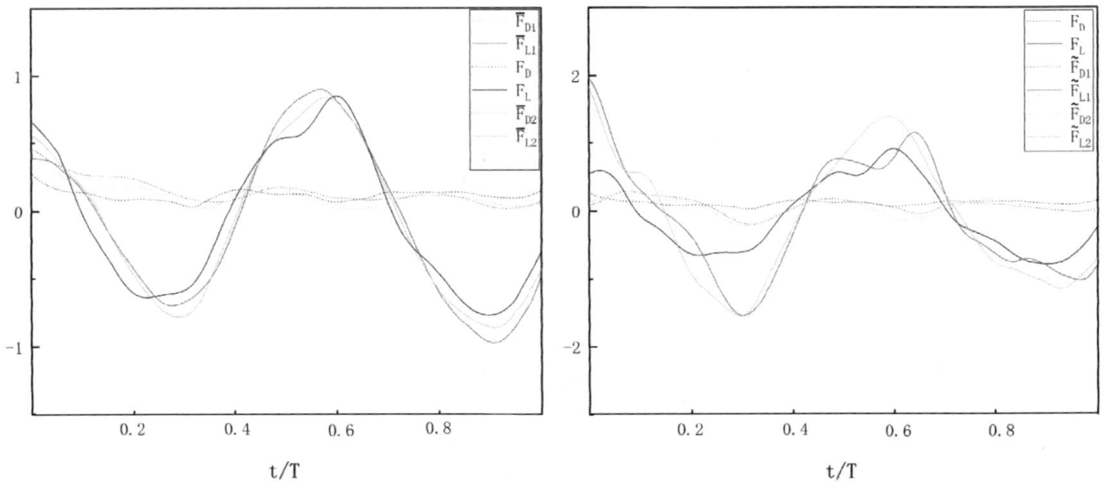

(a) (b)

Fig.2 Time-dependent lift and drag acting on the model

Fig.2 shows the change curve of lift and drag acting on the surface of the model. In the figure, F_D, F_L respectively represent the resistance and lift of a single fish in motion; $\bar{F}_{D1}, \bar{F}_{L1}$ represents the average drag and lift of the fish in the first arrangement; $\bar{F}_{D2}, \bar{F}_{L2}$ represents the average drag and lift of the fish in the second arrangement; $\tilde{F}_{D1}, \tilde{F}_{L1}$ and $\tilde{F}_{D2}, \tilde{F}_{L2}$ respectively represent the resistance and lift of a single model at the middle of the end of the team in the first and second arrangement formations. In (b), it can be seen that the lift generated by the fish at the end of the queue is obviously greater than that of the single fish when moving, but the resistance is basically the same. That is to say, the lagging fish can absorb energy from the tail vortex of the fish in front to provide power for itself. In (a), it is found that the average lift resistance curve of a fish shoal shows the same trend as that of a single fish, because the latter fish absorbs the energy of the forefish and the forefish reduces the lift generated [10].

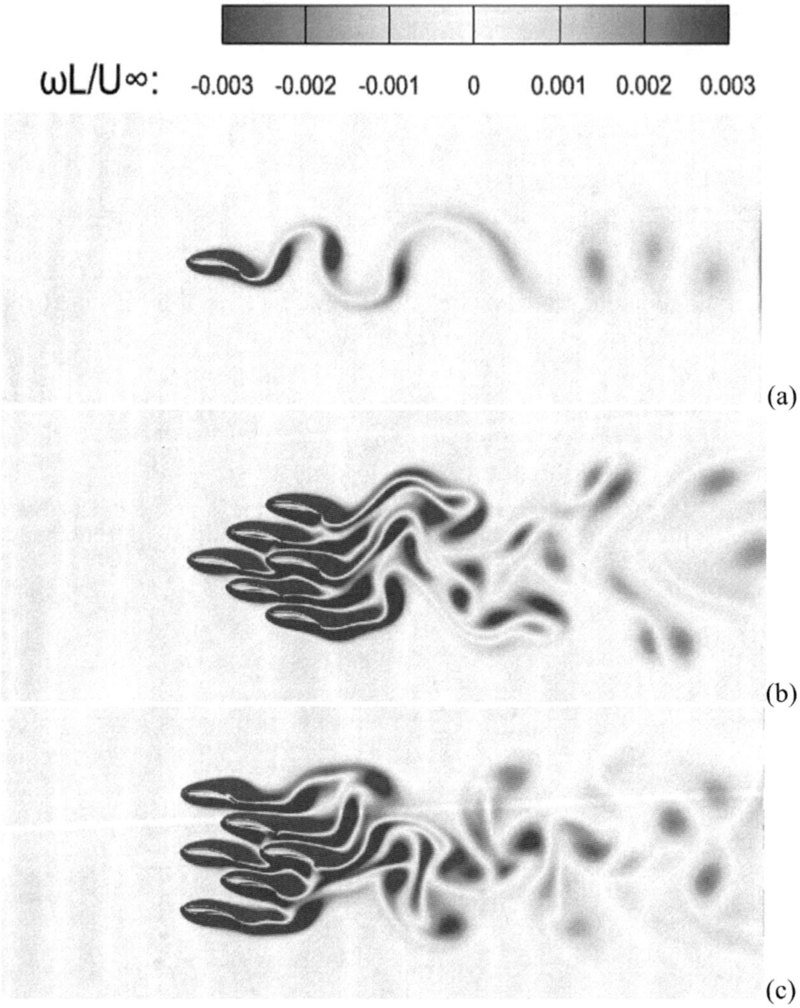

Fig.3 Vortex contour near the swimming model in a uniform flow.

According to the instantaneous vorticity diagram of the flow field in a period of motion, it can be seen that a periodically changing shear layer is produced around the model. The shear layer develops flow separation near the trailing edge of the airfoil, and alternating vortices with opposite signs form a classic Karman vortex street in the wake of the wave wing, as shown in Fig.3(a). (b)(c) shows the overall swimming formation of the two fish schools and the induced vorticity field. The wake vortex produced by the front fish will tilt towards the rear fish and merge with the rear vortex, making it propulsion There is a difference in performance. At this time, the school of fish mainly relies on the wake vortex of the fish directly in front rather than the vortex of the fish on the side to absorb energy.

5. Conclusion

By studying the fish school unit composed of 6 fishes, it is found that the backward fish can produce greater lift, that is, the backward fish can absorb energy from the vortex of the previous school of fish, achieving the effect of energy saving. Therefore, the strong fish should be ranked first when the school of fish is swimming. The fish school as a whole does not achieve the purpose of energy saving, which may be related to the swing frequency of the fish and the distance between each other, not just the formation of the fish. Since the swimming environment of six fish is relatively complex, autonomous swimming under different swing frequencies will be further studied in the follow-up work to better simulate the motion state of fish.

References

[1] D. H. Cushing, F. R. Harden Jones. Why do fish school? NATURE, 1968, 218(5145): 918-920.
[2] J. Krebs. Fish schooling. NATURE, 1976, 264(5588): 701-70
[3] Magnuson JJ. Locomotion by Scombrid Fishes: Hydromechanics, Morphology and Behavior, in Fish Physiology Vol.7: Locomotion, Hoar WS, Randall DJ, eds. New York: Academic, 1978. 239-313
[4] Guo Z , Zheng C , Shi B . Discrete lattice effects on the forcing term in the lattice Boltzmann method[J]. Physical Review E, 2002, 65(4):046308.
[5] Peskin C S . The immersed boundary method[J]. Acta Numerica, 2002:1-36.
[6] Mittal, R. and G. Iaccarino (2005). "Immersed Boundary Methods." Annual Review of Fluid Mechanics 37(1): 239-261.
[7] Huang W X , Tian F B . Recent trends and progress in the immersed boundary method[J]. ARCHIVE Proceedings of the Institution of Mechanical Engineers Part C Journal of Mechanical Engineering Science 1989-1996 (vols 203-210), 2019, 233(23-24):095440621984260.
[8] Yang Ming, Liu Jubao, Yue Qianbei, et al. Numerical simulation on the vortex-induced collision of two side-by-side cylinders. Chinese Journal of Theoretical and Applied Mechanics, 2019, 51(6): 1785-1796 (in Chinese).
[9] Lauder George V. Hydrodynamics of median-fin interactions in fish-like locomotion: Effects of fin shape and movement [J] Physics of Fluids,2020
[10] Wang L, Wu CJ. Research on energy saving mechanism of "channel effect" in fish swimming [J].Chinese journal of theoretical and applied mechani,2011,43(01):18-23 (in Chinese).

Journal of Physics: Conference Series **2271** (2022) 012023 doi:10.1088/1742-6596/2271/1/012023

Stability analysis of expansion and efficiency improvement of bulb tubular turbine

Xiaoyong Tang

(Datang Hydropower Science & Technology Institute, Nanning, Guangxi 530000)

Email: txytcsf@163.com

Abstract. Many bulb tubular turbines were put into operation earlier. The construction of upper and lower cascade power stations leads to the decrease of generating head. Unit efficiency decline can not reach the design output. Large quantities of water were abandoned from the power station. Taking the expansion and efficiency improvement of water turbine generator unit in Bailongtan Hydropower Station as an example. The stability and efficiency improvement before and after expansion are compared and analyzed. The transformation not only solves the problem of high vibration of low waterhead and failure to reach the designed output, but also increases the flow and output of the unit, reduces abandoned water of the power station, improves the utilization rate of hydraulic resources, improves the stability of the unit and ensures the safe operation of the power plant. The result of unit transformation is good. It can be used as a reference for the transformation of similar tubular units.

1. Introduction

Bailongtan Hydropower station is located in the middle reaches of hongshui River at the junction of Du 'an Yao Autonomous County and Mashan County. The hydropower station is equipped with six bulb tubular turbine generators with a single capacity of 32MW, with a total installed capacity of 192MW [1-2]. The Bailongtan hydropower station was built and put into operation earlier, and has been in operation for more than 20 years. The operation conditions of hydropower stations have changed greatly due to the operation of upstream cascade power stations. The maximum waterhead is reduced from 18m to 16m, and the running time of low waterhead is increased obviously. The lower the head is, the greater the vibration is, the more difficult it is to ensure the safe and stable operation, the lower the efficiency of the unit, and the output is severely limited [3-5]. Power plant is forbidden to operate in water head less than 5m for the safe operation of unit. As a result, a large amount of water is abandoned at low head, resulting in serious loss of power generation and reduced economic benefits of power plants. In order to ensure the safe and efficient production of the power plant, new technology will be adopted to expand and improve the efficiency of the No. 3 unit [6-7].

2. The basic situation of the unit transformation

The main object of transformation is the runner and the flow passage components. This reform maintains the runner size unchanged, the main change unit runner room, turbine runner, spindle seal, inner guide ring and other components. By optimizing the runner configuration and adopting advanced technologies such as VOD refining and numerical control machining, the unit output is increased from 32MW to 33MW. Comparison of turbine parameters before and after modification is shown in Table 1.

The new runner consists of a blade, a hub body, an operating mechanism and a drain cone. The hub ratio is reduced from 0.428 to 0.38, and the runner has sufficient strength and stiffness to withstand any maximum pressure, centrifugal force and pressure pulsation that may occur on the runner. Under the operating conditions of the power station, there will be no blade stuck, water inflow, excessive oil leakage or any other abnormal phenomena.

Table 1. Comparison of new and old turbine parameters

Name	Bulb propeller turbine	
	Old turbine	**New turbine**
Runner diameter	6.4m	6.4m
Rated head	9.7m	9.5m
Rated flow	377.5m^3/s	388.4 m^3/s
rated power	32MW	33MW
Rated speed	93.8r/min	93.8r/min
Runaway speed	310r/min	274.51r/min
Rated efficiency	92%	91.92%
Maximum efficiency	95.2%	95.4%
Waterhead	3m~18m	3m~16m

The runner chamber adopts a thick wall structure with sufficient stiffness and strength. According to the transportation conditions, the combination of block welding is adopted. The inner surface profile is similar to the accepted model. In order to avoid clearance cavitation, the flow surface of the runner chamber is made of 06Cr19Ni10 stainless steel plate. The water guide mechanism adopts original parts except the inner ring of guide vane. The inner ring of guide vane is molded by steel plate and processed by numerical control. The upstream and downstream sides are connected with flanges, with reliable leak-stopping device.

The new spindle seal consists of an overhaul seal and a working seal. The seal structure of the main shaft adopts lip seal with reliable performance. Maintenance seal adopts pressurized solid air girdle structure.

3. Stability analysis of unit after modification

Stability tests (variable speed test, variable excitation test and variable load test) were carried out before and after the modification to comprehensively analyze the stability index performance of the unit before and after the expansion and efficiency improvement. Verify whether the modified performance parameters are in line with national industry-related standards and contract guaranteed values. It can be used as reference for expansion and efficiency improvement of similar units.

3.1 Test parameters

Table 2. Stability test parameters before and after modification

Test	Pre-modification test	Post modification test
waterhead /m	11	9
speed (Variable speed test)	60%、70%、80%、 90%、100%	The same as the test before modification
The excitation voltage (Variable excitation test)	20%、40%、60%、 80%、100%	The same as the test before modification
Power/MW (Variable power test)	3、6、9、12、15、18、 21、24、27、30、32	The same as the test before modification

Measuring points (10)	Hydroconductivity pendulum (2), hydroconductivity vibration (3), combined vibration (3), bulb head vibration (2)	The same as the test before modification

3.2 Stability test results

Analysis and comparison of pendulum vibration amplitude data and change trend chart of each measuring point in stability test shows that:

(1) As shown in Table3-Table 5, the stability test results after expansion and efficiency enhancement transformation show that. The vibration and pendulum values of each measuring point are small. It shows that the unit has good alignment and mass distribution, no electromagnetic unbalance and good hydraulic performance. All the stability test data of the modified unit are within the standard allowable range, and all the data are far less than the standard value. It indicates that the new unit has good stability.

Table 3. Pendulum vibration values of unit at variable speed test before and after modification

Measuring point	Before modification (Maximum peak-to-peak value)	After transforming (Maximum peak-to-peak value)	Standard values
Pendulum of guidebearing plus X	267	217	≯420
Pendulum of guidebearing plus Y	248	232	≯420
Horizontal vibration of guidebearing	185	83	≯120
Vertical vibration of guidebearing	46	33	≯120
Axial vibration of guidebearing	227	98	≯100
Horizontal vibration of combined bearing	17	15	≯120
Vertical vibration of combined bearing	5	5	≯120
Axial vibration of combined bearing	49	48	≯100
Horizontal vibration of the bulb head	52	35	≯120
Vertical vibration of the bulb head	19	17	≯120

(2) A comparative analysis of the detection data of each measuring point before and after the transformation shows that most parameters of the unit after the transformation are obviously better than those before the transformation. The result is shown in Figure 1. The vibration data of guidebearing changes obviously before and after modification. Before the reconstruction, the vibration data of guidebearing increases with the increase of the load, and when the load reaches 30MW, the vibration exceeds the standard. The vibration of the modified unit is not affected by load variation.

Table 4. Pendulum vibration values of unit under variable excitation test before and after modification

Measuring point	Before modification (Maximum peak-to-peak value)	After transforming (Maximum peak-to-peak value)	Standard values
Pendulum of guidebearing plus X	250	236	≯420
Pendulum of guidebearing plus Y	241	236	≯420
Horizontal vibration of guidebearing	192	117	≯120
Vertical vibration of guidebearing	48	32	≯120
Axial vibration of guidebearing	168	89	≯100
Horizontal vibration of combined bearing	16	15	≯120

Vertical vibration of combined bearing	5	5	≯120
Axial vibration of combined bearing	49	47	≯100
Horizontal vibration of the bulb head	45	33	≯120
Vertical vibration of the bulb head	19	16	≯120

Table 5. Pendulum vibration values of unit under variable load test before and after modification

Measuring point	Before modification (Maximum peak-to-peak value)	After transforming (Maximum peak-to-peak value)	Standard values
Pendulum of guidebearing plus X	237	245	≯420
Pendulum of guidebearing plus Y	243	248	≯420
Horizontal vibration of guidebearing	147	114	≯120
Vertical vibration of guidebearing	39	22	≯120
Axial vibration of guidebearing	83	83	≯100
Horizontal vibration of combined bearing	21	12	≯120
Vertical vibration of combined bearing	28	22	≯120
Axial vibration of combined bearing	32	46	≯100
Horizontal vibration of the bulb head	42	26	≯120
Vertical vibration of the bulb head	39	40	≯120

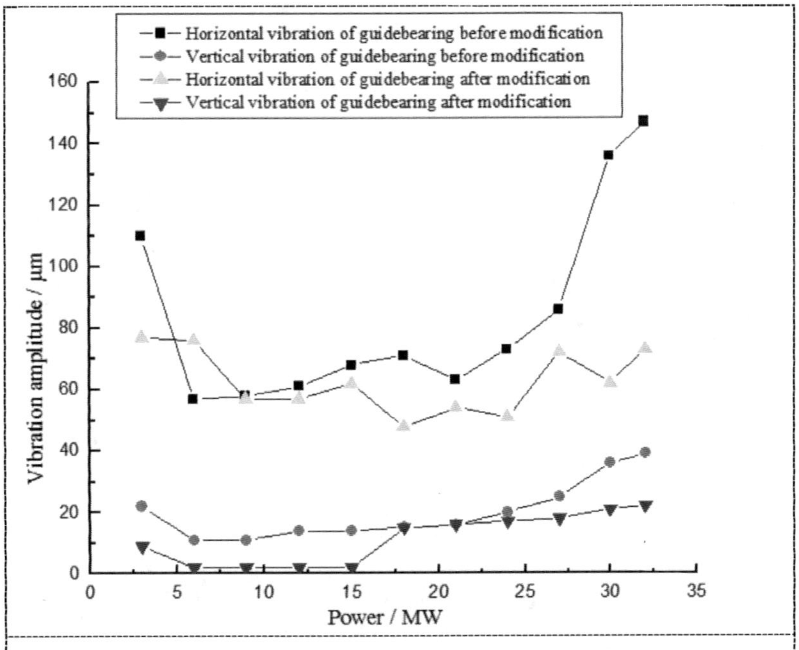

Fig 1. Comparison of stability curves before and after modification

As unit 3 of Bailongtan Power Station has been in operation for more than 20 years, and the water level of upstream and downstream has changed. According to the results of related stability tests, the stability of the unit began to deteriorate, and some key parts of the parameters have exceeded the standard. According to the requirements of relevant regulations, the acceptance test of No.3 unit after transformation is carried out. All the test data are good and meet the requirements of the regulation. The unit meets the acceptance standard. After commissioning and grid connection of the unit, the technical indexes meet the design requirements, and the vibration, swing and temperature of each part are normal. After expansion and efficiency improvement, the new unit can well adapt to the new hydrological conditions, and the overall stability is significantly improved. Even in the range of low head and low load, the vibration pendulum value of each measuring point is far lower than the relevant national standard value.

4. Analysis on the effect of unit expansion and efficiency improvement

On the premise that the runner passage and runner chamber diameter are not changed, the runner structure of unit 3 of power station is optimized and reformed. The overflow of the unit is increased, the design output is increased from 32MW to 33MW, the water abandonment rate of the power station is reduced, and the economic benefit of the power station is improved. The unit adapts to the new hydrologic working condition and the modification effect is good. After modification, the hydraulic performance of the unit is better and the influence of tail water pressure change on the stability of the unit is reduced. The purpose of unit expansion and efficiency enhancement under low head condition is realized. At the same time, the overall stability performance of the unit is optimized, the efficiency and output of the unit are improved. The value of vibration pendulum also decreases obviously compared with that before modification.

Table 6. Variation value of unit efficiency after modification

Head/m	Increase rate overflow after transformation /%	Increase rate of output after transformation /%	Increase rate of efficiency after transformation /%
7.6	12	25	11.5
9.1	10	23	10

After modification, the hub ratio of the runner is reduced from 0.428 to 0.38, which increases the flow area and improves the flow capacity of the runner by 10%~12%. Under the same guide vane opening, the output increases by 10%~25% compared with the original unit. See Table 6. When the water head of Unit 3 is 9.1m, the output of the transformed unit reaches 26MW, which is 5MW more than the original unit's output of 21MW. The output of the unit is improved by 23% and the efficiency is improved by about 10%. When the water head is 7.6m, the output of the modified unit reaches 20MW, which is 4MW higher than the original unit's output of 16MW. The output of the unit is increased by 25%, and the efficiency is increased by about 11.5%. It shows that the unit output is significantly increased and the efficiency is significantly improved after the transformation. It has created good economic and social benefits.

5. Conclusion

(1) The technical transformation of the unit not only solves the problems of large vibration of low head and failure to reach the designed output, but also improves the stability of the unit and eliminates the hidden danger of safety of the original unit. So that the unit can adapt to the situation of constant change of head, especially the situation of increasing running time of low head. It ensures the safe and stable operation of the unit.

(2) Through the expansion and efficiency improvement of the unit, the overflow of the unit is increased and the output of the unit is improved. The unit can achieve the design output under the condition of low head, and the unit's production efficiency and utilization rate of water resources are significantly improved. At 9.1m head, unit output is increased by 23% and efficiency is increased by 10%. At 7.6m head, the unit output is increased by 25% and the efficiency is increased by 11.5%.

The expansion and efficiency improvement of No. 3 unit of Bailongtan Power Plant has achieved good results, eliminating the hidden danger of the original unit, adapting to the new hydrological conditions, improving the stability of the unit, and increasing the output and efficiency of the unit. It can provide a reference for the similar renovation work of bulb tubular turbines.

References

[1] Guangxi Electric Power Industry Survey and Design Institute. Feasibility Study on Turbine Reconstruction of Bailongtan Hydropower Station in Hongshuihe [R]. Nanning: Guangxi Electric Power Industry Survey, Design and Research Institute, 2012.
[2] Ding Wang, Nanwen Chen. Research on Technical Transformation of Large Bulb Tubular Turbine Runner [J]. Hongshui river, 2019, 38 (6): 13 to 16.
[3] Chongzhi Li, Guangyao Zhai. Research and Practice of Turbine Efficiency and Capacity Expansion in the Second Phase of Hong Kong Power Station [J]. China Rural Water and Hydropower, 2018, (6):182-184.
[4] Peng Xue, Xin Wang, Yajuan Tian, et al. Main problems and Solutions of Efficiency and Capacity Expansion of small and medium-sized hydropower stations [J]. China Rural Water and Hydropower, 2014, (2):133-136.
[5] Xifang Chen. Capacity Enhancement and Optimal Operation of Water Turbine Generator Set [M]. Beijing: China Water and Power Press, 2010:1-4.
[6] Long Ma, Wu Zhou, et al. Stability Analysis of Capacity Expansion and Efficiency Enhancement for Unit 3 of Gezhouba Power Station [J]. Large Electric Machine Technology, 2012, 6:44-47.
[7] Xuping Fu, Feng Zheng. Efficiency of Bulb Tubular Turbine [J]. Water and power energy science, 2008, 26(6):132-133.

Calculation Method of Foundation-Subgrade Dynamic Interaction for Large Wind Turbine

Zhifan Han[1a*], Guohua Lu[1], Wen Li[1], Guang Deng[2]

[1] Construction Institute, Guangdong Technology College, Zhaoqing 526000, China

[2] Gong Cheng Management Consulting Co., Ltd., Guangzhou,510000, China

[a]*hanzfhanzf@163.com

Abstract: Taking a 3MW large wind turbine foundation as the research object, the dynamic interaction model of wind turbine foundation and subgrade is established. By intercepting the subgrade in a limited range and setting spring-damper elements at the boundary of the subgrade, the semi-infinite space of the subgrade is simulated. In order to facilitate the use of structural designers, the calculation methods of the stiffness coefficient and damping coefficient of the spring-damper element are simplified. The simplified calculation formulas for the stiffness coefficient and damping coefficient of the spring-damper unit are obtained when the intercepted subgrade range is 5 times and 10 times the wind turbine foundation.

1. Introduction

Wind power is a clean form of power generation with little impact on the ecological environment. In recent years, the world's wind power generation has grown rapidly, and wind power has become the fastest growing renewable energy technology[1]. The installed capacity of wind power in China ranks first in the world, which relies mainly on onshore wind power[2]. The power generation cost of onshore wind power is lower than that of offshore wind power, and is comparable to that of hydropower[3]. It is expected that by 2040, the growth rate of China's onshore wind power installations will maintain a relatively high growth rate. The single-unit capacity of newly built onshore wind turbines is mainly 3MW, but with the advancement of technology, the single-unit capacity will further increase in the future. Large-scale onshore wind turbines mainly use circular reinforced concrete to expand the foundation, and the foundation diameter of 3MW wind turbines is basically greater than 20m.Onshore wind turbines are mainly concentrated in the mountains and hills with high altitudes, where the terrain is complex, and the construction is inconvenient. During the construction process, the wind turbine foundation costs a large amount of money and time. However, the dynamic interaction between the fan foundation and the subgrade is not considered in the design of the fan foundation, and the results obtained in the dynamic analysis often greatly deviated from the actual situation, resulting in an overly conservative design.

For the study of the interaction between the foundation and the subgrade of large wind turbines, Zaaijer[4] used the stiffness matrix to replace the subgrade to simplify the dynamic model of the wind turbine foundation, but there was a high error in the dynamic calculation of some wind turbines. Achmus et al.[5] used the degradation stiffness model to represent the influence of subgrade on the foundation of the wind turbine, but the model needs further verification in practical application, and it is recommended to adopt a conservative method in practical design. Ke et al.[6] considered the foundation-subgrade interaction by arranging mass springs and dampers on the fan foundation and subgrade interface. In practical engineering, engineers intercept a limited range of subgrades and impose fixed constraints on the boundaries of the intercepted subgrades. The intercepted subgrades are calculated together with the fan foundation to consider the fan foundation-subgrade interaction[7-8]. These methods do not regard the subgrade as a semi-infinite space, and do not consider the wave propagation and energy dissipation in the subgrade semi-infinite space, and the calculated results deviate greatly from the actual situation.

In view of this, this paper takes a large onshore 3MW wind turbine in Guangdong Province as an example, establishes a calculation model considering the dynamic interaction between the foundation and the subgrade of the wind turbine, and proposes a simplified method that is convenient for structural engineers to calculate.

2. Establishment of a large wind turbines foundation structure and dynamic interaction model

This paper is based on a 3MW large wind turbine foundation in Guangdong Province. The wind turbine foundation adopts a circular expansion foundation with a diameter of 23.4m and a height of 4.2m. The diameter of the foundation column is 8.0m, depth of foundation is 4m, and the concrete strength grade is C40. The design load combination of wind turbine foundation under extreme working conditions is horizontal force F_{xy}=1673.2kN vertical force F_z=4126.1kN, resultant moment M_{xy}=143554$kN{\cdot}m$, safety factor is taken as 1.0, and the load combination is applied to the top surface of the foundation.

$$v = v_0 + v_t \tag{1}$$

When calculating wind loads, the total wind speed v is usually regarded as the superposition of the average wind speed v_0 and the fluctuating wind speed v_t:

$$F = F_0 + F_a \sin(2\pi t/T) \tag{2}$$

The handling of semi-infinite subgrades is critical when modeling wind turbine foundation-subgrade dynamic interactions. The most widely used method is intercepting a limited range of subgrades and imposing artificial boundaries on the subgrade boundaries, which is easy for structural designers to understand and easy to implement in calculation software. The large-scale wind turbine foundation-subgrade calculation model is shown in Figure 1. Under this situation, the size of the intercepted subgrade range and the setting of artificial boundaries will affect the calculation accuracy and efficiency, which are the two key factors for the application of this method.

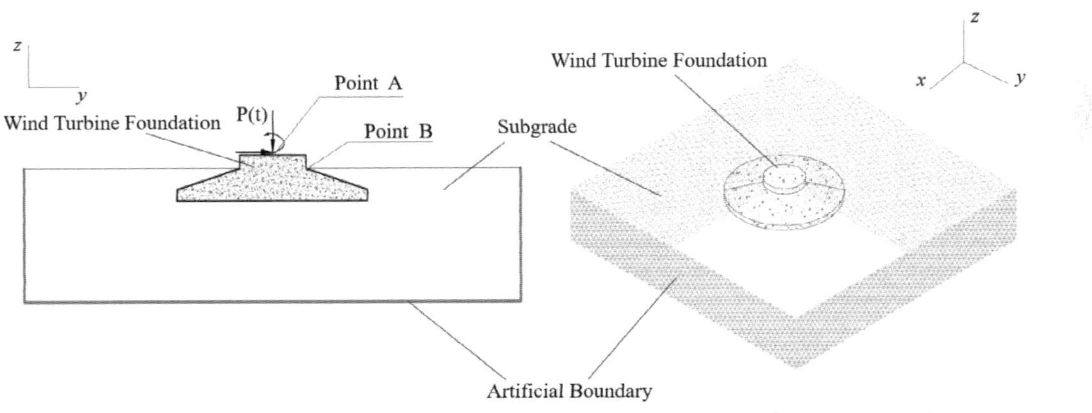

Fig.1 Calculation model of large-scale wind turbine foundation-subgrade dynamic interaction

$$\sigma(r,t) = -\rho C\dot{u}(r,t) - Ku(r,t) \tag{3}$$

In the design and calculation, simply setting the artificial boundary in Fig. 1 as a fixed boundary cannot accurately simulate the subgrade semi-infinite space, and the wave will cause reflection after propagating to the boundary and affect the calculation result. Using a stress-type artificial boundary, the boundary stress can be expressed by velocity and displacement at the artificial boundary node, as follows:

$$\sigma(r,t) = -\rho C\dot{u}(r,t) - Ku(r,t) \tag{4}$$

Among them, r is the distance from the artificial boundary point to the point source.

After analyzing formula (4), the velocity term coefficient $-\rho C$ can be equivalent to the damper on the boundary node, and the displacement term coefficient $-K$ can be equivalent to a linear spring. By setting the spring-damper on the artificial boundary element, the process of dynamic waves passing through artificial boundaries is simulated. Based on the time domain expression of the artificial boundary displacement wave, the subgrade semi-infinite domain is assumed as an elastic body, and the physical and geometric equations are introduced to obtain the normal stress and shear stress of the boundary nodes:

$$\sigma_N = -\frac{(\lambda+2G)}{r(1+br/al)} u_N(r,t) - \rho c_p \dot{u}_N(r,t) \tag{5}$$

$$\sigma_T = -\frac{G}{r(1+br/al)} u_T(r,t) - \rho c_s \dot{u}_T(r,t) \tag{6}$$

ρ is the subgrade density; λ is the Lame coefficient; G is the shear modulus; c_p is the longitudinal wave velocity; c_s is the shear wave velocity; a is the number of scattering wavelets; b is the number of plane wavelets; l is the dimension Coordination factor.

The dimensionless parameter $A = br/al$, $A \in [0, +\infty)$ is introduced as the adjustment coefficient of the artificial boundary stiffness. Considering the incidence of mixed wavelets, the dimensionless parameter B is introduced as the correction coefficient.

Parameters $\alpha = E(1-\mu)/[(1-2\mu)(1+\mu)]$ and $\beta = (1-2\mu)/(2-2\mu)$ are introduced, where μ is Poisson's ratio. The stiffness and damping coefficients of the boundary node spring-damper element are:

Normal direction N of boundary node:

$$K_N = \frac{\alpha}{r}\frac{1}{(1+A)}$$

$$C_N = B\ (\alpha\rho)^{1/2} \tag{7}$$

Tangent direction T of boundary node:

$$K_T = \frac{\alpha}{r}\frac{1}{(1+A)}\beta$$

$$C_T = B\ (\alpha\beta\rho)^{1/2} \tag{8}$$

The parameter r is the distance from the wave scattering source to the artificial boundary node, which can be simplified as the vertical distance from the basic geometric center to the corresponding artificial boundary in the actual engineering calculation. The values of A and B are related to the propagation characteristics of the wave and the incident angle. According to the existing experimental results, in the mean linear elasticity subgrade, when the interception range of the subgrade is 10 times the foundation[9], the optimal values are A is 0.8, and B is 1.1 , and the stiffness and damping coefficients of the node physical components can be further simplified as: $K_N = \alpha/18$, $C_N = 1.1\sqrt{\alpha\rho}$, $K_T = \alpha\beta/18$, $C_T = 1.1\sqrt{\alpha\beta\rho}$.

3. Calculation results and analysis of wind turbine foundation-subgrade dynamic interaction

Structural designers set up spring-damper elements at the intercepted subgrade boundary nodes through the simplified calculation formulas of stiffness and damping coefficient, which can simulate the semi-infinite space of subgrade. As shown in Figure 1, the fan foundation-subgrade is modeled by the finite element software ADINA, and the observation point A and the observation point B are set at the center of the top surface of the foundation and the variable section respectively. The subgrade interception range is 10 times the foundation size, the subgrade boundary of work condition 1 adopts a fixed boundary, the subgrade boundary of work condition 2 adopts a stress-type artificial boundary, and the boundary nodes are set with spring-damper elements. The material parameters of subgrade and wind turbine foundation are shown in Table 1. The design load combination of wind turbine foundation under extreme conditions is loaded on the top surface of the foundation according to formula (2). The dynamic calculation time is 6s, and each calculation step is 0.01s, a total of 600 steps.

Table 1 Material parameters of wind turbine foundation and subgrade

material parameters	subgrade	wind turbine foundation
elastic modulus ($N \cdot m^{-2}$)	2.5×10^8	3.25×10^{10}
density ($Kg \cdot m^{-3}$)	2000	2500
Poisson ratio	0.2	0.25

Through calculation, the Y-direction acceleration time-history curves of the two working conditions at points A and B are shown in Figure 2, where the abscissa is the time, and the ordinate is the acceleration.

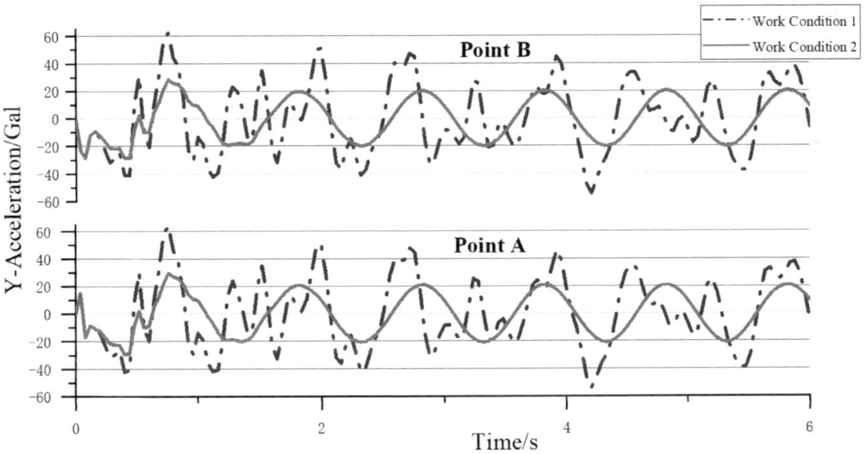

Fig.2 Y direction acceleration time-history curve of point A and point B under different artificial boundary conditions

It can be seen from Figure 2 that the acceleration time-history curves of points A and B are similar, and the fixed boundary produces wave reflection, which influences the accuracy of the calculation results. This makes the calculation results too large, thus cannot truly reflect the wind turbine foundation-subgrade dynamic interaction and results in an overly conservative design. Although it is simple and easy to set the fixed boundary for the subgrade boundary, the deviation of the calculation result is large, and it is not suitable for the design of the wind turbine foundation. The structural designer intercepts the subgrade 10 times the size of the wind turbine foundation, and applies spring-damper elements at the subgrade boundary nodes. The values of K and $\boldsymbol{C}$ are calculated according to the simplified formula, which are only related to the material parameters of the subgrade. This calculation method is simple and efficient, and it is a practical method.

The size of the intercepted subgrade has a great influence on the calculation efficiency. Reducing the intercepted subgrade size can greatly reduce the amount of calculation. In actual engineering, designers often use a subgrade that is 5 times the size of the foundation size for calculation. When the intercepted subgrade range is 5 times the size of the foundation size, and the same calculation method as in working condition 2 is used, the calculation result is smaller than the actual value, and the calculation accuracy is affected. The reason for this result may lie in the value of the damper elements. It can be seen from equations (7) and (8) that the value of $\boldsymbol{C}$ has nothing to do with the size of the intercepted subgrade. The intercepted subgrade size changes but the value of $\boldsymbol{C}$ does not change. Therefore, when a subgrade of 5 times the size of the wind turbin foundation is used for calculation, it is necessary to adjust the parameter B in the $\boldsymbol{C}$ value calculation formula. Under this condition, 1.1 may not be the optimal value of B.

Based on the above analysis, the modeling is carried out according to working condition 2. The interception range of subgrade is 5 times the size of the wind turbine foundation. Through calculation, the acceleration time-history curve and the velocity time-history curve of observation point A in working conditions 2-5 are drawn in Fig. 3 and Fig. 4 .

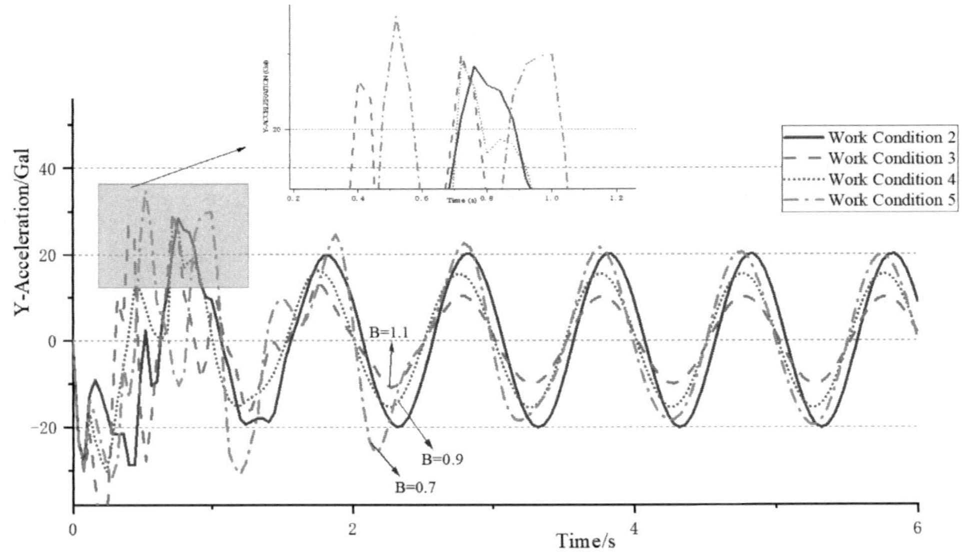

Fig.3 Y direction acceleration time-history curve of point A in working conditions 2-5

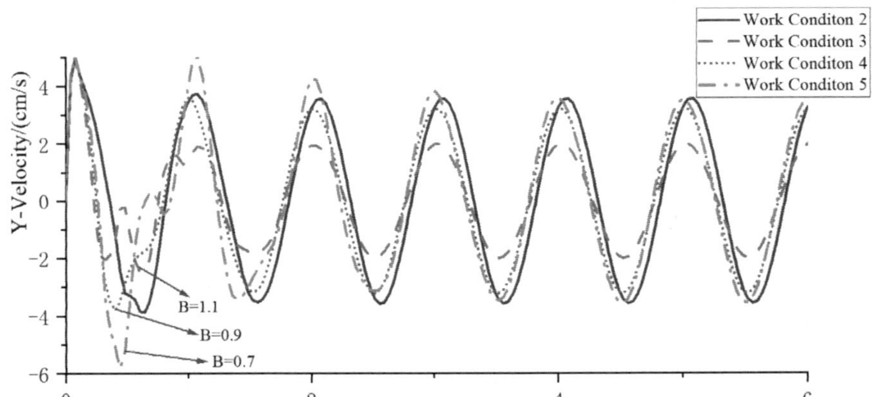

Fig.4 Y direction time-history curve of velocity of point A in working conditions 2-5

Based on the analysis of Fig. 3 and Fig. 4, it can be seen that when the intercepted subgrade range is 5 times the size of the wind turbine foundation, with the change of the value of parameter B, the acceleration and velocity show a similar variation trend. When B=1.1, the calculation result is too small. When B is 0.7, the calculation result is too large, especially the maximum deviation is large. When B is 0.9, the calculation results are in good agreement with the cut-off subgrade range which is 10 times the size of the wind turbine foundation. It can be seen that when the intercepted subgrade range is 5 times the size of the wind turbine foundation, the stiffness and damping coefficients of the spring-damping elements on the artificial boundary nodes can be simplified as: $K_N = \alpha/9$, $C_N = 0.9\sqrt{\alpha\rho}$, $K_T = \alpha\beta/9$, $C_T = 0.9\sqrt{\alpha\beta\rho}$.

4. Conclusion

By intercepting a limited range of subgrades and setting artificial boundaries at the subgrade boundaries, a wind turbine foundation-subgrade dynamic interaction model is established, and numerical simulation calculations are performed. By comparing the calculation results, the following conclusions can be obtained:

(1) It is simple and easy to use a fixed boundary for the artificial boundary, but the calculation results are too large and cannot truly reflect the dynamic interaction between the wind turbine foundation and the subgrade.

(2) Using a stress-type artificial boundary and setting up spring-damper elements at the subgrade boundary can effectively simulate the subgrade semi-infinite space, which can be implemented in most computing software.

(3) When the interception range of subgrade is 10 times of the wind turbine foundation , the stiffness and damping coefficient of spring-damper elements can be simplified as: $K_N = \alpha/18$, $C_N = 1.1\sqrt{\alpha\rho}$, $K_T = \alpha\beta/18$, $C_T = 1.1\sqrt{\alpha\beta\rho}$. When the intercepted subgrade range is 5 times the size of the wind turbine foundation, the stiffness and damping coefficients of the spring-damper elements can be simplified as: $K_N = \alpha/9$, $C_N = 0.9\sqrt{\alpha\rho}$, $K_T = \alpha\beta/9$, $C_T = 0.9\sqrt{\alpha\beta\rho}$.

References

[1] Chen Z, Guerrero J M, Blaabjerg F. (2009). A review of the state of the art of power electronics for wind turbines. IEEE Transactions on Power Electronics, 24(8), 1859-1875.
[2] Global Wind Energy Council. Global wind report 2021[R]. Belgium, 2021.
[3] International Energy Agency, Nuclear Energy Agency, Organisation for Economic Co-operation and Development. Projected Costs of Generating Electricity 2020[R]. France, 2020.
[4] Zaaijer, M. (2006). Foundation modelling to assess dynamic behaviour of offshore wind turbines. Applied Ocean Research, 28(1), 45-57.
[5] Achmus, M., Kuo, Y., & Abdel-Rahman, K. (2009). Behavior of monopile foundations under cyclic lateral load. Computers and Geotechnics, 36(5), 725-735.
[6] Ke, S., Wang, S., Cao, J., Zhao, L., & Ge, Y. (2015). Analysis on wind-induced responses of large wind power structures considering soil-structure interaction. China Civil Engineering Journal, (2), 18-25.
[7] Wu, S., Yang, H., Yu, Y., & LI,D.(2011). The Three dimensional Finite Element Analysis of the Foundation for Wind Turbo generator Units of Liziqing Wind Farm. Yunnan Water Power, 26(2), 3.
[8] Xu,X., Liu,Z., & Zhang D. (2012). Study on Ground Mechanical Characteristics of Tower Foundation of Wind Turbine with Capacity More Than One Megawatt. China Water Power, 38(12), 3.
[9] Han, Z., Chen, G., & Wu, X. (2021, December). Simplified analysis of dynamic machine foundation-subgrade dynamic interaction. In International Conference on Intelligent Equipment and Special Robots(ICIESR 2021) (Vol. 12127, pp. 436-441). SPIE

Static characteristics of high-speed bubble-oil-lubricated spiral groove thrust bearing

Feida Zhou, Shuyun Jiang*

School of Mechanical Engineering, Southeast University, NanJing, 210000, China

*Corresponding author: Shuyun Jiang

E-mail address: jiangshy@seu.edu.cn (S. Jiang)

Abstract. In this paper, the effect of bubble oil on the performance of spiral groove thrust bearing is Investigated. A theoretical model for thrust bearing is established based on bubble-oil two-phase flow considering centrifugal effect and viscosity-temperature relationship. The static characteristics of thrust bearing are analyzed. The result shows that both of viscosity-temperature effect and centrifugal effect can reduce the load capacity and friction torque. Meanwhile, with the increase of aeration level in bubble oil, the capacity, friction torque and maximum temperature rise of spiral groove thrust bearing are lower.

1. Introduction

Oil film lubrication has become the most common lubrication method of hydrodynamic journal bearing due to its high viscosity and excellent lubrication characteristics. However, the wear power consumption of oil-film lubricated increase sharply owing to oil film shearing at high speed, leading to large thermal deformation and bearing failure sometimes. In recent years, a new lubrication technology using bubble-oil mixture as lubricating medium has been put forward to solve the problem of high yield heat in high-speed journal bearing.

The rheological properties of bubble oil have been studied theoretically and experimentally. Many researchers have conducted theoretical analysis on the influence of static characteristics, where shows the bearing capacity increase with bubble oil, the position of pressure center is also offset[1,2]. A new viscosity and density models of aeration oil has been developed considering the influence of bubble surface tension, the result shows that aeration oil can improve capacity theoretically[3-5], while has little effect experimentally. They ignore the effect of temperature on performance. Some aeration oil viscosity experiment devices are developed which aim to study the equivalent viscosity model of aeration oil. It is found the bearing capacity and volume flow rate are reduced at high speed, but the equivalent viscosity of aeration oil increases significantly at low shearing rates[6,7].

Most of research study on theory and experiment in common journal bearing. However, the divergence area of journal bearing will introduce additional cavitation bubbles inevitably, effect the state of bubble oil. In addition, few literatures consider the thermal factor in the study of bubble-oil lubricated performance, but the influence of temperature on the lubrication characteristics of bubble oil cannot be ignored.

Therefore, the spiral groove thrust bearing is selected in this paper. Based on the lubrication characteristics of bubble oil, the Reynolds equation and energy equation are derived considering centrifugal effect. The influence of temperature and centrifugal effect is analyzed on thrust bearing, and the static characteristics of bubble-oil-lubricated spiral groove thrust bearing is explored.

2. Theoretical model

2.1. Reynolds equation in polar coordinates

Figure 1 shows the schematic diagram of outwardly pumping spiral groove thrust bearing, and the liner shape is logarithmic helical groove, which can be expressed as

$$r = r_b e^{\theta cot\beta} \tag{1}$$

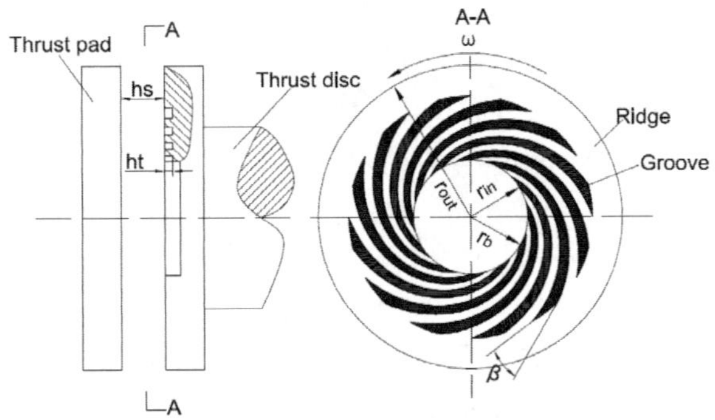

Figure 1. Schematic diagram of spiral groove thrust bearing

The density and viscosity of bubble oil vary with the internal pressure and temperature of the bearing in high speed. Therefore, a compressible Reynolds equation needs to be established. For spiral groove thrust bearing, the centrifugal force of lubrication also has a great influence. Based on the above facts, the following assumptions are set for the establishment of Reynolds equation:

1) Bubbles are discrete in bubble-oil mixture without contact with each other.
2) Ignore non-Newtonian fluid properties of bubble oil in the film thickness.
3) The heat conduction effect of thrust bearing is not considered.

Based on the assumptions, the stable Reynolds equation for compressible fluid is derived with centrifugal effect in polar coordinates.

$$\frac{\partial}{\partial r}\left(\frac{\rho r h^3}{12\mu}\frac{\partial p}{\partial r}\right)+\frac{\partial}{\partial \theta}\left(\frac{\rho h^3}{12\mu r}\frac{\partial p}{\partial \theta}\right)=\frac{h\omega r}{2}\frac{\partial p}{\partial \theta}+\frac{\partial}{\partial x}\left(\frac{h^3}{12\mu}\rho^2 {u_m}^2\right) \tag{2}$$

Where u_m is defined as the average velocity along the film thickness.

2.2. Energy equation

According to the hydrodynamic lubrication theory, the energy equation of variable viscosity compressible lubrication can be express as:

$$\frac{c_v}{r}q_\theta\frac{\partial T_m}{\partial\theta}+c_v q_r\frac{\partial T_m}{\partial r}=\mu\frac{\omega^2 r^2}{h}+\frac{h^3}{12\mu}\left\{\left(\frac{\partial p}{r\partial\theta}\right)^2+\left(\frac{\partial p}{\partial r}-\frac{\rho {u_m}^2}{r}\right)^2\right\}-p\left(\frac{\partial}{r\partial\theta}\frac{q_\theta}{\rho}+\frac{\partial}{r\partial r}\left(\frac{r q_r}{\rho}\right)\right) \tag{3}$$

Where, q_θ is circumferential mass flow, q_r is on behave of radial mass flow, T_m is defined as the average temperature over the film thickness.

The viscosity-temperature relationship is described as

$$\mu = \mu_0 e^{-a(T-T_0)} \tag{4}$$

where μ_0 is the initial viscosity, a is a parameter determined by empirical formula.

2.3. Reynolds equation and energy equation in boundary-fitted coordinates

The polar coordinate system (r,θ) is transformed into boundary-fitting coordinate system (ζ,η) in order to improve the accuracy of solution in the calculation of spiral groove thrust bearing. The coordinate transformation relation is as follows

$$r = \zeta, \ \ \theta = \eta + f(\zeta), \ \ f(\zeta) = \frac{1}{cot\beta} ln\,(\frac{\zeta}{\zeta_b}) \tag{5}$$

The Reynolds equation after coordinate transformation is

$$\frac{\partial}{\partial\zeta}\left(\zeta\frac{\rho h^3}{\mu}\frac{\partial p}{\partial\zeta}\right) + \frac{\partial}{\partial\eta}\left((\frac{1}{cot^2\beta}+1)\frac{\rho h^3}{\zeta\mu}\frac{\partial p}{\partial\eta}\right) - \frac{\partial}{\partial\zeta}\left(\frac{1}{cot\beta}\frac{\rho h^3}{\mu}\frac{\partial p}{\partial\eta}\right) - \frac{\partial}{\partial\eta}\left(\frac{1}{cot\beta}\frac{\rho h^3}{\mu}\frac{\partial p}{\partial\zeta}\right) = 6\omega\zeta\frac{\partial}{\partial x}(\rho h) + \frac{\partial}{\partial\zeta}\left(\frac{h^3\rho^2 {u_m}^2}{\mu}\right) - \frac{1}{\zeta cot\beta}\frac{\partial}{\partial\eta}(\frac{h^3\rho^2 {u_m}^2}{\mu}) \tag{6}$$

The film thickness of spiral groove thrust bearing is expressed as

$$h(\zeta,\eta) = \begin{cases} h_s & \text{ridge area} \\ h_s + h_t & \text{groove area} \end{cases} \tag{7}$$

The boundary condition of the Reynolds equation are as

$$\left.\begin{array}{c} p|_{\zeta=0} = p_0 \\ p|_{\zeta=\zeta_{out}} = p_0 \\ p(\eta) = p(\eta + 2\pi) \end{array}\right\} \tag{8}$$

Similarly, coordinate transformation of energy with bubble oil is

$$\frac{c_v}{\zeta}q'_\eta\frac{\partial T_m}{\partial\eta} + c_v q'_\zeta(\frac{\partial T_m}{\partial\zeta} - \frac{1}{\zeta cot\beta}\frac{\partial T_m}{\partial\eta}) = \mu\frac{\omega^2\zeta^2}{h} + \frac{h^3}{12\mu}\left\{\left(\frac{\partial p}{\zeta\partial\eta}\right)^2 + \left(\frac{\partial p}{\partial\eta} - \frac{1}{\zeta cot\beta}\frac{\partial p}{\partial\eta} - \frac{\rho {u_m}^2}{\zeta}\right)^2\right\} - p(\frac{\partial}{\zeta\partial\eta}\frac{q'_\eta}{\rho} + \frac{\partial}{\zeta\partial\zeta}\left(\frac{\zeta q'_\zeta}{\rho}\right) - \frac{1}{\zeta^2 cot\beta}\frac{\partial}{\partial\eta}\left(\frac{\zeta q'_\zeta}{\rho}\right)) \tag{9}$$

The boundary conditions for Energy equation are

$$\left.\begin{array}{c} T|_{\zeta=0} = T_0 \\ T(\eta) = T(\eta + 2\pi) \end{array}\right\} \tag{10}$$

2.4. Equivalent model of bubble oil

The equivalent viscosity and density of bubble oil are affected by the aeration level, and results on the viscosity theoretical model of bubble-oil two-phase flow are not unified. Therefore, viscosity model is often measured by experiments. In this paper, the viscosity model of experimental results of An Qi[7] is adopted

$$\frac{\eta_b}{\eta_0} = 1 - 0.01\beta\frac{p}{p_0} \tag{11}$$

The equivalent density model of bubble oil can be expressed as the average density, its dimensionless density as

$$\bar{\rho} = \frac{\rho}{\rho_{oil}} = \frac{1}{1+\lambda}(1 + \lambda\frac{\bar{p}}{\bar{T}}) \tag{12}$$

Where λ is defined as the ratio of bubble volume to oil volume, $\bar{T}$ is dimensionless temperature, $\bar{p}$ represents dimensionless pressure.

2.5. Calculation of static characteristics

The bearing pressure distribution can be obtained by using finite difference method, and then capacity can be calculated by integrating the pressure on the bearing surface.

$$W_P = \int_0^{2\pi}\int_{r_{in}}^{r_{out}} prdrd\theta \tag{13}$$

Friction occurs between the thrust pad and the fluid surface of spiral groove thrust bearing, the general expression of friction is as follow

$$F_f = -\int_0^{2\pi}\int_{r_{in}}^{r_{out}} \tau rdrd\theta \tag{14}$$

Then the expression of the friction torque is

$$M_f = -\int_0^{2\pi}\int_{r_{in}}^{r_{out}} \tau r^2 dr d\theta \tag{15}$$

3. Results and discussions

The calculated geometrical parameters of spiral groove thrust bearing are shown in Table 1.

Table 1. Geometric parameters of spiral groove thrust bearing

Item	Value
Inner radius $r_{in}(mm)$	7.5
Base circle radius $r_b(mm)$	7.5
Outer radius $r_{out}(mm)$	20
Groove number N	12
Groove depth $h_t(\mu m)$	40
Groove angle $\beta(deg)$	70
Viscosity of pure oil $\mu(pa.s)$	0.02
Density of pure oil $\rho(kg/m^3)$	870
Pressure of supply $p_{in}(MPa)$	0.05
Initial temperature of bubble oil T_0(°C)	20
Specific heat of bubble oil $c_v(J/(kg.k))$	2000
Bearing rotating speed $n(rpm)$	10000

Figure 2 shows the variation of bearing capacity and friction torque with aeration level under different models. As can be seen, with the increase of aeration level, the capacity and friction torque of thrust bearing decreases, and both temperature and centrifugal effect affect the bearing capacity. The reason for this phenomenon is that the presence of bubbles can significantly reduce the viscosity and density of the bubble oil, thence friction torque is smaller with high aeration level. In addition, Compared with temperature and centrifugal effect, the bearing capacity and friction torque increases when only temperature or centrifugal effect is considered. That can be explained that viscosity of bubble oil decreases with the rise of temperature, leading to the decline of capacity and friction torque. Centrifugal effect can also reduce bearing capacity, which will increase the radial flow rate. Therefore, the flow through the radial oil film becomes more, the circumferential flow of lubricating fluid becomes smaller, which reduces the capacity. It also can be seen that the friction torque and capacity are higher without considering temperature or centrifugal effect, and the viscosity-temperature effect has greater impact than the centrifugal effect.

Figure 3 shows the variation changes law of load capacity, friction torque and maximum temperature rise of spiral groove thrust bearing with oil film thickness with different aeration level. As expected, the static characteristics decrease with the increase of film thickness. The static characteristics of pure oil are obviously higher than those of the bubble oil. In addition, with the increase of film thickness, the difference of lubrication performance with different aeration level, which can be explained that when the film thickness is large, the decrease of hydrodynamic effect has a greater impact on the static characteristics than the effect of aeration level.

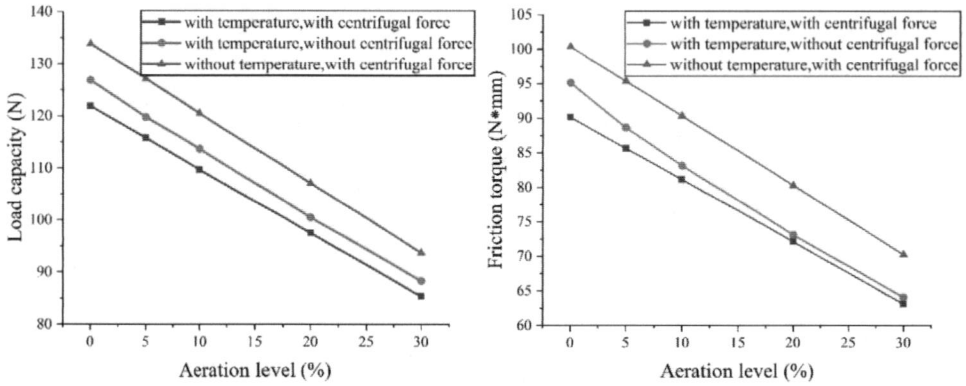

Figure 2. Comparison of load capacity and Friction torque under different conditions

4. Conclusion

Based on the viscosity-temperature effect, the Reynolds equation and energy equation with bubble oil spiral groove thrust bearing are constructed in combination with the viscosity and density model of two-phase flow. The influence of aeration level on the static characteristics is analyzed. The result shows that:

1) The viscosity-temperature effect and centrifugal effect will significantly reduce the viscosity of bubble oil, reduce the static characteristics of thrust bearings, and the influence of viscosity- temperature effect is greater than that of centrifugal effect.

2) Under the same film thickness, the capacity with low aeration level is larger, and the friction torque and maximum temperature rise of the bearing decrease with the increase of aeration level.

3) When the oil film thickness reaches a certain range, the influence of aeration oil on static characteristics gradually decreases.

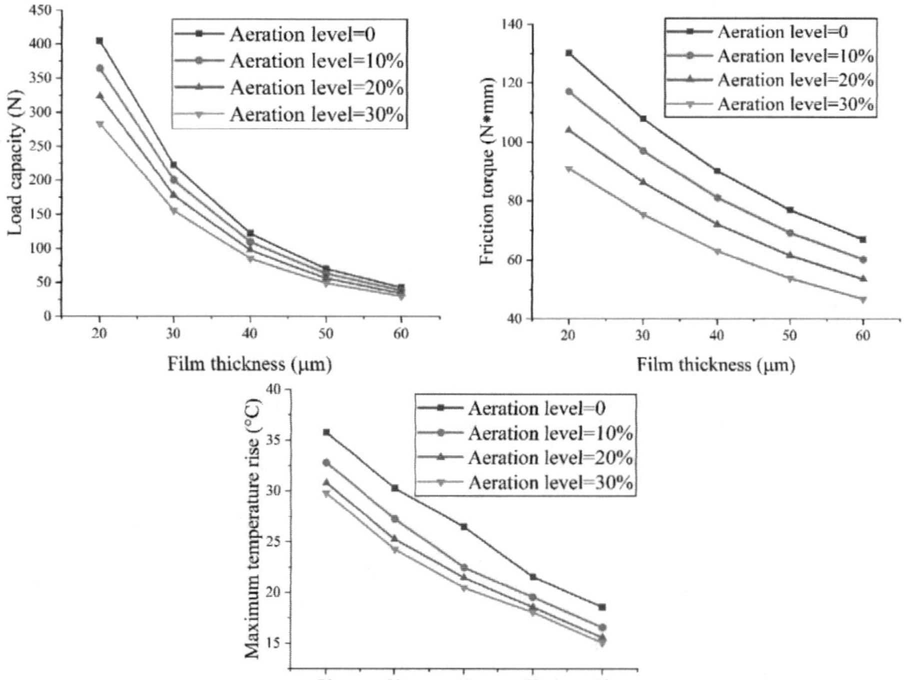

Figure 3. Influence of aeration level on static characteristics

References

[1] Abdel-Latif L A., Peeken H., Benner J.(1985) Thermohydrodynamic Analysis of Thrust-Bearing With Circular Pads Running on Bubbly Oil. ASME J of Trib., 107: 527-537.
[2] Tonder K. (1977) Effect of Gas Bubbles on Behavior of Isothermal Michell Bearings. Trans.asme J.of Lub.tech., 99: 354-358.
[3] Nikolajsen J L. (1999) Viscosity and Density Models for Aerated Oil in Fluid-Film Bearings. ASME Transactions., 42: 186-191.
[4] Nikolajsen, Jorgen L. (1999) The Effect of Aerated Oil on the Load Capacity of a Plain Journal Bearing. Tribology Transactions., 42: 58-62.
[5] Goodwin M J, Dong D. (2007) Theoretical and experimental investigation of the effect of oil aeration on the load-carrying capacity of a hydrodynamic journal bearing. Journal of Engineering Tribology., 221: 779-786.
[6] Yang H R, Wang X L, Zhang Y Y. (2014) Study on the bubbly lubrication of journal bearings at various shear rates and temperatures. Tribology International., 71: 132-139.
[7] An Q, Zhou Y, Quan Y. (1997) Study on the viscosity properties of bubbly oil and the static characteristics of journal bearing lubricated with bubbly oil. Wear., 213: 159-164.

Simulation research of single phase half wave controllable rectifier circuit based on MATLAB

Jingwei Cui[1]*

[1]School of Automatic Control,Liuzhou Railway Vocational and Technical College,Liuzhou, Guangxi,China

* 327845890@qq.com

Abstract—The function of the rectifier circuit is to convert AC electrical energy into DC electrical energy to provide energy for DC electrical equipment. The input end of the single phase rectifier circuit is a single phase AC power supply. The working principle of the single phase rectifier circuit is related to its load, the load is different, and its working process is also different. In this paper, MATLAB 9.5 software is used to simulate the single phase half wave controllable rectifier circuit under different load working conditions. In this paper, MATLAB9.5 is used to simulate the single-phase half-wave controllable rectifier circuit, study the working process under different load conditions,and analyze the working principle of the rectifier circuit, which is the research of the single phase half wave controllable rectifier circuit in practical industry and application provide a very important reference.

1. Introduction

Rectification genus in electric power four major transformations medium basic type,the rectifier circuit can convert AC power into DC power, DC power provides power to DC electrical equipment [1-3].Rectifier circuits have very important applications in industry, transportation,power systems and other fields.Rectifier circuits can be classified according to a variety of methods.According to the number of AC input phases, they can be divided into multi-phase circuits and single phase circuits. Single phase rectifier circuits include single phase half wave controllable rectifier circuits,single phase fully-controlled bridge-type controllable rectifier circuits,single phase bridge-type half-controlled rectifier circuits, and so on. The AC side of the single phase rectifier circuit is a single phase power supply. The loads of the single phase half wave controllable rectifier circuit in practical applications are resistive load, resistive-inductive load with freewheeling diode, and resistive-inductive load [4-6].

The thyristor is the core component in the single phase half wave controlled rectifier circuit, and generally works in the switching state. The thyristor is a semi-controlled power electronic technology component, which can use control signals to control the thyristor to turn on,but it cannot control the thyristor to turn off. When analyzing the working principle of a single-phase rectifier circuit, the thyristor is generally regarded as an ideal component. When the thyristor is in the conducting state,the voltage difference between the anode and the cathode of the thyristor is zero, and when the thyristor operates in the blocking condition, its leakage current is equal to zero.Under ideal conditions, the turn-on process and turn-off process of the thyristor in the circuit are completed instantaneously. The thyristor acts as a switching device in the circuit. This article will use MATLAB9.5 simulation platform to build a single phase half wave controllable rectifier circuit, and analyze the influence of different loads on the circuit operation through simulation [1].

2. Single phase half wave controllable rectifier circuit

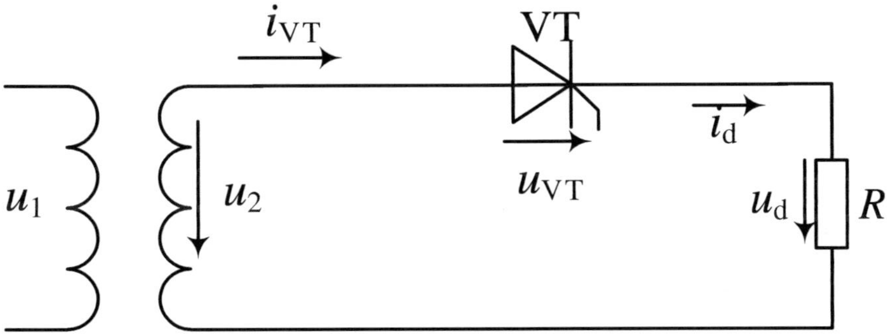

Fig.1 Single phase half wave controllable rectifier circuit (resistive load)

Fig.1 is the electrical principle structure diagram of single-phase half-wave controllable rectifier circuit with resistive load. In practical applications,the load of the resistance heating furnace and the electroplating device has a resistance characteristic,and its circuit structure is simple. u_1 and u_2 are AC power sources, T is the transformer (the transformer has the function of voltage conversion and electrical isolation), VT is a thyristor, and the load is R.Assume that the control angle of the thyristor is α,the conduction angle of the thyristor when it is working is θ, the forward voltage experienced across the load is u_d, the forward voltage experienced between the anode and cathode of the thyristor is u_{VT}, the current at the thyristor point is i_{VT}, the current at the load in the circuit is i_d.

The single phase half wave controllable rectifier circuit is simulated and studied by MATLAB 9.5/Simulink simulation software, and the working process of the rectifier circuit under different loads is analyzed [4]. The working process of the rectifier circuit with different loads is mainly analyzed. Modules such as AC power supply, thyristor, oscilloscope, signal generator and load are mainly used in the simulation. Extract Scope, AC Voltage Source,Thyristor, Voltage Measurement, Current Measurement, Powergui, Pulse Generator, Series RLC Branch, Demux in MATLAB respectively.

3. Matlab simulation of single phase half wave controllable rectifier circuit

3.1. Single phase half wave controllable rectifier circuit (resistive load)

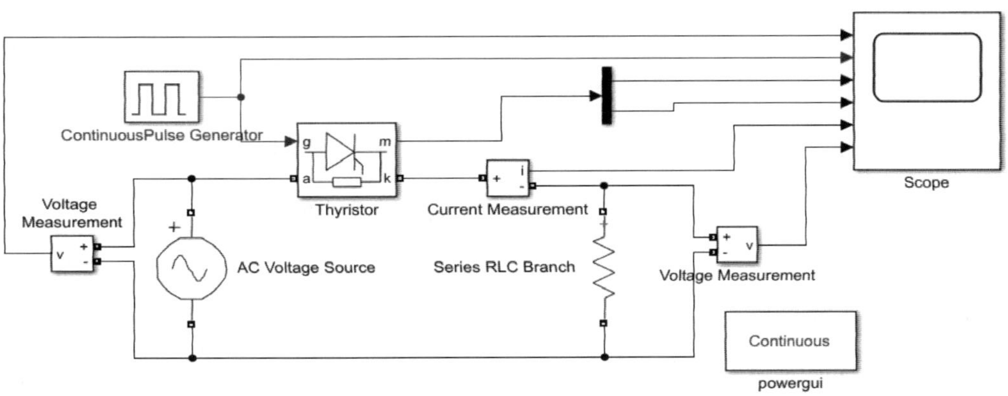

Fig.2 Single phase half wave controllable rectifier circuit (resistive load)

Fig.2 is a simulation model built according to the principle structure diagram. A single phase half wave controllable rectifier circuit under resistive load operation is established in Simulink, and then the parameters of each component in the simulation circuit are set.

Journal of Physics: Conference Series **2271** (2022) 012026 doi:10.1088/1742-6596/2271/1/012026

Table 1 Block parameter setting table in Simulink

The name of the component	Parameter settings	
AC Voltage Source	Peak amplitude (V):150	Frequency (Hz):50
Thyristor	Resistance Ron(Ohms) :0.001	Forwardvoltage Vf (V) :0.8
	Snubber resistance Rs (Ohms) : 1000	Snubber capacitanceCs (F) : 100e-9
Continuous Pulse Generator	Amplitude:10	Period (secs):0.02
	Pulse Width (% of period):5	Phase delay (secs):0.0025
Series RLC Branch	Resistance (Ohms):10	
Simulation/Model Configuration Parameters/Solver	Start time:0.0	stop time:0.05
	Solver selection type:Variable step	Solver:ode23tb (stiff/TR-BDF2)

After the parameter setting is completed, the circuit simulation starts. The Scope waveform is as follows:

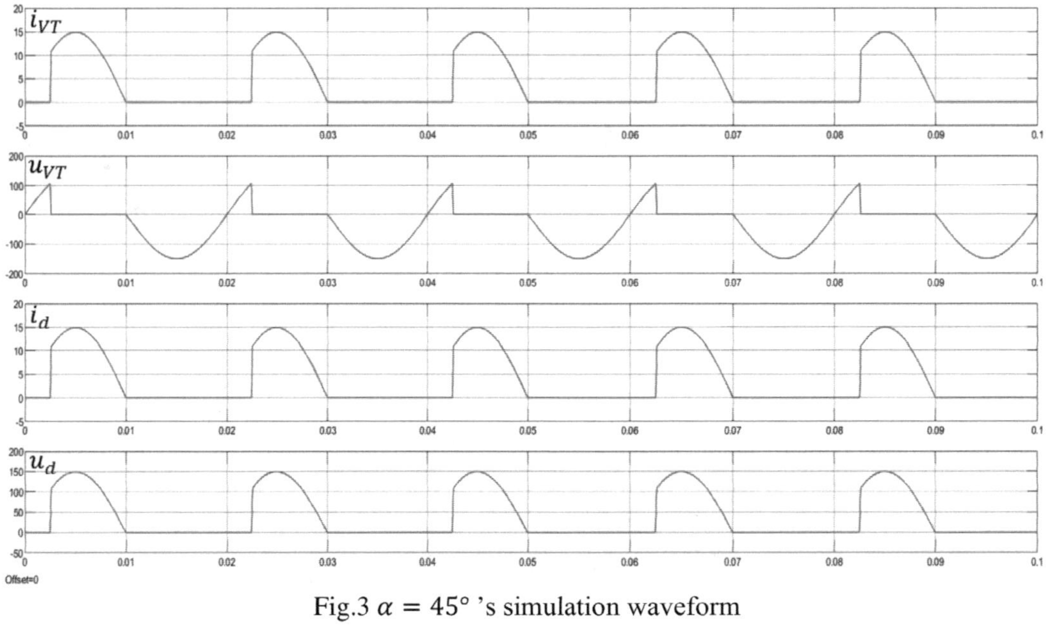

Fig.3 $\alpha = 45°$'s simulation waveform

The role of the thyristor in the circuit is a switching device, if the thyristor wants to be turned on, two conditions must be met at the same time. The first condition is that the anode and the cathode of the thyristor need to withstand a forward voltage, and the second condition is that there is a trigger pulse between the gate and the cathode of the thyristor. From the simulation waveform in Fig.3, it can be obtained, We analyze the first duty cycle of the circuit, there are three working states of the thyristor. The first stage is that the two ends of the thyristor are only withstand the forward voltage but there is no trigger pulse between the gate and the cathode, the thyristor cannot meet the conduction condition, and the entire closed circuit is in an open state, $i_d = 0$, $u_d = 0$, $i_{VT} = 0$, $u_{VT} = u_2$; In the second stage, when the thyristor meets the conduction conditions, the thyristor changes from the off state to the on state. The entire circuit is a series closed loop. Under ideal conditions, the thyristor plays a role similar to that of a wire, and its two ends bear the voltage is $u_{VT} = 0$, $i_{VT} = i_d$, $u_d = u_2$. The state of the thyristor in the third stage changes from on to off, when u_2 is equal to zero, at this time in the circuit, $u_{VT} = 0$, $i_d = i_{VT} = 0$, $u_d = 0$.

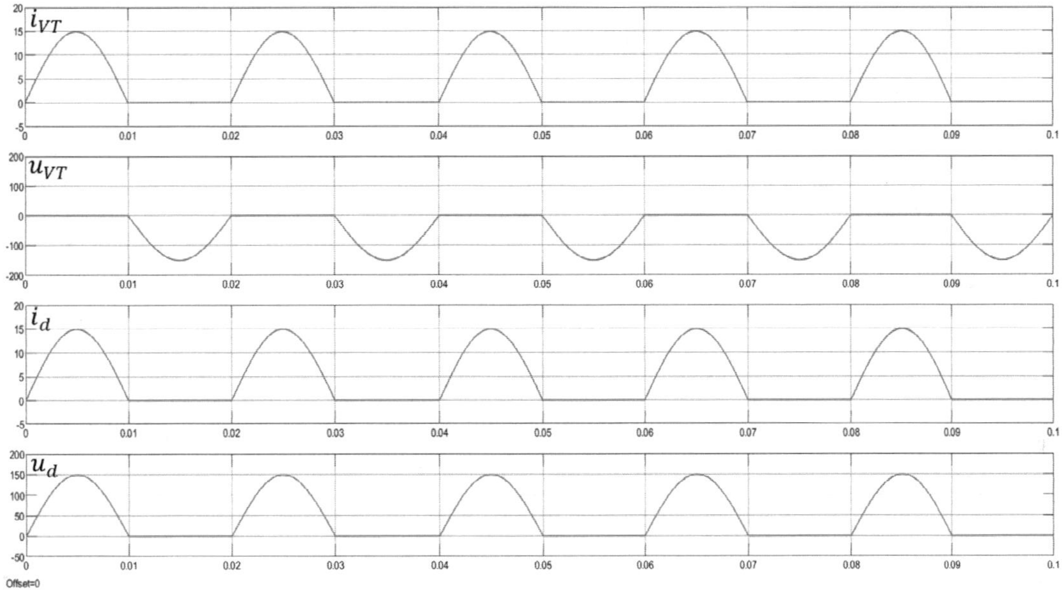

Fig.4 $\alpha = 0°$'s simulation waveform

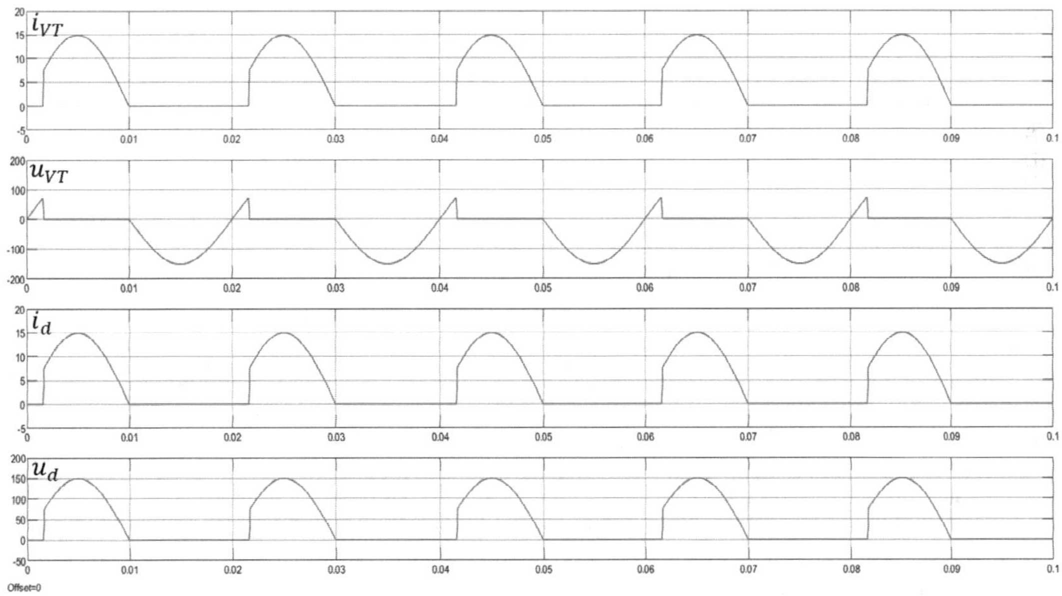

Fig.5 $\alpha = 30°$'s simulation waveform

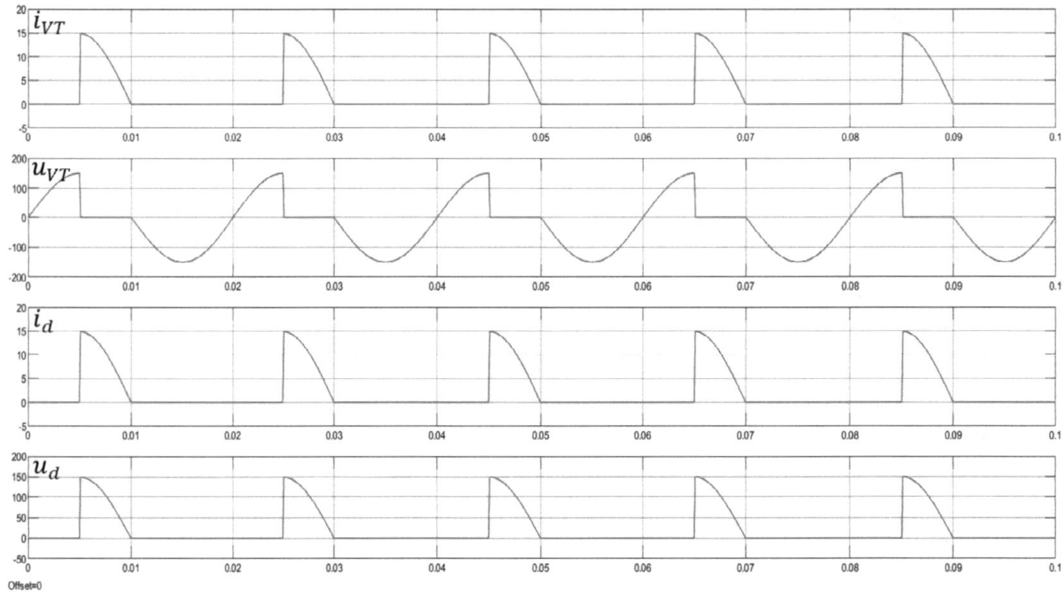

Fig.6 $\alpha = 90°$'s simulation waveform

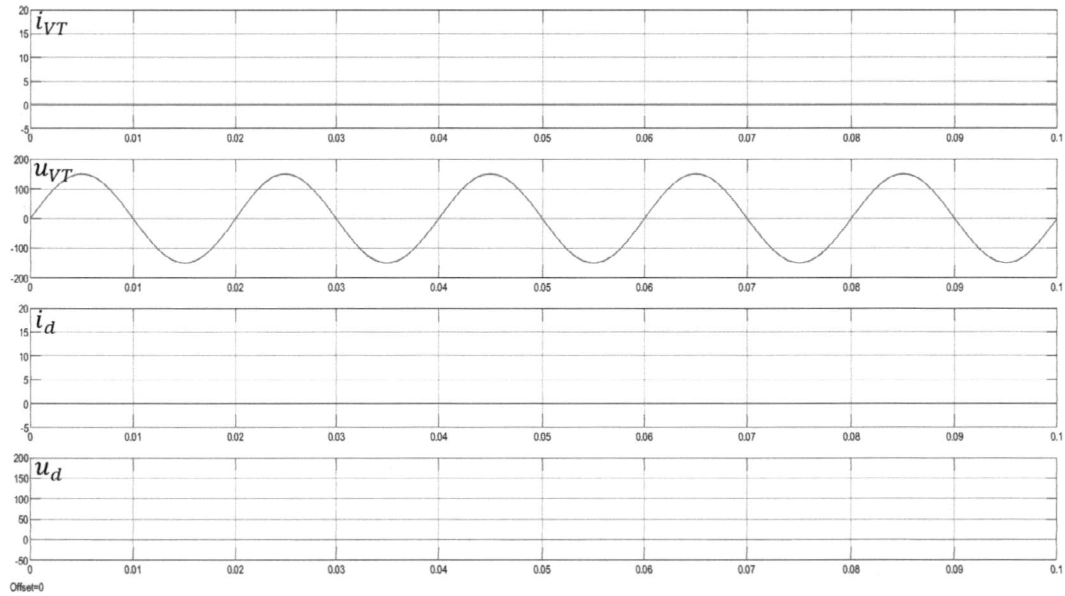

Fig.7 $\alpha = 180°$'s simulation waveform

Fig.4, Fig.5, Fig.6 and Fig.7 show the simulated output waveforms of the circuit at different control angles, respectively. The angle range of the single-phase half-wave controllable rectifier circuit (resistive-inductive load) is 0°~180°. Observe the simulation output waveform and find that the waveforms of u_d and i_d are similar.

$$U_d = \frac{1}{2\pi}\int_{\alpha}^{\pi} \sqrt{2}U_2 \sin \omega t d(\omega t) = \frac{\sqrt{2}U_2}{2\pi}(1 + \cos \alpha) = 0.45U_2 \frac{1+\cos \alpha}{2} \quad (1)$$

$$I_d = \frac{U_d}{R} \quad (2)$$

U_d is the average value of the DC output voltage in the circuit, I_d is the average value of the DC output current in the circuit.Through the formula, it can be found that the size of U_d is related to the size of α, $\alpha = 0°$, the value of U_d is the largest, $U_d = 0.45U_2$, the value of $I_d$is the largest;$\alpha = 180°$, U_d has the smallest value, $U_d = 0$,$I_d = 0$. The value range of α is 0°~180°, As the value of α gradually increases, the value of U_d gradually decreases.

When single phase half wave controllable rectifier circuit is applied in practice, most of its load types exist at the same time as resistance and inductance, and the resistance and inductance are in the form of series connection.

3.2. Single phase half wave controllable rectifier circuit (resistive inductive load)

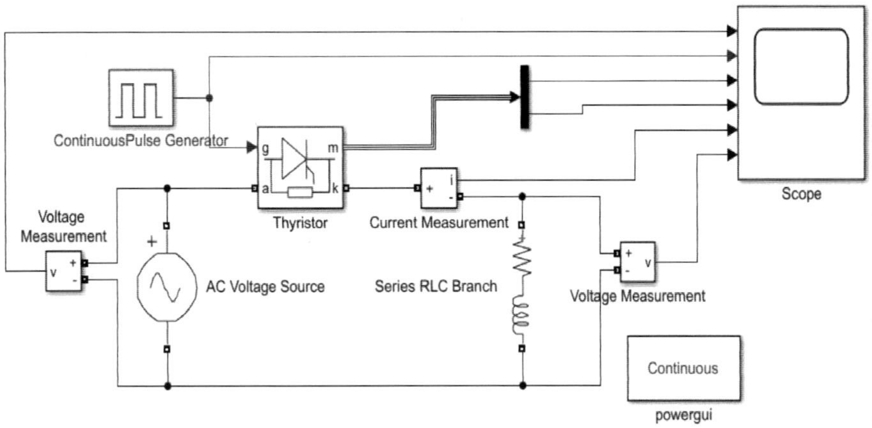

Fig.8 Single phase half wave controllable rectifier circuit (resistive inductive load)

Fig.8 is a simulation model built according to the principle structure diagram. Set the parameters in the circuit, Resistance(Ohms):10, Inductance(H):0.01.

The simulation waveforms of the single phase half wave controllable rectifier resistive inductive load circuit are as follows:

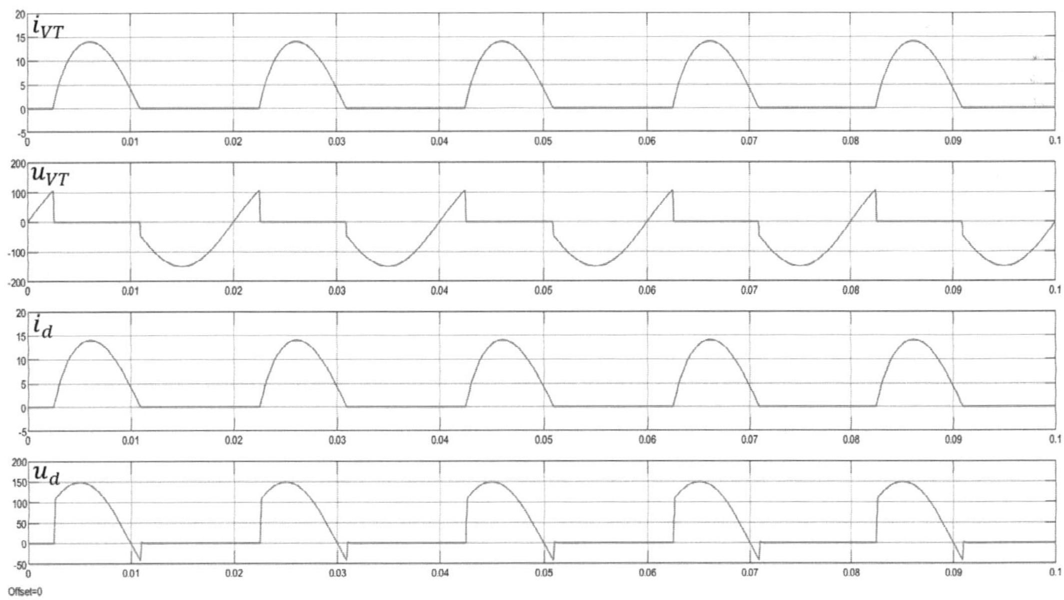

Fig.9 $\alpha = 45°$'s simulation waveform

It is found from Fig.9 that when the conduction condition of the thyristor is not met, the rectifier circuit is in an open state,at this moment, $u_d = 0$, $i_d = i_{vt} = 0$, $u_{VT} = u_2$; When the thyristor is turned on, $u_{VT} = 0$, $i_{VT} = i_d \neq 0$, with the gradual increase of u_2, the voltage across the load is $u_d = u_2$, the current i_d in the circuit gradually increases from zero. When u_d is the maximum value,because the inductance has an obstructive effect on the change of the current flowing through the inductance, i_d does not reach the maximum value. When u_2 gradually decreases, u_d gradually decreases. After i_d in the circuit reaches the peak value, it gradually decreases. $u_2 = 0$, due to the effect of inductance, $i_{VT} = i_d \neq 0$, the thyristor is still on, at this time, u_d and i_d are not zero, and the existence of inductance makes the thyristor turn on for a longer period of time, and the waveform of u_d has a negative part.

$$U_d = U_{dR} + U_{dL} = \frac{1}{2\pi}\int_{\alpha}^{\alpha+\theta} u_R\, d(\omega t) + \frac{1}{2\pi}\int_{\alpha}^{\alpha+\theta} u_L\, d(\omega t) \tag{3}$$

$$U_{dL} = \frac{1}{2\pi}\int_{\alpha}^{\alpha+\theta} u_L\, d(\omega t) = \frac{1}{2\pi}\int_{\alpha}^{\alpha+\theta} L\frac{di_d}{dt} d(\omega t) = \frac{\omega L}{2\pi}\int_{0}^{0} di_d = 0 \tag{4}$$

$$U_d = \frac{1}{2\pi}\int_{\alpha}^{\alpha+\theta} U_R d(\omega t) \tag{5}$$

According to formula (5), the value of U_d is related to the size of α and θ. As the angle of α gradually increases, the negative area under the x axis of the waveform of u_d gradually increases, when the positive area on the x axis is equal to the negative area on the x axis, $U_d = 0$, $\theta = 360° - 2\alpha$, at this time, the value of U_d has nothing to do with the size of α. When the single phase half wave controllable rectifier circuit works with a resistive and inductive load, the inductance makes the value of U_d smaller, $U_d = 0$, when the load cannot satisfy the actual work, a diode needs to be added to the circuit at this time.

3.3. Single phase half wave controllable rectifier circuit (resistive and inductive load with freewheeling diode)

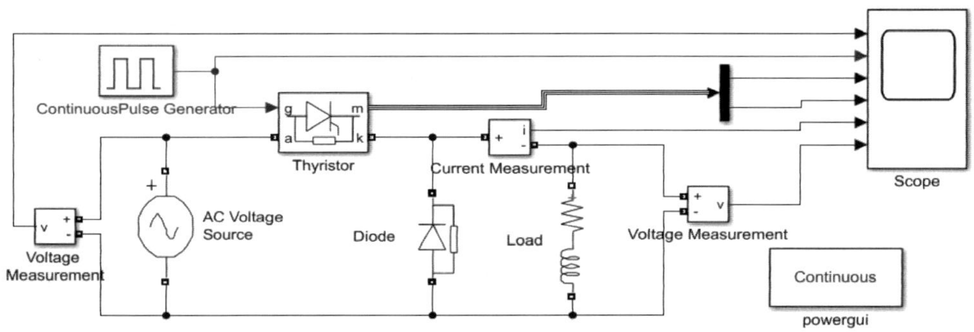

Fig.10 Single phase half wave controllable rectifier circuit (resistive and inductive load with freewheeling diode)

Fig.10 is a simulation model built according to the principle structure diagram. Set the parameters Resistance(Ohms):10. Inductance (H):0.01.

Simulate the circuit, and the output waveform is as follows:

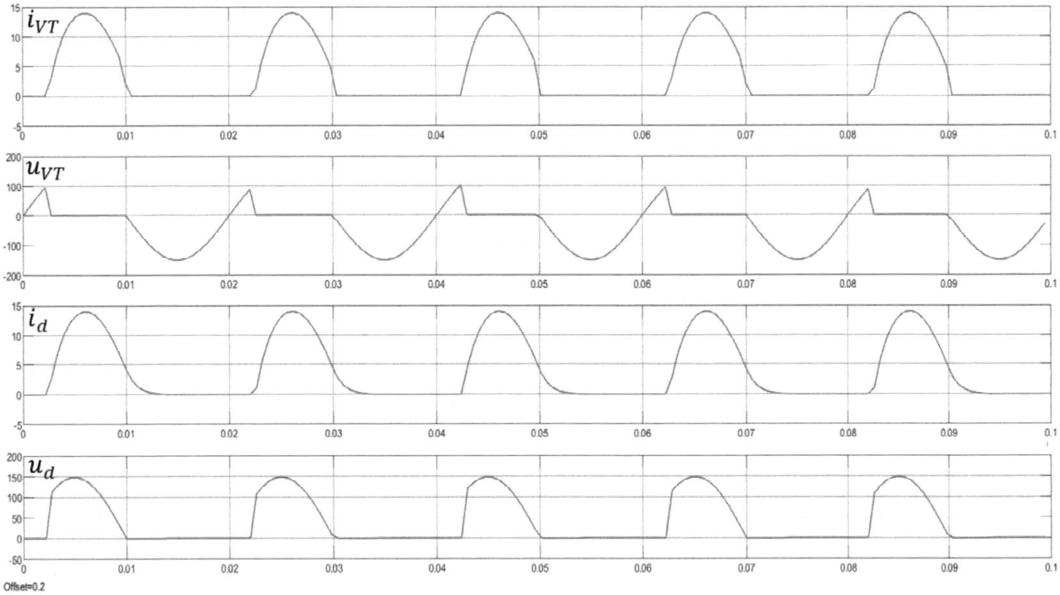

Fig.11 $\alpha = 45°$'s simulation waveform

Keep the other parameters unchanged, only change the inductance parameter, Inductance (H):0.1. Simulate the circuit, and the output waveform is as follows:

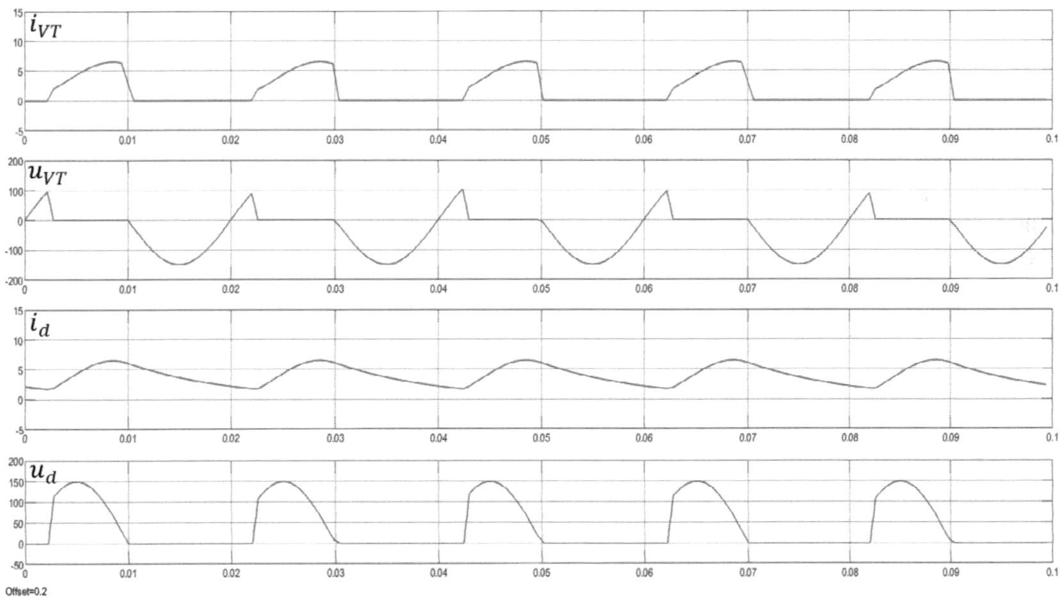

Fig.12 $\alpha = 45°, L = 0.1H$'s simulation waveform

Fig.11 and Fig.12 show the circuit simulation waveforms when the other parameters of the circuit are the same and the inductance is different. The inductance affects the working process of the circuit.u_d has a waveform in the positive half of u_2 and no waveform in the negative half of u_2. Inductance L stores energy, if L is large enough and the load is a large inductive load, it can make i_d have a waveform in a complete cycle and the waveform is continuous and the waveform is approximately a horizontal line. The size of L affects the waveforms of i_d and i_{VT}, while the waveform of u_d does not change much. When u_2 negative half cycle, the working state of the diode changes from off to on, at this time, due to the reverse voltage between the anode and the cathode of the thyristor,it is turned off.$u_{VT} = u_2$, $u_d = 0$. In a cycle, when u_d is at the maximum, i_d is not at the maximum but in the rising phase, when $u_d = 0$, i_d is not zero.

4. Conclusion

This article uses MATLAB9.5/Simulink toolbox to simulate the working process of a single phase half wave controllable rectifier circuit, and simulates the circuit with different loads. Based on the above simulation waveform results, the following conclusions are drawn:

(1)The simulation result found that the load is different, and the parameter waveforms in circuits such as u_d and i_d are different.

(2) Under the working conditions of resistive inductive load and resistive inductive load with freewheeling diode, the size of L and α also affect the operation and application of the circuit. When the resistive inductive load has a freewheeling diode, the circuit solves the problem that U_d decreases when the resistive inductive load works.

Acknowledgments

This work was financially supported by the project of Liuzhou Railway Vocational Technical College: "Development of Power Electronics Training Device"(No.2021-KJA06).

References

[1] Wei,L.M.,Yang,K.,Lv,X.Y.(2016)Simulation Analysis of Single-phase Half-wave Controlled Rectifier Circuit Based on MATLAB/Simulink. Northern Architecture,1(02):72-74.
[2] Wang,H.J.(2020)Application of MATLAB Simulink in circuit experiment.Laboratory Science ,23(06):80-84.
[3] Li,D.S.(2020)Application of Virtual Simulation Technology in the Teaching of Power Electronics Technology in Application-Oriented Universities. Popular Science & Technology,22(10):105-108.
[4] Chen,Z.X.,Li,Z.C.(2021)Simulation of thyristor single-phase half-wave controlled rectifier circuit based on MATLAB. Electric Engineering,(24):186-188.
[5] Zhang,Y.F.(2020)Application of Simulation Technology Based on Matlab in the Teaching of Power Electronic. Journal of Tianjin Sino-German University of Applied Sciences,(05):38-42.
[6] Gong,J.(2020)Power system fault analysis based on MATLAB.Electronic Design Engineering, 28(04):176-179+184.

Journal of Physics: Conference Series **2271** (2022) 012027 doi:10.1088/1742-6596/2271/1/012027

Research on on-line monitoring data cleaning of transformer oil chromatogram based on machine learning and neural networks

Shoubao Liu[1,2*], Yanyang Dang[2] and Ruijian Li[3]

[1] Datang Hydropower Science & Technology Research Institute Co., Ltd, Nanning, Guangxi 530007, China

[2]Tebian Electric Apparatus Stock Co. Ltd, Changji, Xinjiang, 652300, China

[3]School of microelectronics and communication engineering, Chongqing University, Chongqing, 400030, China

[*]Corresponding author's e-mail: 627909585@qq.com

Abstract. Transformer oil chromatogram on-line monitoring device is a widely used sensing device to monitor the operation state of transformer. Due to the lack of hardware verification means, the on-line chromatographic data generally has the problem of drift. In actual operation, it is mostly used as the basis for trend judgment, but can not reflect the effective value of chromatographic measurement. In this paper, the LSTM neural networks is used to correct the on-line monitoring data of transformer oil chromatogram. The off-line test data combined with neural networks data cleaning technology are used to repeatedly feed back and calibrate the on-line data, so that the corrected chromatographic on-line number is steadily close to the off-line detection data, which greatly improves the availability and accuracy of on-line chromatographic data. The research results of this paper have important reference significance for the data cleaning of on-line monitoring and measurement of electrical equipment.

1. Introduction

Transformer is a complex device, which plays a very important role in power plants, substations and other power facilities. Due to the complex structure and long-term withstand of high voltage and electric power, the transformer is likely to have various hidden defects. Transformer oil color spectrum analysis is an important means, which can find the latent faults such as damp, partial overheating and low-energy discharge in the transformer by analyzing various gas components and their changes in the transformer insulating oil[1-2]. At present, there are two main methods of transformer oil chromatogram analysis: on-line monitoring and off-line detection. The accuracy of off-line detection is high, but cumbersome sampling and testing are required, and the real-time performance is also poor; Online monitoring and sampling is convenient, but due to the difficulty of monitoring equipment calibration, the test data after long-term work is often not accurate enough and can only reflect the general trend[3-6]. Therefore, based on the off-line detection data, using data processing technology to correct the on-line monitoring oil chromatogram data into more accurate and reliable data is very important for the effective utilization of oil chromatogram on-line monitoring data.

Machine learning has unique advantages in finding patterns. It is based on neural networks, obtains knowledge, establishes a model through a series of algorithms, and automatically finds the relationship between things. After the model training is completed, the machine can correct the on-line monitoring data of transformer oil chromatogram data obtained by it, so as to make it more in line with the actual situation, which not only retains the advantages of convenient and fast on-line monitoring, but also ensures the authenticity and reliability of the data.

This paper extracts the on-line monitoring data of Transformer Oil Chromatogram in a power plant, and modifies the on-line monitoring data by regularly carrying out off-line testing. The specific way is to input the on-line gas concentration data into LSTM neural networks for training, so that the neural networks can find a model that can transform the on-line monitoring data into close to the off-line monitoring data. Since the off-line monitoring data is more accurate, Therefore, it can further enable its output to modify the on-line monitoring data of transformer oil color spectrum, make the on-line data close to the model of the real situation, and then better reflect the real situation of transformer oil color spectrum through the on-line data.

2. LSTM neural networks

The training of LSTM neural networks adopts back propagation algorithm[7-8]. It mainly includes three steps. One is to calculate the output value of each neuron forward, and the other is to calculate the error value of each neuron backward δ, Third, calculate the gradient of each weight according to the error term[9-10]. That is, first calculate the input test data layer by layer, and calculate the error value. The error value propagates from the output layer to the input layer, and adjust its weight layer by layer. After the input signal is calculated layer by layer through the hidden layer, the output result is generated. When the output results are in good agreement with the expected results, the training ends. If the difference between the output result and the expected result is unacceptable, enter the reverse calculation process. According to the formula, the neural networks derives the parameters (weights, coefficients, etc.) of the previous layer, and divides the error equally into each unit of each layer, so that the function of the neural networks is gradually close to the real function. After all the test data go through the above cycle process, the model can make the calculation results closest to the expected results. The trained model can correct the data. Therefore, a complete LSTM networks has input layer, output layer and hidden layer, as shown in Figure 1.

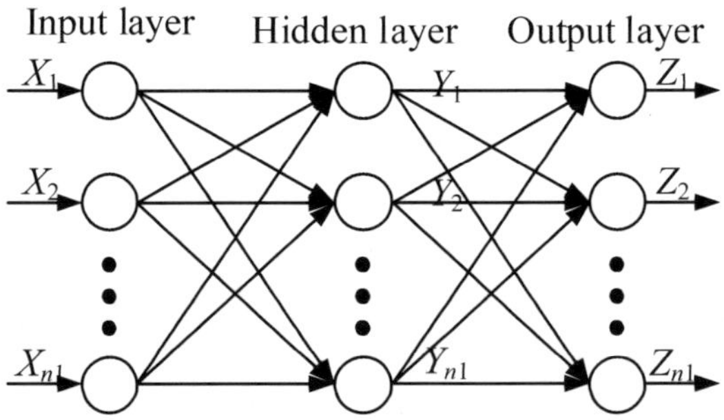

Figure 1. Three layer structure of LSTM networks.

There are actually two states in the LSTM hidden layer, one is the long-term storage state C and the other is the current state H. These two states are controlled by three gates, namely three functions. One is the forgetting gate, which controls how much information from the previous cell state is retained into the cell state of this layer; One is the input gate, which controls how much current input information enters the unit state of this layer; One is the output gate, which controls how much information about the unit status of this layer is output. The long-term state can store valuable data of a long period of layers before, and the current state reflects the results of the interaction between the current input and long-term storage, as shown in Figure 2.

Journal of Physics: Conference Series **2271** (2022) 012027 doi:10.1088/1742-6596/2271/1/012027

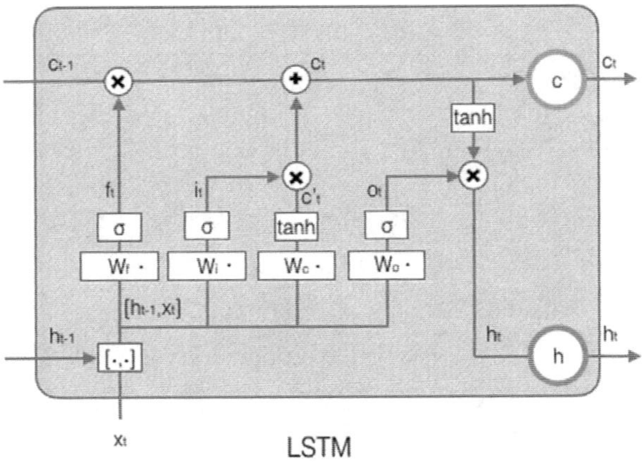

Figure 2. LSTM algorithm control structure.

According to figure 2, the forward calculation expansion is described as follows:

The previous cell state c_{t-1} is input into the hidden layer, and the previous layer output h_{t-1} is input into the hidden layer.

Note activation function $\sigma(x)=\frac{1}{1+e^{-x}}$ The weight matrix is W. Then the forgetting door is calculated as follows:

$$f_t = \sigma(W_f \cdot [h_{t-1}, x_t] + b_f) \tag{1}$$

Square brackets indicate connecting two vectors. Similarly, the input gate is calculated as follows:

$$i_t = \sigma(W_i \cdot [h_{t-1}, x_t] + b_i) \tag{2}$$

The unit status $\tilde{c}_t$ of the current input is calculated according to the last output and this input:

$$\tilde{c}_t = \tanh(W_c \cdot [h_{t-1}, x_t] + b_c) \tag{3}$$

Where $\tanh$ is a hyperbolic tangent function.

$$\tanh(x)=\frac{e^x - e^{-x}}{e^x + e^{-x}} \tag{4}$$

Therefore, the cell state c_t at the current time can be calculated:

$$c_t = f_t \circ c_{t-1} + i_t \circ \tilde{c}_t \tag{5}$$

$\circ$ represents Hadamard product, also known as matrix multiplication by elements, that is, multiply the elements at the corresponding positions of two matrices with the same structure.

After obtaining the above states, the output gate can be calculated:

$$o_t = \sigma(W_o \cdot [h_{t-1}, x_t] + b_o) \tag{6}$$

Finally, the final output of this layer is calculated as follows:

$$h_t = o_t \circ \tanh \tag{7}$$

Equations 1 to 7 are the forward calculation formulas of LSTM neural networks.

LSTM neural networks reverse calculation, set t time error term δ_t is:

$$\delta_t \overset{def}{=} \frac{\partial E}{\partial h_t} \tag{8}$$

The time error term of t-1 is

$$\delta_{t-1}^T = \frac{\partial E}{\partial h_{t-1}} = \frac{\partial E}{\partial h_t}\frac{\partial h_t}{\partial h_{t-1}} = \delta_t^T \frac{\partial h_t}{\partial h_{t-1}} \tag{9}$$

The formula that the error term is transmitted forward to any k time can be obtained by using the full derivative formula and partial derivative formula:

$$\delta_k^T = \prod_{j=k}^{t-1} \delta_{o,j}^T W_{oh} + \delta_{f,j}^T W_{fh} + \delta_{i,j}^T W_{ih} + \delta_{\tilde{c},j}^T W_{ch} \tag{10}$$

A formula that transfers errors to the next level.

$$\begin{aligned}\frac{\partial E}{\partial net_t^{l-1}} &= \frac{\partial E}{\partial net_{f,t}^{l}}\frac{\partial net_{f,t}^{l}}{\partial x_t^l}\frac{\partial x_t^l}{\partial net_t^{l-1}} + \frac{\partial E}{\partial net_{i,t}^{l}}\frac{\partial net_{i,t}^{l}}{\partial x_t^l}\frac{\partial x_t^l}{\partial net_t^{l-1}} + \frac{\partial E}{\partial net_{\tilde{c},t}^{l}}\frac{\partial net_{\tilde{c},t}^{l}}{\partial x_t^l}\frac{\partial x_t^l}{\partial net_t^{l-1}} \\ &\quad + \frac{\partial E}{\partial net_{o,t}^{l}}\frac{\partial net_{o,t}^{l}}{\partial x_t^l}\frac{\partial x_t^l}{\partial net_t^{l-1}} \\ &= \delta_{f,t}^T W_{fx} \circ f'(net_t^{l-1}) + \delta_{i,t}^T W_{ix} \circ f'(net_t^{l-1}) + \delta_{\tilde{c},t}^T W_{cx} \circ f'(net_t^{l-1}) \\ &\quad + \delta_{o,t}^T W_{ox} \circ f'(net_t^{l-1}) \\ &= (\delta_{f,t}^T W_{fx} + \delta_{i,t}^T W_{ix} + \delta_{\tilde{c},t}^T W_{cx} + \delta_{o,t}^T W_{ox}) \circ f'(net_t^{l-1})\end{aligned} \tag{11}$$

When dealing with the gradient of weight, the gradient of each parameter at time t shall be calculated first, and then added, that is:

$$\begin{aligned}\frac{\partial E}{\partial W_{oh}} &= \sum_{j=1}^{t} \delta_{o,j} h_{j-1}^T \\ \frac{\partial E}{\partial W_{fh}} &= \sum_{j=1}^{t} \delta_{f,j} h_{j-1}^T \\ \frac{\partial E}{\partial W_{ih}} &= \sum_{j=1}^{t} \delta_{i,j} h_{j-1}^T \\ \frac{\partial E}{\partial W_{ch}} &= \sum_{j=1}^{t} \delta_{\tilde{c},j} h_{j-1}^T\end{aligned} \tag{12}$$

The gradients of the offset terms should also be added, i.e.:

$$\begin{aligned}\frac{\partial E}{\partial b_o} &= \sum_{j=1}^{t} \delta_{o,j} \\ \frac{\partial E}{\partial b_i} &= \sum_{j=1}^{t} \delta_{i,j} \\ \frac{\partial E}{\partial b_f} &= \sum_{j=1}^{t} \delta_{f,j} \\ \frac{\partial E}{\partial b_c} &= \sum_{j=1}^{t} \delta_{\tilde{c},j}\end{aligned} \tag{13}$$

Finally, the weight gradient of each door with respect to input x:

$$\begin{aligned}\frac{\partial E}{\partial W_{ox}} &= \frac{\partial E}{\partial net_{o,t}}\frac{\partial net_{o,t}}{\partial W_{ox}} = \delta_{o,t} x_t^T \\ \frac{\partial E}{\partial W_{fx}} &= \frac{\partial E}{\partial net_{f,t}}\frac{\partial net_{f,t}}{\partial W_{fx}} = \delta_{f,t} x_t^T \\ \frac{\partial E}{\partial W_{ix}} &= \frac{\partial E}{\partial net_{i,t}}\frac{\partial net_{i,t}}{\partial W_{ix}} = \delta_{i,t} x_t^T \\ \frac{\partial E}{\partial W_{cx}} &= \frac{\partial E}{\partial net_{\tilde{c},t}}\frac{\partial net_{\tilde{c},t}}{\partial W_{cx}} = \delta_{\tilde{c},t} x_t^T\end{aligned} \tag{14}$$

The on-line monitoring data of transformer oil chromatogram to be corrected is input into the trained model to obtain the corrected data.

After the model training is completed, another set of data to be corrected is input into the model, that is, the corrected data can be obtained.

3. Oil chromatogram online monitoring data correction process

In order to correct the on-line monitoring data of oil chromatography with off-line test data, the following methods are adopted in this paper.

(1) Obtain the online monitoring data of historical transformer oil chromatogram and offline monitoring data of transformer oil chromatogram, and integrate and preprocess them.

Collect historical online and offline monitoring data, remove some irrelevant information and error information, and preliminarily integrate the data set used for training.

(2) The preprocessed data is further processed into a training set.

The data set is vectorized, that is, the online monitoring data and offline monitoring data are integrated into ordinal pairs to facilitate the reading of neural networks and further remove some unreasonable data.

(3) The training set is input into LSTM neural networks for training, so that its output can modify the model of on-line monitoring data of transformer oil chromatogram.

The processed training set, that is, the vectorized data containing on-line and off-line oil chromatographic data, is input into the neural networks for training. Because the off-line data is relatively close to the real situation, the neural networks can find the relationship between the on-line monitoring data and the off-line monitoring data, and the output model can correct the on-line monitoring data of transformer oil chromatogram to be more in line with the real situation to a certain extent.

(4) The on-line monitoring data of transformer oil chromatogram to be corrected is input into the trained model to obtain the corrected data.

After the model training is completed, another set of data to be corrected is input into the model, that is, the corrected data can be obtained.

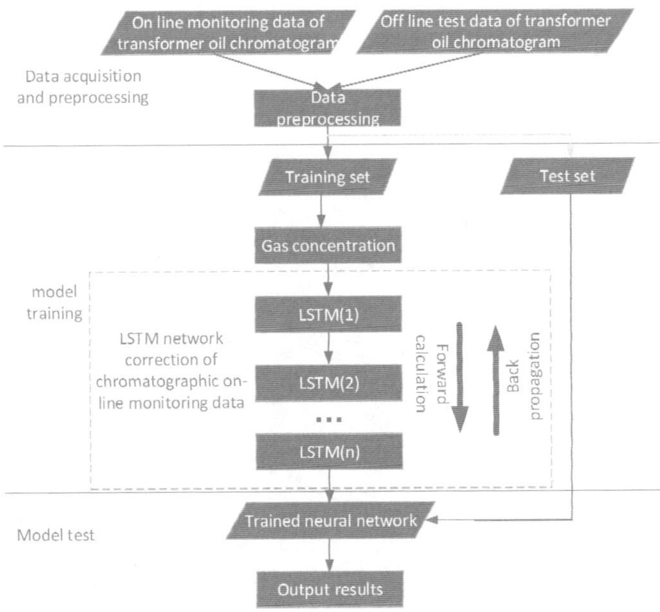

Figure 3. Flow chart of oil chromatography online monitoring data correction.

4. Correction of on-line monitoring data of oil chromatogram of a transformer

Taking the on-line monitoring data of a 500kV transformer oil color spectrum as the analysis object, this paper takes 13 consecutive off-line sampling tests in 94 days, and the on-line monitoring frequency of oil color spectrum is set to once every two days. The on-line monitoring data are corrected repeatedly through the off-line test data, so as to verify the effectiveness of LSTM neural networks in correcting the on-line monitoring data of transformer oil color spectrum.

As shown in Figure 4, the off-line test values, on-line monitoring values and corrected values after cleaning of hydrogen, methane and total hydrocarbon gas contents in the transformer oil chromatogram data are given.

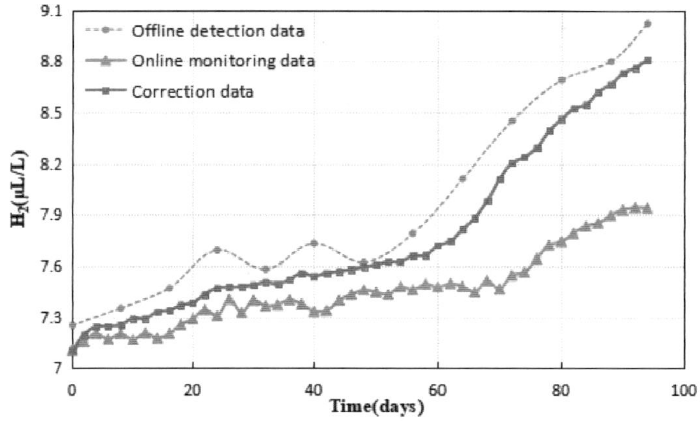

(a) hydrogen content

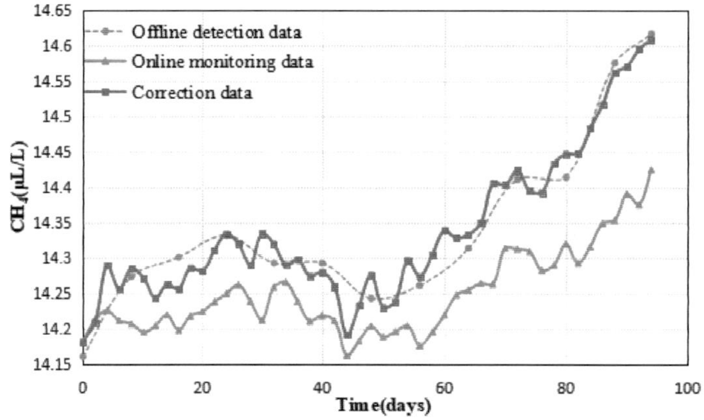

(b) methane content

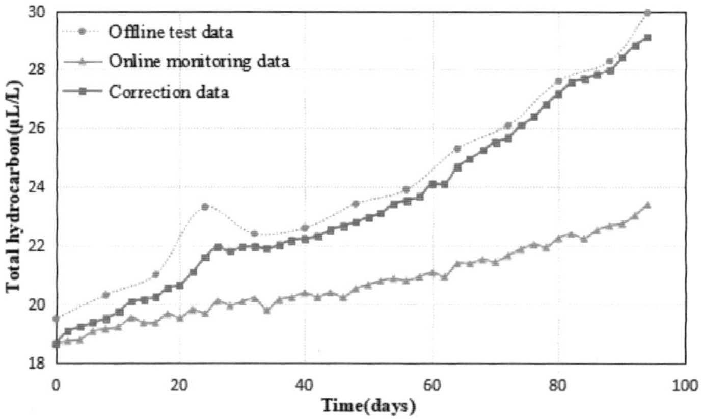

(c) total hydrocarbon content

Figure 4. Correction of on-line monitoring data of transformer oil chromatogram.

It can be seen from the above application that at the beginning of correction, the gap between offline data and corrected data is large, while the gap between corrected data and online monitoring data is small. With the increase of off-line test data, the accuracy of the corrected data after multiple feedback and correction of LSTM neural networks continues to improve and gradually approaches the real value. In this case, the gap between the on-line monitoring device and the real value gradually expands. The calibration of oil chromatogram data provided in this paper can effectively reflect the actual number of chromatography, so as to change the disadvantage that online chromatography can only reflect the change of trend.

5. Conclusion

In this paper, the oil chromatogram is used to modify the on-line monitoring data by using the off-line test data, and the on-line monitoring data is modified by LSTM neural networks, which effectively makes up for the deficiency of the on-line monitoring data in reflecting the real value of chromatography. The main conclusions are as follows:

(1) The on-line monitoring data of oil chromatogram is difficult to meet the absolute correctness of the test data due to the problems of the device itself, but it has reference significance in ttrend analysis of oil chromatogram.

(2) When using data processing methods such as neural networks to process objects with unclear internal mechanism and many influencing factors, it has good adaptability and can save complex programming and calculation process.

(3) Through the off-line test of oil chromatography and LSTM neural networks, the on-line monitoring data are corrected. The corrected data can better approach the real value, which plays an important role in improving the accuracy and availability of on-line data.

References

[1] Feng Guoji, Chen Wenhong, Zeng Li. (2022) Discussion on fault restart treatment method of oil chromatogram on-line monitoring device. J. Electromechanical information, 04: 54-56.
[2] Yin hang, Wang Lei, Meng Tao. (2022) Transformer oil color spectrum fault detection method based on improved neural networks algorithm. J. Jilin Electric Power, 50(01):14-18.
[3] Sha Weiyan, Li xiuguang, he Ninghui. (2022) Transformer fault diagnosis method based on oil color spectrum time-frequency domain information and residual attention networks. J. Power Grid and Clean Energy, 38 (01): 66-75.
[4] Xu Yunshui. (2021) Automatic pretreatment technology of characteristic gas components in transformer oil. J. High voltage Apparatus, 57(12): 218-224.
[5] Yang Li, Zhao Jing, Zhang Dewen. (2021) Influence analysis of low temperature environment on on-line monitoring device of dissolved gas in transformer oil. J. Heilongjiang Electric Power, ,43 (06): 530-532.
[6] Li Kunpeng. (2021) Research on fault diagnosis method of oil immersed transformer based on deep learning. D. North China Electric Power University.
[7] Liu Di, Qian Hui, Wang Feng. (2022) Reconstruction of distorted signal with finite innovation rate based on LSTM feature extraction. J. Acta Electronica Sinica, 50 (01):217-225.
[8] Li Lei, Ni Fusheng, Jiang Shuang. (2022) Velocity prediction model of dredged pipeline based on LSTM. J. Automation and Instrumentation, 37 (02): 86-90.
[9] Li Chunlan, Wang Jing, Shi Zhai. (2022) Research on extraction method of electric shock current based on vmd-lstm. J. JOURNAL OF HUNAN UNIVERSITY (NATURAL SCIENCE EDITION), 49 (02): 149-159.
[10] Leu van Kien, Sun Yu, Chen Lijuan. (2022) Research on fault diagnosis method of pneumatic circuit based on bidirectional LSTM. J. Computer and digital engineering, 50 (02): 367-372.

Zoning Study on Influence Intensity of Cao'e River Sluice on River Network Drainage Based on 2D Numerical Model

Junfeng Yu[1],Weijin Chen[2,a],Zhihao Fang[3,4], Dongfeng Li[4], Zihao Li[4], Le Yang[4]

[1]Shaoxing Water Conservancy and Hydropower Project Quality and Safety management Center, Shaoxing,31200, China
[2]Shaoxing No.1 Water Conservancy Ecological Construction Co., Ltd, Shaoxing,31200, China
[3]Zhejiang Ocean University, School of Naval Architecture and Maritime, Zhoushan 216022, China
[4]Key Laboratory for Technology in Rural Water Management of Zhejiang Province Zhejiang University of Water Resources and Electric Power, Hangzhou 310018, China

[a] Coresponding authors: chenweijin2011@126.com

Abstract: Shaoxing City is located in the central and northern part of Zhejiang Province, generally showing the topographic trend of high in the South and high in the North. The South is mountainous and the North is plain. There are many rivers in the river network, which are staggered and complex. A two-dimensional mathematical model for flood control and drainage of river network in Shaoxing plain is established, the hydrodynamic influence of Cao'e River gate on river network in different regions is studied, and the characteristics of each region are analyzed. According to the zoning of the impact intensity of Cao'e River sluice on river network hydrodynamic force, six obvious zoning from strong to weak are obtained, and the characteristics of each region are analyzed. It has reference value for comprehensively formulating sluice dispatching operation procedures, changing river network hydrodynamic force distribution and other non-engineering countermeasures, and provides decision-making basis for river network comprehensive management and hydrodynamic force utilization.

1. Introduction

Flood control and drainage regulation can be mainly divided into pre discharge, flood control and waterlogging removal and scouring and silting regulation. The research on the sluice pre discharge is of great significance to the research of flood control and drainage [1]-[2]. Although many water flow projects have been built in Cao'e River Basin, flood control and waterlogging drainage are still the focus of water conservancy work. Make full use of the pre-discharge of Cao'e River sluice, doing a good job in flood control and drainage is the basis of economic and social development [3]. The above goal of waterlogging drainage can be achieved through engineering countermeasures and non-engineering countermeasures. It is of great significance to study the zoning of the impact intensity of Cao'e River sluice on river network drainage [4]-[5]. Mathematical models have been widely used in regional division and classification. However, in the study of Shaoxing river network, there are more one-dimensional mathematical models [6]-[7] and less two-dimensional mathematical models. Using two-dimensional mathematical model to study Shaoxing river network is of great significance [8] - [10]. In this paper, a two-dimensional numerical model is established to study the zoning of river network, and the corresponding countermeasures can be further formulated by analyzing the zoning results.

2. Two dimensional hydrodynamic mathematical model of river network

2.1. Basic equation of two-dimensional model of water flow in river network

Flow continuity equation:

$$\frac{\partial z}{\partial t}+\frac{\partial (hu)}{\partial x}+\frac{\partial (hv)}{\partial y}=0 \tag{1}$$

X. Y-direction flow equation:

$$\frac{\partial u}{\partial t}+u\frac{\partial u}{\partial x}+v\frac{\partial u}{\partial y}-fv+g\frac{\partial z}{\partial x}+g\frac{u\sqrt{u^2+v^2}}{c^2h}=\lambda\Delta u \tag{2}$$

$$\frac{\partial u}{\partial t}+u\frac{\partial v}{\partial x}+v\frac{\partial \beta v}{\partial y}+fu+g\frac{\partial z}{\partial y}+g\frac{v\sqrt{u^2+v^2}}{c^2h}=\lambda\Delta v \tag{3}$$

Where: x and y are spatial coordinates; z is the water level; t is the time; u，v is the velocity component in X direction and Y direction respectively; h is the water depth; f is Coriolis force, $f = 2W\sin\Phi$; W is the rotation velocity of the earth; Φ is the latitude of the earth; c is Xie Cai coefficient; g is gravitational acceleration.

Definite solution condition:

Initial conditions: give the initial terrain, velocity, water depth, etc. of each calculation grid node

Boundary conditions: the flow process at the inlet of a given model; discharge process of given water level at outlet.

2.2. 2D model validation

（1）Study area

Shaoyu plain belongs to the Xiaoshao canal water system of Cao'e River Basin. The North is a plain area, crisscrossed by reticulated rivers and numerous lakes. Shaoxing has the unique river and ocean power of river network flow movement in this area. Relying on this characteristic, the development and utilization of water resources, flood control, drainage and pollution control of the plain river network can be realized. The West of the basin is Dongxiaojiang River; Cao'e River in the East; Kuaiji mountains in the South; In the North are the estuaries of Cao'e River and Qiantang River. The water flows from west to east and from south to north, mainly through Xinsanjiang gate and Mashan gate and then into Cao'e River. In this calculation, almost every river channel and lake in the river network water area is selected to establish the model. The selection range of the model and the six horizontal and five vertical backbone rivers, the model range and boundary conditions are shown in Fig. 1.

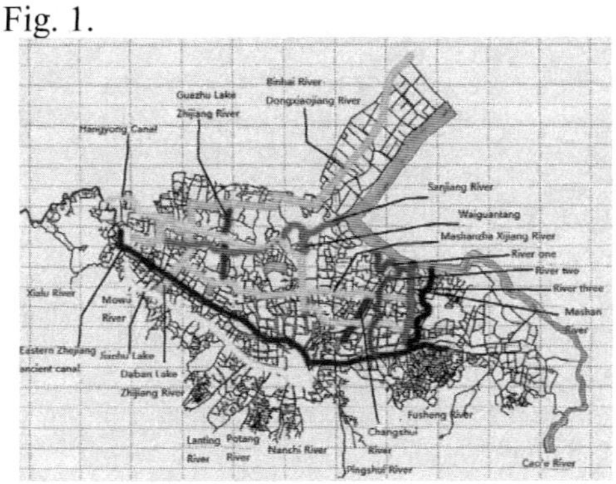

Figure.1 Model scope and boundary conditions

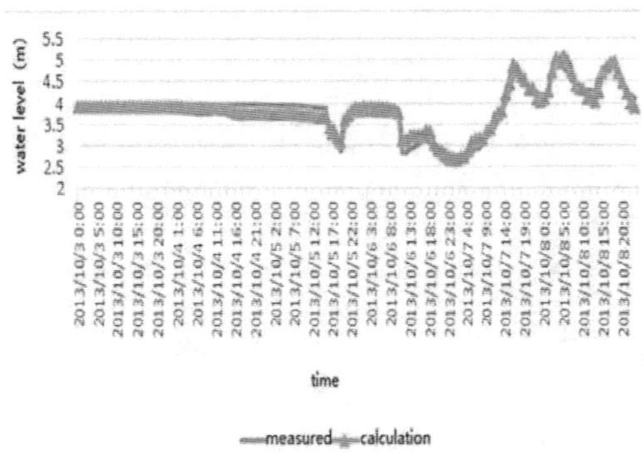

Figure.2 Comparison of measured and calculated water level process in Mashan sluice

（2）Model validation

Typhoon Fitow in 2013 is selected as the reference flood for parameter calibration calculation. The fitting results of calculated water level and simulated water level of each reference station under the selected parameters are shown in Table.1, and the simulation results of water level process are shown in Fig. 2 to 4. The measured values are in good agreement with the calculated values, and the drainage evolution calculation method and proposed parameters are reasonable .

Table.1 Calculation results of parameter calibration of typhoon Fitow model

water level station	error range / m	relative error range /%	average error / m	average relative error /%
inside Mashan gate	-0.17~0.24	-7.05~9.67	0.06	2.51
Qianqing	-0.26~0.23	-7.01~4.77	-0.05	-2.67
Gaobu	-0.27~0.22	-6.95~4.64	-0.06	-2.77

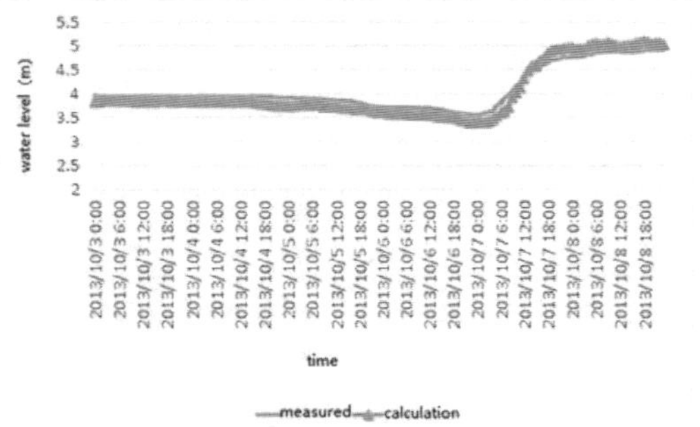

Figure.3 Comparison between measured and calculated water level process of Qianqing station

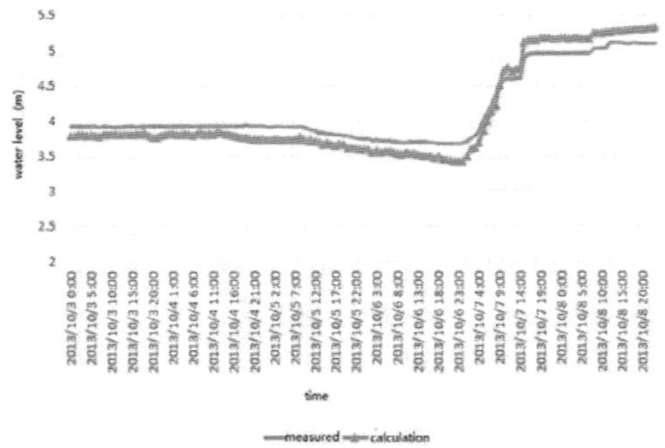

Figure.4 Comparison between measured and calculated water level process in Gaobu

3. Study on hydrodynamic impact zoning of Cao'e River Network after opening the sluice

3.1. Calculation conditions

The division of hydrodynamic impact strength of river network after the opening of Cao'e River sluice is studied, that is, the division of the hydrodynamic impact intensity of the river network with the opening of the Cao'e River Sluice and the opening of the sluice along the left bank of the Cao'e River is only considered. The normal water level of the initial condition is set at 3.9m in the river channel of the whole river network. The control water level of Cao'e River sluice is set at 2.5m, The flow calculation of river network waters is analyzed.

3.2. Calculation and analysis

According to the operation data of Cao'e River gate, the data of 12 hours after the opening of Cao'e River gate are selected for analysis. According to the hydrodynamic calculation results of river network, the water level isoline at typical time is selected for analysis. Fig. 5 shows the contour distribution of river network water level when the gate is opened for 13 hours. The numbers in the figure are the water level at this position. The more dense the marked data is, the greater the variation range of water level near here. It can be seen from the isoline distribution of river network water level in Fig. 5 that after 12 hours, the impact of Cao'e River sluice opening on river network water level is getting smaller and smaller from near to far, and has been reduced by 0.1m at the farthest position. It can be seen from the water level contour distribution map of the whole river network that there are six distinct areas in the whole river network according to the different water levels.

The water level variation range of the most affected area is 2.5m-3.0m, which is referred to as the first area for short. According to different locations, this area can be divided into three areas. The three areas are: (1) one is in the water area of Cao'e River channel, which affects the water area from Cao'e River gate to Mashan gate, with an influence range of about 22.6km, and is named area 11; (2) Xinsanjiang River channel, The impact range is from the intersection of Xinsanjiang River and Cao'e River to the intersection of the upstream of Xinsanjiang River and G92 highway, nearly 3.7km, which is named zone 12; (3) the left bank of Xinsanjiang River below the intersection of Xinsanjiang River and G92 highway and the river network water area on the left bank of Cao'e River are named zone 13.

The variation range of water level in the area with strong influence is 3.0m-3.3m, which is referred to as the second area for short. It is mainly distributed in four areas (1) the area from Mashan gate to neem gate, with a range of about 1695m and a water surface gradient of 0.5%%, which is named zone 21; (2) Caoe River water area from neem gate to Huilian gate, with a range of about 410m and a water surface gradient of 0.15%%, is named zone 22; (3) the water area of Cao'e River from Huilian gate to the entrance of Cao'e River of Hangzhou Ningbo canal, with a range of about 10500m and a water surface gradient of 0.063%%, is named zone 23; (4) the water area from the entrance of Cao'e River of Hangzhou Ningbo canal to the confluence gate of Cao'e River of Xiaocao canal is about 7600m, and the water surface gradient is 0.046%%. It is named zone 24.

The variation range of water level in the area with strong influence is the river network area enclosed by the 3.5m contour line from the right bank of Xinsanjiang sluice, the left bank of Cao'e River and the left bank of Xiaocao canal, which is referred to as the third area for short. Several drainage gates on the left bank of Cao'e River have a great impact on it.

The variation range of water level in the area with strong influence is 3.5m-3.6m, which is referred to as the fourth area for short. This area is divided into two parts (1) the isoline on the left bank of Xinsanjiang sluice is between 3.5m and 3.6m. The shape of this part of river network is similar to inverted trapezoid. The distance between 3.5m-3.6m isolines close to Xinsanjiang River channel is relatively small and the distance is relatively large. The area of this part is relatively small, (2) the contour line on the right bank of Xinsanjiang sluice is between 3.5m and 3.6m. The shape of this part of the river network is similar to the regular trapezoid. The distance between the contour lines of 3.5m-3.6m close to the Xinsanjiang River is relatively small and the distance is relatively long. This part is far from the left bank of Cao'e River, with little impact, but the river network area is large.

The variation range of water level in the area with relatively weak impact is 3.6m-3.7m, which is referred to as the fifth area for short. This part of the area is far from the left bank of Cao'e River and the end of Xinsanjiang sluice. It is less affected by the opening of Cao'e River sluice. The area is large.

The variation range of water level in the area with weak influence is 3.7m-3.8m, which is referred to as the sixth area for short. This part of the area is farthest from the left bank of Cao'e River and the end of Xinsanjiang sluice, and is least affected by the opening of Cao'e River sluice. It can be seen that such a belt area in the northwest and southeast direction is relatively large.

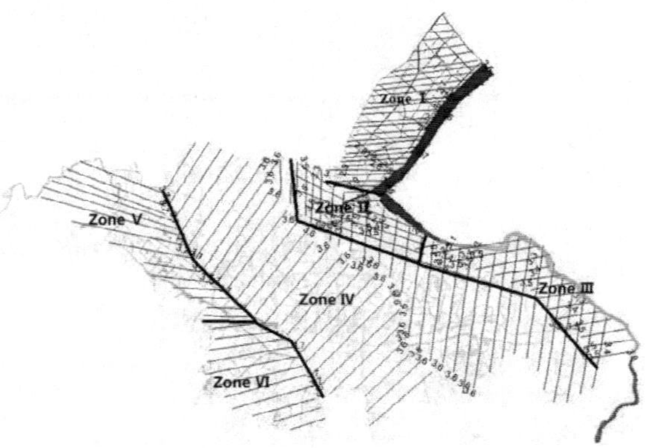

Figure.5 Velocity distribution at typical time

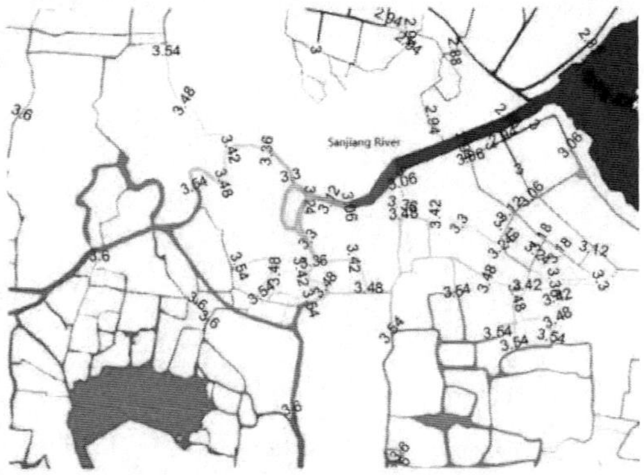

Figure.6 Water level distribution of river network near the downstream end of Sanjiang River

4. Significance of impact intensity zoning to engineering countermeasures and non engineering countermeasures in river network regulation planning

The planning and design of waterlogging drainage projects should make full use of these areas, especially the first and second areas, through planning and design of river project, pump station sluice project and other engineering countermeasures, the river network floods in the West, South and Southeast of the fourth, fifth and sixth regions are discharged into the first, second and third regions to reduce the river network water level and meet the requirements of river network drainage. In terms of river network drainage, the sluice on the left bank of Cao'e River also plays an important role in river network drainage. In particular, within the range of nearly 4.5km downstream of Xinsanjiang sluice, the river water level is below 3.15m, and the water level in this reach decreases from 3.1m to 2.85m, and the water surface gradient is 0.73%. See Fig. 6 water level distribution map of river network near the downstream end of Sanjiang River. This shows that Sanjiang has strong drainage capacity. Under the same other conditions, the flood near the river network is preferentially discharged into the river channel 4.5km downstream of Sanjiang River.

5. Conclusion

The impact intensity of Cao'e River sluice on river network drainage has six obvious divisions. The impact of Cao'e River sluice on river network hydrodynamic force is getting weaker and weaker from near to far. In addition to making full use of the water area of Cao'e River for drainage, the 4.5km river at the downstream end of the three rivers is also an important area for drainage of river network. The planning and design of river network drainage project should give priority to the drainage of river network flood into the waters of the first and second areas.

Acknowledgments

This research was supported by the Joint Funds of the Zhejiang Provincial Natural Science Foundation of China （No.LZJWZ22E090004,No. LZJWZ22C030001） and the Funds of Water Resources of Science and Technology of Zhejiang Provincial Water Resources Department, China(No.RB2115)，（No.RC2040），the Funds Key Laboratory for Technology in Rural Water Management of Zhejiang Province，the National Key Research and Development Program of China(No.2016YFC0402502)，the National Natural Science Foundation of China(51979249).

References

[1] Ministry of water resources of the people's Republic of China. (1995) Technical management regulations for sluice. China water resources and Hydropower Press, Beijing.
[2] W.G. Yu. (2011) Guidelines for the preparation of control and operation plans for large and medium-sized sluices in Zhejiang Province. Zhejiang water conservancy Estuarine Research Institute, Hangzhou.
[3] F. Chen. (2019) Impact analysis of pre discharge on flood disaster of plain river network. Zhejiang water conservancy science and technology.,47(03): 24-27 + 37.
[4] W.G.Yu. (2015) Research on Sluice dispatching. Zhejiang water conservancy science and technology.,43(05): 76-79 + 89.
[5] F. Zhou, H.F.Sheng. (2013) 2013 control and operation plan report of Cao'e River Sluice. Zhejiang water resources and Hydropower Survey and Design Institute, Hangzhou.
[6] Y. Wu, L.Q. He. (2019) Discussion on constructing hydrodynamic model of Cao'e River flow area in Shaoxing City. Zhejiang water conservancy science and technology.,47(03): 21-23+31.
[7] F. Chen, X.L. Ying. (2016) Research and application of river network model in Shaoxing Plain. Journal of Zhejiang University of water resources and hydropower. ,28(05): 43-47
[8] Y.Y. Tong, D.F. Li. (2016) Nie Hui Study on hydrodynamic and plane two-dimensional mathematical model of river network in Datian Plain. Journal of Zhejiang University of water resources and hydropower.,28(01): 14-17
[9] F.X. Yang, P.J. Zhang, W.J. Chen, D.F. Li, H.W. Zhang. (2015) Analysis of hydrodynamic

two-dimensional numerical model of river network and lakes of Qingshui project in Keqiao main urban area. Journal of Zhejiang University of water resources and hydropower. ,27(03): 18-21
[10] B.J. Liang, F.X. Yang, D.F. Li, H.W. Zhang. (2015) Hydrodynamic analysis of clean water project in Keqiao main urban area of Shaoxing. Journal of Zhejiang University of water resources and hydropower. ,27 (01): 16-20

Study on Geese Array Effect and Optimal Layout of Herringbone PV array

Guiyuan Liu [1a*], Junhong Hu [2b], Jiang Liu [1c]

[1] Xinjiang Water Resources and Hydropower Planning and Design Administration, Urumqi, China

[2] Tarim University, Aral, China

[a*]18894006520@163.com, [b]18294490922@163.com, [c]27200869@qq.com

Abstract—Layout parameters play a significant role in wind loads of PV array. In view of this, wind loads of the herringbone PV array composed of 9 panels under five array angles (30°,40°,45°,50°,60°), five installation angles (30°,40°,45°,50°,60°) and five array distances (1.5m,1.75m,2m,2.25m,2.5m), are investigated by using the SST k-ω turbulence model and unsteady RANS approach. Then laws of wind pressure subjected to parameters are evaluated by drawing the contour and boxplot. Results show that: in the construction of herringbone photovoltaic panels, array angle is preferably not greater than 45°, installation inclination angle is not greater than 50°, and optimal array distance is between 1.75m and 2.5m. It is expected to provide theoretical basis and scientific guidance for layout of the herringbone array and sand control in desert areas.

1. Introduction

Photovoltaic stations in Gobi area, as a new way to effectively improve the status of desertification [1], can not only utilize the abundant local solar energy resources, but also change the soil moisture and local microclimate [2-3], make crop growth well [4]. Therefore, it is of great importance to find the optimal layout of PV array to further improve the ecological environment and desertification control in the Gobi area.

Over the past decades, many scholars have done a lot of researches on ground mounted arrays. Hassan et al. [5] analyzed the wind pressure distribution of photovoltaic panels in different wind directions. Agarwal et al. [6] found that the maximum and minimum values of the windward lift coefficient appeared when inclination is 60° and 90° respectively; Jubayer et al. [7] studied the wind pressure of PV panels at different windward angles, and found that the largest force is at 0° and 180° while the moment is at 45° and 135°. Shademan et al. [8] found that proper longitudinal spacing can minimize lift coefficient. Workamaw et al. [9] found that the wind pressure reduced seriously in row 2. Wind reduction characteristics of photovoltaic panels under different heights, inclination angles and spacing arrangements were analyzed by Zhou Wei et al. [10].

Due to lack of study on wind pressure of photovoltaic array. Attempts have been made in this paper to determine pressure coefficients on PV panels subjected to various layout parameters. Conclusions are helpful to reasonably guide the construction of photovoltaic power station in the Gobi area and better improve the ecological environment in arid areas.

2. Models and Methods

2.1. Geometric modelling

Numerous studies show that the installation angle, array angle and distance play a decisive role in wind pressure carried by the photovoltaic array. In order to avoid the scaling effect that peak load obtained by numerical simulation is smaller than the measured value, full scale model is adopted [11]. Inspired by the goose array effect proposed by Weimerskirch et al. [12], the photovoltaic panels are arranged in a Y hell shape in this study. Fig 1 shows the layout parameters of photovoltaic panels. Size of 9 panels is 2m(l)×1m(b)×0.05m(t), with a constant height 0.2m above ground. The experiment is divided into three groups, each group has only one variable: group 1, array angle was taken as 30°,40°,45°,50° and 60° respectively; group 2, installation inclination was 30°,40°,45°,50° and 60° respectively; group 3, array distance was taken as 1.5m, 1.75m, 2m, 2.25m and 2.5m respectively, other parameters remain unchanged.

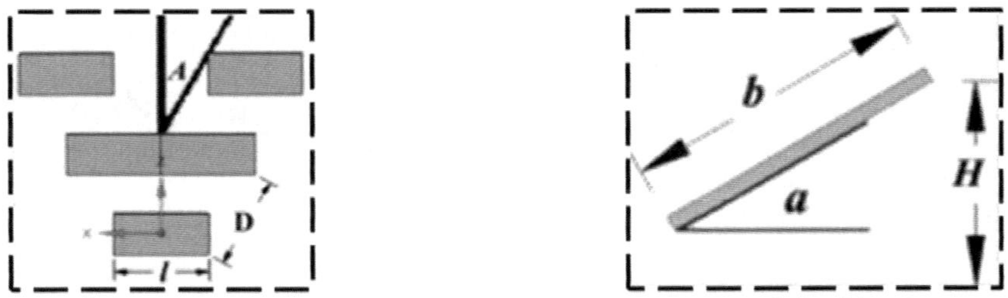

（a） vertical view （b） side view

Fig 1 Schematic diagram of photovoltaic panel layout parameters

2.2. Numerical Simulation

Computational domains are created refer to Franke et al. [13] "Best practice guideline for the CFD simulation of flows in the urban environment". For photovoltaic panel, adapt to size transition by dividing unstructured mesh, locally optimized near the boundary with the help of boundary layer grid and local size function. Referring to Bitsuamlak et al. [14] research on inclined plate, SST K-ω turbulence model is employed. Inlet velocity is constant at 10m/s and the free pressure outlet. The time step is estimated by Courant Number. With the increase of the number of iterations, the residuals of the monitoring variables and the lift-drag coefficient tend to be stable, calculation converged.

3. Results and Analysis

To minimize wind loads of herringbone PV array, response laws that array geese effect and wind pressure distribution characteristic subjected to arrangement parameters are discussed respectively by drawing wind pressure contour and boxplot.

3.1. Contour

In order to analyse the distribution of the wind pressure, each photovoltaic panel was divided into 15 monitoring areas by the equal spacing method. Fig 2 shows the contours of rows 1-5 for each array angle. The distribution of wind pressure is non-uniform, and the maximum wind pressure of each row appears on the alary connection on the outer side of the windshield wall; the maximum wind pressure of the 2nd-5th row moves inward with the increase of the array angle. Wind pressure carried by row 1 is much greater than others, which means goose array effect appeared. With the increase of array angle, the wind pressure difference between rows gradually decreases, goose array effect weakens.

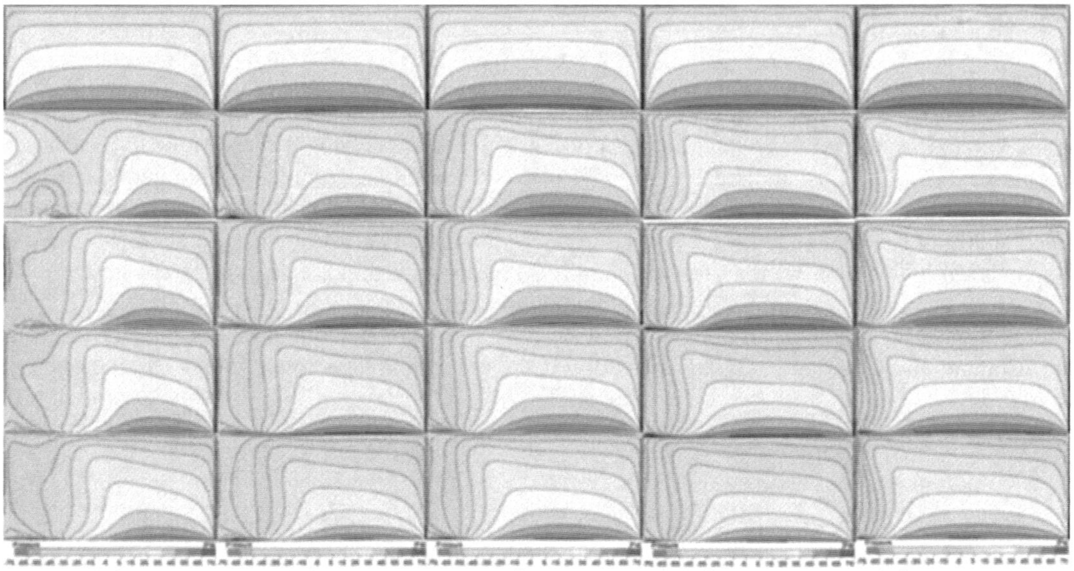

（a）A=30° （b）A=40° （c）A=45° （d）A=50° （e）A=60°

Fig 2 Contours of rows 1-5 for each array angle

3.2. Response law of wind pressure to layout parameters

In order to analyse the response law of the wind pressure of the wind barrier to the layout parameters, the classical statistical analysis of the wind pressure at the monitoring points is carried out. Fig 3-5 are the box diagrams of rows 1-5 for each array angle, installation inclination angle and array distance.

Fig 3 shows that at the same array angle, average wind pressure decreases with the increase of the number of rows, and the amplitude reduction of the second row is particularly significant, which is consistent with the research conclusion of Workamaw et al. [15] on single row photovoltaic panels. Box height first decreases and then increases with the increase of the number of rows, and reaches the minimum in the second row.

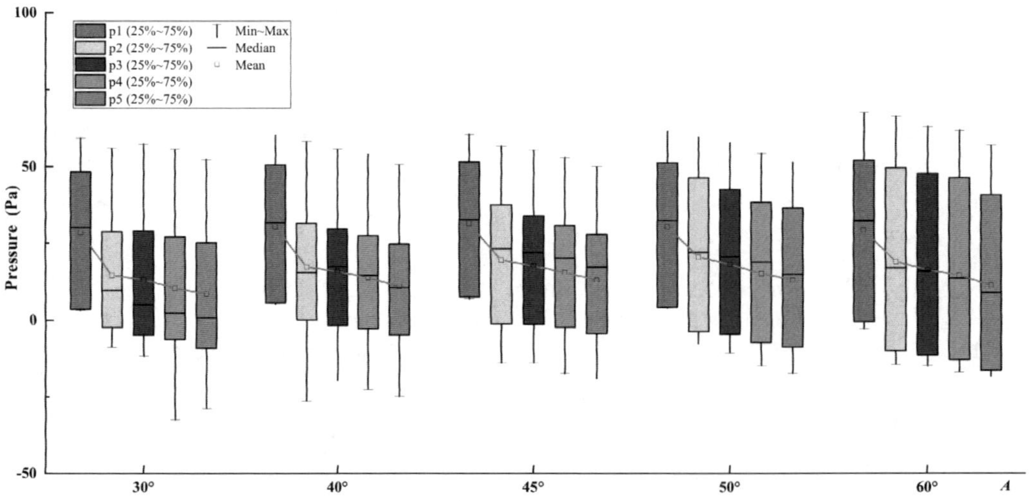

Fig 3 Box diagram of rows 1-5 for each array angle

With the increase of array angle, average wind pressure first increases and then decreases. For different array angles, box height of row 1 tends to be the same, box height of rows 2-5 first decreases and then increases with the increase of array angle, with a minimum at 40° and a maximum at 60°.

Fig 4 shows that with the increase of installation angle, average wind pressure of each row is 60° > 45° > 50° > 40° > 30°. Box height of the first row reaching peak at 50°, box height of rows 2-5 first decreases and then increases, reaching minimum at 40° and the maximum at 60°.

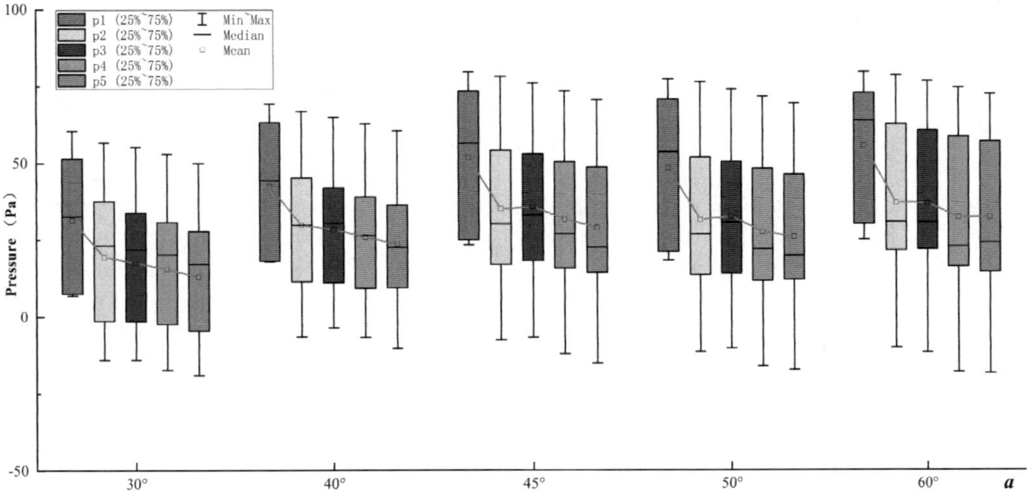

Fig 4 Box diagram of rows 1-5 for each inclination angle

Fig 5 shows that for the same array distance, average wind pressure decreases with the increase of the number of rows, and the reduction amplitude first increases and then decreases with the increase of array distance. The goose array effect first strengthens and then weakens with the increase of array distance, reaches the peak at 1.75m.

With the increase of the array spacing, the average wind pressure of each row first decreased significantly and then increased slightly, and both reached the minimum value when the array distance was 1.75m, and reached the maximum value when the array distance was 1.5m; box height of each array distance first decreases and then increase, reaching the peak when array distance is 1.5m. for rows 1 and 3, box height is the smallest when the array distance is 1.75m; for row2 ,row 4 and row 5, box height is the smallest when array distance is 2m.

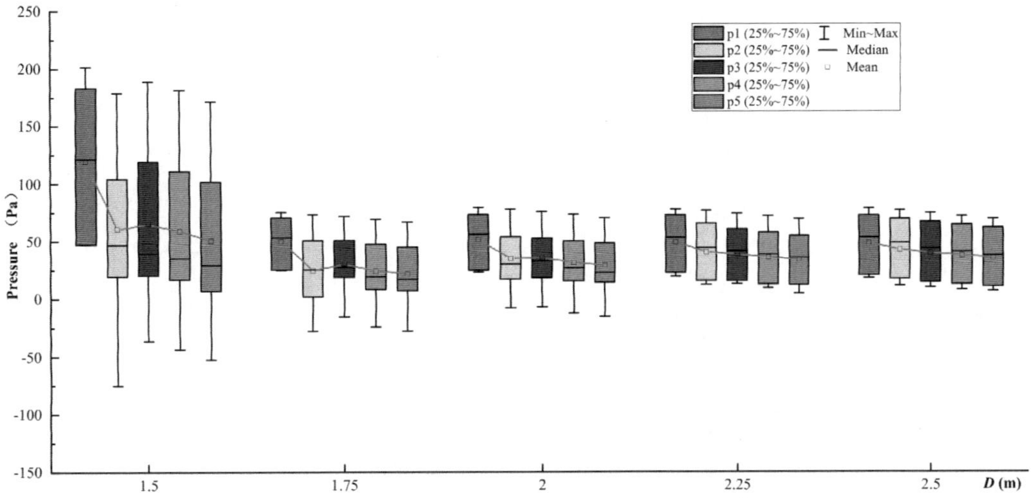

Fig 5 Box diagram of rows 1-5 for each array distance

4. Conclusion

Through comparative analysis, it is confirmed that the herringbone PV array has a geese array effect, the single factor analysis method is further used to explore the wind pressure dependent on layout parameters, following are conclusions:

(1) Wind pressure of the first row is much greater than behind rows, which means geese array effect appeared, local reinforcement method can be used to improve load absorption of herringbone PV array.

(2) The wind pressure of the herringbone array are subject to the array angle, installation inclination and array distance. As far as the average wind pressure is concerned, it increases and then decreases by rows with the increase of the array angle; it shows 60°>45°>50°>40°>30° with different installation inclinations; with the increase of the array distance, it decreased significantly and then increased slightly, reaching a peak at 1.5m and a valley at 1.75m. As far as the coefficient of variation is concerned, box height of the 2-5 rows decreases and then increases by rows with different array angles or inclinations, reaching a minimum at 40° and a maximum at 60°; box height of each row all reach their peak at 1.5m with different array distances.

(3) In the actual construction of the herringbone array, array angle and installation inclination are preferably not more than 45° and 50° respectively, while the array space is recommended between 1.75 and 2.5m.

References

[1] Wang Q, Wang F L, Liu S J, et al. (2020). The Spatial-Temporal Variation Characteristics and Ecological Significance of Solar Radiation Utilization by Photovoltaic Industry Analogy to Desert Vegetation: Taking Gansu Hexi Corridor as an Example[J]. Ecological Economy, 36(1): 79-85.

[2] Yin D, Ma L, Qu J, et al. (2017). Effect of Large Photovoltaic Power Station on Microclimate of Desert Region in Gonghe Basin[J]. Bulletin of Soil and Water Conservation, 37(3): 15-21.

[3] Yuan S P, Zhang Z S, Dang T H, et al. (2018). Comparative Study of Growth Status of Three Plant Measures and Its Ecological Functions in Photovoltaic Power Station of Mu Us Sandy Land[J]. Research of Soil and Water Conservation, 25(2): 235-239.

[4] Gao X Q, Yang L W, Lv F, et al. (2016). Effect of PV farm on soil temperature in golmud desert area[J]. Acta Energiae Solaris Sinica, 37(6):1439-1445.

[5] Irtaza H, Agarwal A. (2018). CFD Simulation of Turbulent Wind Effect on an Array of Ground-Mounted Solar PV Panels[J]. Journal of the Institution of Engineers, 99(2): 205-218.

[6] Agarwal A, Irtaza H, Zameel A. (2017). Numerical study of lift and drag coefficients on a ground-mounted photo-voltaic solar panel[J]. Materials Today-Proceedings, 4(9): 9822-9827.

[7] Jubayer C M, Hangan H. (2014). Numerical simulation of wind effects on a stand-alone ground mounted photovoltaic (PV) system[J]. Journal of Wind Engineering and Industrial Aerodynamics, 134: 56-64.

[8] Shademan M, Barron R M, Balachandar R. (2014). Numerical simulation of wind loading on ground-mounted solar panels at different flow configurations[J]. Canadian Journal of Civil Engineering, 41(8): 728-738.

[9] Warsido W P, Bitsuamlak G T, Barata J, et al. (2014). Influence of spacing parameters on the wind loading of solar array[J]. Journal of Fluids & Structures, 48: 295-315.

[10] Zhou W, He B, Cai J , et al. (2018). Wind Load Characteristics and Reduction Analysis of a Structural System of Photovoltaic Power Station[J]. Structural Engineers, 34(2): 86-94.

[11] Aly A M. (2016). On the evaluation of wind loads on solar panels: The scale issue[J]. Solar Energy, 135: 423-434.

[12] H W. (2001). Energy saving in flight formation[J]. Nature, 6857(413): 697-698.

[13] Franke J. Best practice guideline for the CFD simulation of flows in the urban environment[M]. Hamburg, Germany:Cost Office Brussels, 2007.

[14] Bitsuamlak G T, Dagnew A K, Erwin J. (2010). Evaluation of wind loads on solar panel modules using CFD[C]. Fifth International Symposium on Computational Wind Engineering (CWE), Chapel Hill, NC, May, 23-27.
[15] Warsido W P, Bitsuamlak G T, Barata J, et al. (2014). Influence of spacing parameters on the wind loading of solar array[J]. Journal of Fluids and Structures, 48: 295-315.

Research on the Reflection Characteristics of the Shock Wave Boundary in the Numerical Simulation of Underwater Explosion

Houlin Fang[1, a*], Dezhi Zhang[1, b*], Quanshi Guo[1, c], Qiang Lu[1, d], Liangyong Zhang[1, 2, e], Ao Li[1, f], Yunzhe Liu[1, g]

[1]Northwest Institute of Nuclear Technology, Xi'an 710024, China

[2]College of Meteorology and Oceanography, National University of Defense Technology, Changsha 410073, China

[a*]fanghoulin@nint.ac.cn, [b*]zhangdezhi@nint.ac.cn, [c]guoquanshi@nint.ac.cn, [d]luqiang@nint.ac.cn, [e]zhangliangyong@nint.ac.cn, [f]liao@nint.ac.cn, [g]liuyunzhe@nint.ac.cn

Abstract—A two-dimensional axisymmetric model of the underwater explosion is established in this paper. After verification by empirical formula, the variation law of the reflection coefficient of the shock wave with different boundary sizes under different boundary conditions is studied. The results show that the peak pressure of the near-field shock wave is relatively close to the empirical formula. The calculated value of the far-field shock wave peak pressure is smaller than the empirical value, and refining the mesh can make the near-field shock wave peak pressure closer to Cole's empirical formula; when the "Flow-out" boundary condition is set, with the increase of the proportional boundary size, the boundary pressure decreases and the reflection coefficient increases. When the proportional boundary size reaches 5, the reflection coefficient of the boundary is the same as that without the "Flow-out" boundary, which is close to 90%.

1. Introduction

The commonly used large-scale finite element software for numerical simulation of underwater explosions include LS-DYNA, ABAQUS, DYTRAN, and AUTODYN. By comparing the application of four numerical calculation software in underwater explosion simulation, it is concluded that AUTODYN has the most comprehensive functions in underwater explosion simulation, with better calculation stability and obvious advantages [1].

These finite element software are equipped with ALE 2D and 3D solvers, which can be used to solve highly nonlinear dynamic problems of solids, fluids, gases, and interactions, and have important applications in the simulation of underwater weapon power and numerical simulation of ship anti-explosive shock value [2]. Since both underwater explosion and detonation products and water belong to fluids, compared with the Lagrange description method, the Euler method is more convenient and effective to describe [3]; mapping techniques of some software can map the one-dimensional spherical symmetry calculation results to the two-dimensional axisymmetric solver, thus solving the problems of too small mesh size and too long calculation time, especially for the working conditions where the explosion source has a certain distance from the free surface, the bottom or the structure in the water [4]. Based on these software, many scholars have carried out a lot of beneficial work. Hu, Y.T. et al. [5] used the AUTODYN one-dimensional method to accurately simulate the near-field underwater explosion shock waves and bubbles and other characteristic parameters, while the far-field shock wave peak pressure calculation value is smaller than the empirical value. Xiao, Q.P. [4] further studied the influence of AUTODYN mesh density and equation of state on the simulation results, and they found that higher mesh density can obtain better accuracy of pressure and impulse at the same time and the far-field shock wave is more accurate in SHOCK equation of state simulation; Liu, K.Z. et al. [2] in order to characterize the underwater explosion near-field high-frequency shock waves, suggested dividing the mesh according to the criterion of sampling at least 100 times within the peak pressure decay time constant, while using a smaller artificial viscosity coefficient and low-pass filtering; Zhang, W. et al. [3] studied the dynamic characteristics of underwater explosion bubbles near the wall and the influencing factors of bubble shape, maximum bubble radius, pulsation period, and jet time. Jiang, T. et al. [6] calculated and analyzed the rupture and plastic deformation of the pressure hull and non-pressure hull of the typical compartment of a double-hull submarine under different blast distance conditions. Jia, X.Z. et al. [7] comprehensively considered the hydrostatic pressure gradient and boundary conditions including free surface, sidewall, and bottom to calculate the bubble pulsation process of underwater explosion in limited waters. Although these finite element software have been used for a long time, there are a few references to study the boundary reflection effect of shock waves. In this paper, a two-dimensional axisymmetric model of the underwater explosion is established, and the "Flow-out" boundary conditions are set to simulate the free field or reduce the boundary reflection of shock waves. The variation law of the shock wave reflection coefficient of different boundary sizes under the "Flow-out" boundary condition is studied to provide technical support for the study of the shock wave effect of underwater explosions.

2. Numerical Calculation Model

Numerical calculation model uses 1g TNT charge, establish a two-dimensional axisymmetric model, with x-axis for symmetry axis, the calculation domain for the rectangular Euler domain, x-direction size for y-direction size 2 times, and the burst center located in the center of symmetry (origin). The rectangle is set up with three sides of the boundary condition "Flow-out", and the sampling point "Gauge 1" is set at a distance d from the boundary, as shown in Figure 1.

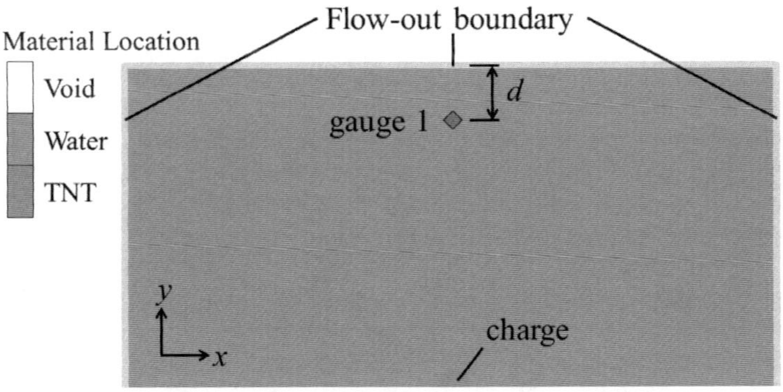

Fig. 1 Two-dimensional axisymmetric model

2.1. *Verification of Calculation Model*

The pressure of the explosive detonation product is described by the JWL equation of state. Generally, there are two sets of parameters in the JWL equation of state of TNT, which are respectively from LEE FINGER (1973) and DOBRATZ (1981). According to the calculation experience [2], the calculation results of the latter parameters are closer to empirical calculations.

The water medium adopts a polynomial state equation. The state equation is different when the medium is under pressure and tension. The internal energy of the medium is considered, which is related to the hydrostatic pressure. If the water medium is considered under atmospheric pressure, the specific internal energy of the water medium is 361.875 J/ kg. Xiao, Q.P. et al. [4] suggested that the SHOCK equation of state should be used in the water medium when simulating mid- and far-field shock waves. The artificial viscosity coefficient adopts the default value, that is, the coefficients of the first-order term and the second-order term are 0.2 and 1, respectively. The attenuation law of shock wave peak pressure with scaled distance under different mesh conditions is calculated and compared with Cole's empirical formula [8]. It can be seen from Figure 2 that the near-field shock wave peak pressure is closer to the empirical formula, while the far-field shock wave peak pressure is smaller than the empirical value, which is the same as the conclusion drawn by Hu, Y.T. et al. [5]; finer meshes can make the near-field physical model closer to the real situation, the calculation accuracy higher, and the peak pressure of the near-field shock wave closer to Cole's empirical formula, but has little effect on the peak pressure of the mid- and far-field shock wave. This proves that the calculation results of the two-dimensional axisymmetric model established in this paper are reliable and can be used for subsequent research.

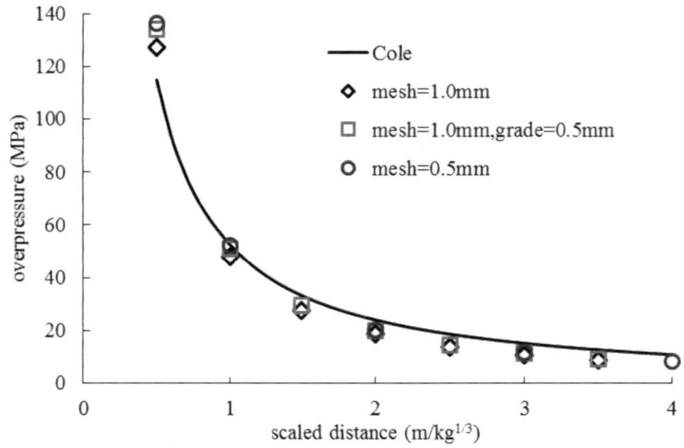

Fig. 2 Attenuation law of shock wave overpressure peak value with scaled distance under different mesh divisions

2.2. Design of calculation conditions

In order to study the variation law of the shock wave reflection coefficient with different boundary sizes under the "Flow-out" boundary condition, the design conditions are shown in Table 2, and the scaled distance between the collection point and the boundary is kept unchanged. In groups 1-9, the boundary size is incremented by 50 mm, and the 10th group is the comparison group of the 9th group, with the scaled distance unchanged, mainly to verify whether the calculation results satisfy the similarity law.

Table 1 Design calculation conditions

No.	Charge (g)	Boundary dimension R_b (mm)	Simulation region $x \times y$ (mm)	Gauge point y (mm)
1	1	100	200×100	50
2	1	150	300×150	100
3	1	200	400×200	150
4	1	250	500×250	200
5	1	300	600×300	250
6	1	350	700×350	300
7	1	400	800×400	350
8	1	450	900×450	400
9	1	500	1000×500	450
10	8	1000	2000×1000	900

3. Computational Results and Analysis

The typical shock wave pressure-time history curve and the pressure nephogram of the sampling point obtained from the calculation are shown in Figure 3. Figure 3(a) contains three curves, namely, the "Flow-out" boundary condition, no "Flow-out" boundary condition (i.e., rigid bounds), and approximately no boundary, in which a larger computational domain is created. From the pressure-time history curve, it can be clearly seen that there is no difference between the "Flow-out" boundary and the approximately no boundary. The former has obviously reflected pressure peaks, the latter has a smooth decay, and the "Flow-out" boundary condition curve is between the two. There is a certain reflected pressure peak in between; it can also be seen from the pressure nephogram that there is obvious shock wave reflection when the "Flow-out" boundary is not set, while the shock wave reflection is weakened when the "Flow-out" boundary is set.

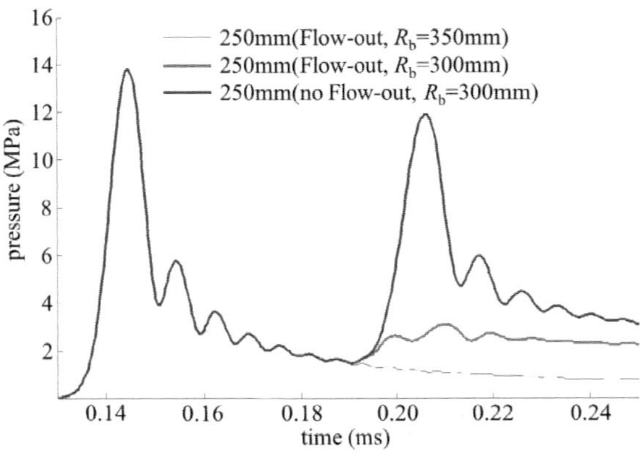

(a) Time history curve of shock wave pressure with different boundary conditions

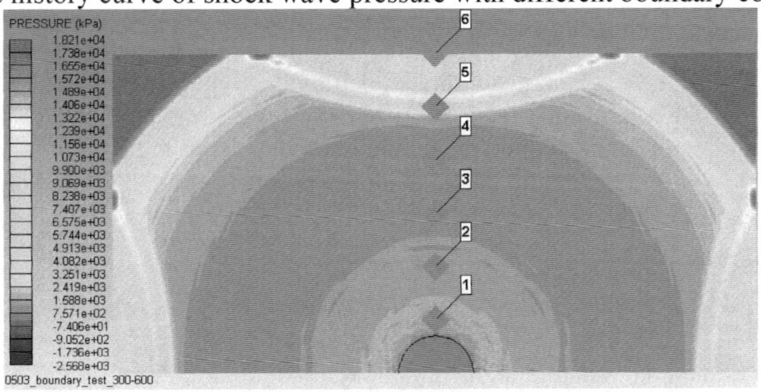

(b) No "Flow-out" boundary pressure

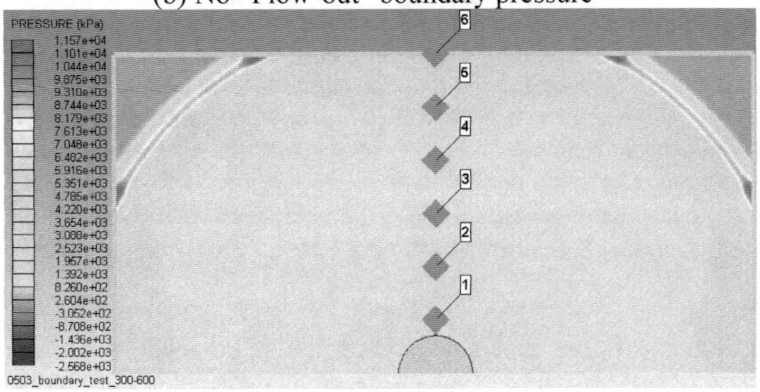

(c) "Flow-out" boundary pressure

Fig. 3 Typical shock wave pressure-time history curve and pressure nephogram

The "Flow-out" boundary shock wave reflection intensity is set directly related to the boundary size. When the calculated boundary size reaches 500 mm, the pressure-time history curve of the "Flow-out" boundary shock wave and the boundary curve without "Flow-out" are set basically completely coincident, as shown in Figure 4, at which the "Flow-out" boundary condition fails.

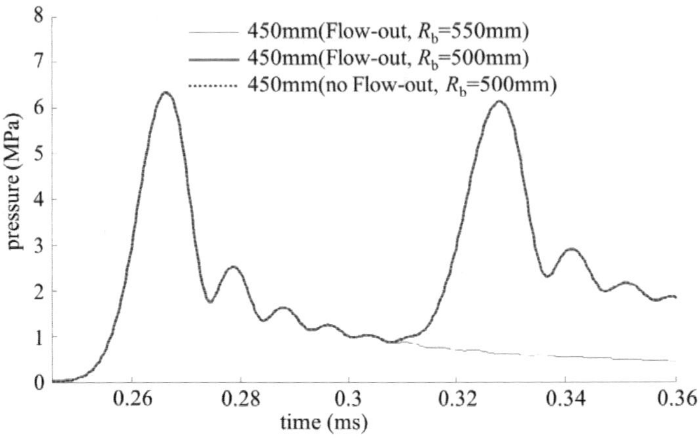

Fig. 4 Time history curve of shock wave pressure with boundary size of 500 mm

In order to quantitatively study the shock wave boundary reflection coefficient under different boundary sizes (here, the pressure peak gain caused by boundary reflection is mainly considered), the reflection coefficient is defined as follows:

$$r = \frac{p_{\mathrm{ar}} - p_0}{p_{\mathrm{ai}}} \tag{1}$$

Where p_{ar} is the peak value of reflected pressure, p_{ai} is the peak value of incident pressure, and p_0 is the pressure value at the moment of peak reflected pressure when there is no reflection. From the comprehensive results, the variation law of the boundary reflection coefficient with the proportional boundary size or boundary pressure is obtained, as shown in Fig. 5, where the boundary pressure is taken as the pressure at the corresponding ratio distance under the approximate no boundary condition.

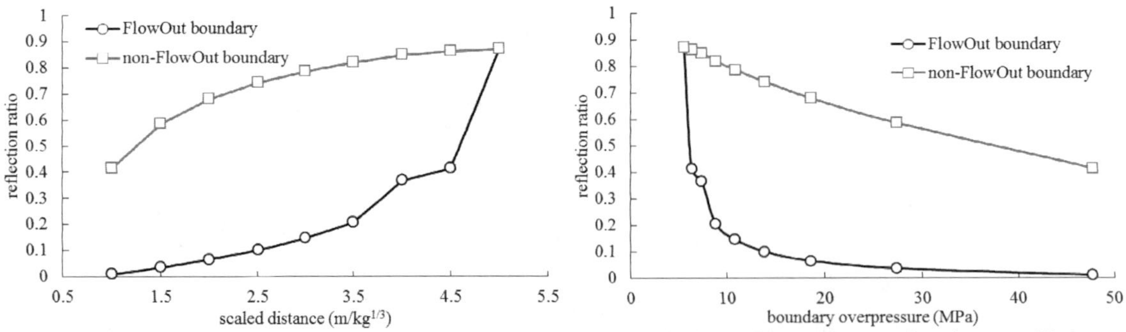

(a) Variation of boundary reflection coefficient with proportional boundary size (b) Variation of boundary reflection coefficient with boundary pressure

Fig. 5 Variation of boundary reflection coefficient with proportional boundary size or boundary pressure

It can be seen from the figure that after setting the "Flow-out" boundary, as the size of the proportional boundary increases, the boundary pressure decreases, and the reflection coefficient increases. When the size of the proportional boundary reaches 5, the boundary reflection coefficient is the same as that of no "Flow-out", close to 90%. Similarly, when the "Flow-out" boundary is not set, as the size of the proportional boundary increases, the reflection coefficient also increases; however, the increase rate of the boundary reflection coefficient differs between the two settings, with the "Flow-out" boundary increasing at a gradual rate and no "Flow-out" boundary increasing at a slow rate. The trend of this variation law is obtained from the analysis of the shock wave pressure-time history curve shown in Figure 6.

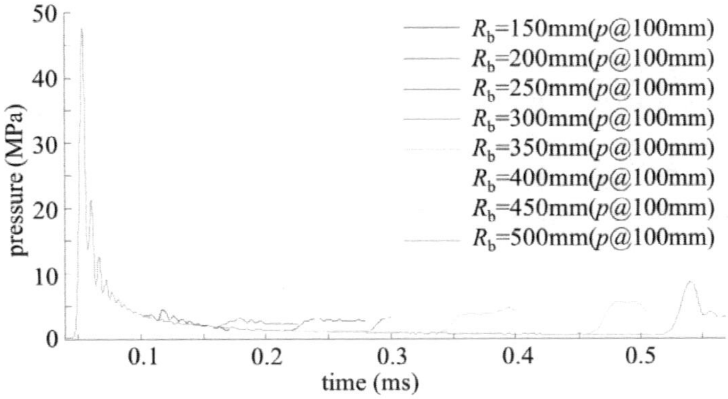

Fig. 6 Time history curve of shock wave pressure with different boundary sizes

4. Conclusion

In this paper, a two-dimensional axisymmetric model of the underwater explosion is established, the calculation model is verified by empirical formulas, and the influence of mesh size is analyzed. Based on this, the variation law of shock wave reflection coefficients for different boundary sizes under Flow-out boundary conditions is studied, and the following conclusions were drawn.

(1) The near-field shock wave peak pressure calculated by the two-dimensional axisymmetric model is relatively close to the empirical formula, and the far-field shock wave peak pressure value is smaller than the empirical value; refining the mesh can make the near-field shock wave peak pressure closer to Cole's empirical formula, however, it has little effect on the peak pressure of the shock wave in the medium and far-field.

(2) The time-history curve of the shock wave reflection pressure at the "Flow-out" boundary is set to be between that under no "Flow-out" boundary and the approximately-no boundary; under the "Flow-out" boundary condition, with the increase of the proportional boundary size, the boundary pressure decreases and the reflection coefficient increases. When the proportional boundary size reaches 5, the boundary reflection coefficient is the same as the boundary without the "Flow-out" set, which is close to 90%.

References

[1] Huang, X.Z., Wang, Z.J. (2013) Numerical Study of Underwater Explosion Bubble Pulse. Explosive Materials, 42: 19–23.
[2] Liu, K.Z., Xu, G.G., Xin, C.L., Yang, Z.L., Qin, J. (2009) Research on numerical simulation in the underwater explosion by AUTODYN. Blasting, 26: 18–21.
[3] Zhang, W., Yue, Y.W., Zhang, A.M., Sun, L.Q. (2012) Numerical simulation of underwater explosion bubble interactions with a solid boundary based on AUTODYN. Chinese Journal of Ship Research, 7: 23–30.
[4] Xiao, Q.P., Chen, W.H., Jia, X.Z., Wu, T., Rao, G.N. (2009) Numerical study of underwater explosion shock wave based on AUTODYN. Ship Science and Technology, 31: 38–43.
[5] Hu, Y.T., Jia, X.Z., Rao, G.N., Liu, J.C. (2009) Numerical study of underwater explosion shock wave and bubble pulse. Ship Science and Technology, 31: 134–140.

[6] Jiang, T., Wang, G.Q., Zhan, F.M., Zhou, F.Y. (2015) Impact damage analysis of typical submarine compartment subjected to underwater blasting based on AUTODYN. Explosive Materials, 44: 61–64.
[7] Jia, X.Z., Wang, J.L., Gao, Z., Guo, W., Jin, P.G. (2015) Numerical simulation of bubble pulse of underwater explosion in confined water area. Ship Science and Technology, 37: 31–34.
[8] Cole, R.H. (1948) Underwater explosions. Princeton University Press.

Numerical Simulation of Hydraulic Characteristics of Spillway Tunnel with Gradually Expanding Outlet and Lower outlet height

Chengfa Deng [1、2], Jincheng Du [3]

[1]Zhejiang Institute of Hydraulics & Estuary, Hangzhou, Zhejiang, China;

[2] Zhejiang Water Conservancy and Hydropower College, Hangzhou, Zhejiang,China;

[3]Taizhou Huangyan District Changtan reservoir affairs center, Taizhou, Zhejiang,China

*Corresponding author's e-mail: 343531932@qq.com

Abstract: Using a practical and efficient overall three-dimensional turbulent numerical model of the spillway tunnel, the numerical simulation of the hydraulic characteristics of the spillway tunnel with the gradually expanding outlet and lower outlet height is carried out. The calculation results show that when the gate is fully open, there is pressure flow in the whole tunnel of the spillway tunnel, and the flow state is stable; when the gate is not fully open and the water level is low, the flow state in the tunnel is unstable, and the free and pressure flow occur alternately. When the gate is at a small opening, the flip bucket raises the water level at the outlet, and the lower outlet height causes insufficient air supply in the cave, which is prone to cavitation. The lift distance is small, and it is easy to scour the slope of the hole. The lower outlet height at the outlet of the spillway tunnel further exacerbates the complexity of the hydraulic characteristics in the back arc section, which should be paid attention to during design and construction, and it is necessary to carry out hydraulic model test research.

1. Introduction

Spillway tunnel is a building used to release water. According to the shape of the water flow, it can generally be divided into pressure tunnels and non-pressure tunnels. During the operation of a pressurized tunnel, it should be ensured that the tunnel body is fully pressurized. When the open flow and pressure flow alternate operation occurs in the pressure tunnel, it is easy to cause cavitation damage, which brings hidden dangers to the safety of spillway tunnel [1]. In order to evaluate the rationality of the structural design of the spillway tunnel, domestic scholars mostly use the method of numerical simulation for research and analysis [2-5]. At the same time, many scholars have also carried out related researches on different outlet types of spillway tunnels [6-8].

Numerical simulation is widely used in the simulation calculation of hydraulic characteristics of spillway tunnels, but there are few studies on the flip flow and lower outlet height of spillway tunnels. In this paper, a gradually expanding outlet and lower outlet height of the spillway tunnel is taken as the research object, and the RNG k-ε turbulence model is used to carry out numerical simulation research on the tunnel and outlet flow pattern under different conditions of gate opening and water level. Provide technical support for the design and operation of similar projects.

2. Computational model

2.1. governing equation

The three-dimensional flow control equations of the flow field in the spill tunnel are the continuity equation and the momentum equation:

$$\frac{\partial u_i}{\partial x_i}=0 \tag{1}$$

$$\frac{\partial u_i}{t}+u_j\frac{\partial u_i}{\partial x_j}=-\frac{1}{\rho}\frac{\partial P}{\partial x_i}+\frac{\partial}{\partial x_j}\left(2vS_{ij}-\overline{u_i^{'}u_j^{'}}\right) \tag{2}$$

where：

$$S_{ij}=\frac{1}{2}\left(\frac{\partial u_i}{\partial x_j}+\frac{\partial u_j}{\partial x_i}\right) \tag{3}$$

$$\overline{u_i^{'}u_j^{'}}=v_t\left(\frac{\partial u_i}{\partial x_j}+\frac{\partial u_j}{\partial x_i}\right)-\frac{2}{3}k\delta_{ij} \tag{4}$$

where u_i is the component velocity in the *i* direction; $u_i^{'}$ is the pulsation velocity in the *i* direction; *P* is the pressure; S_{ij} is the strain rate tensor; $\overline{u_i^{'}u_j^{'}}$ is the Reynolds stress tensor; ρ is the fluid density; *v* is the dynamic viscosity; v_t is the turbulent viscosity; *k* is the turbulent kinetic energy; δ_{ij} is Kronecker symbol （δ_{ij}=1， $i=j$， δ_{ij}=0， $i\neq j$）。

2.2. RNG k-ε equation

RNG k-ε governing equation:

$$\frac{\partial k}{\partial t}+u_j\frac{\partial k}{\partial x_i}=\tau_{ij}\frac{\partial u_i}{\partial x_j}+\frac{\partial}{\partial x_j}\left[\frac{1}{\rho}\left(\mu+\frac{\mu_t}{\sigma_k}\right)\frac{\partial k}{\partial x_j}\right]-\varepsilon \tag{5}$$

$$\frac{\partial \varepsilon}{\partial t}+u_j\frac{\partial \varepsilon}{\partial x_i}=C_{1\varepsilon}\frac{\varepsilon}{k}\tau_{ij}\frac{\partial u_i}{\partial x_j}+\frac{\partial}{\partial x_j}\left[\frac{1}{\rho}\left(\mu+\frac{\mu_t}{\sigma_k}\right)\frac{\partial \varepsilon}{\partial x_j}\right]-C_{2\varepsilon}\rho\frac{\varepsilon^2}{k} \tag{6}$$

where *k* is the turbulent kinetic energy, $k=\frac{1}{2}\overline{u_i^{'}u_j^{'}}$; ε is the turbulent kinetic energy dissipation,

$\varepsilon = v\overline{\left(\frac{\partial u_t^{'}}{\partial x_k}\right)\left(\frac{\partial u_t^{'}}{\partial x_k}\right)}$; μ_t is the eddy viscosity coefficient, $\mu_t = C_\mu \rho k^2 / \varepsilon$; τ_{ij} is the Reynolds stress, $\tau_{ij} = \frac{\mu_t}{\rho}\left(\frac{\partial u_i}{\partial x_j} + \frac{\partial u_j}{\partial x_i}\right) - \frac{2}{3}\rho k \delta_{ij}$; the model constants are C_μ=0.085, $C_{1\varepsilon}$=1.42, $C_{2\varepsilon}$=1.68, $\sigma_k = \sigma_\varepsilon$=0.7179。

3. Project overview and model establishment

3.1. Project Overview

A reservoir is a large (2) type project with a design flood level of 39.39m and a check flood level of 43.01m, and the total storage capacity of 732 million m^3. Spillway tunnel is a circular hole with pressure. The length of the cave is 277m, and the bottom elevation of the inlet is 23.00m. The water inlet is divided into 2 holes, each hole has a net height of 5m and a width of 2.5m. The bottom elevation of the outlet is 20.55m, and the elevation of the bottom ridge is increased from 20.55m to 23.13m at the deflecting flow. The layout of the spillway tunnel is shown in Figure 1.

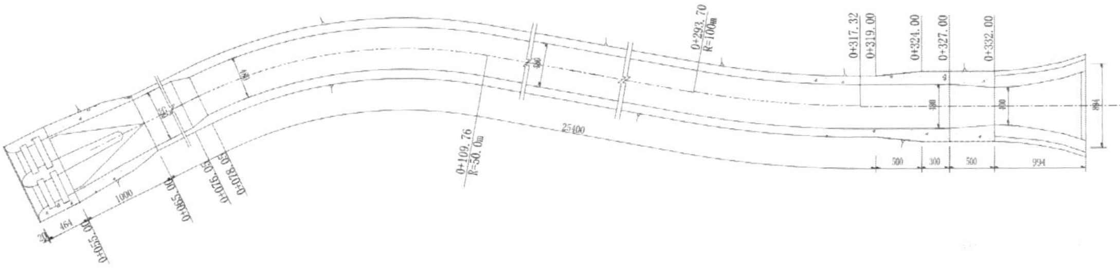

Figure 1. Spillway tunnel floor plan

3.2. computational model

The spillway tunnel model adopts the three-dimensional solid structure obtained by the Boolean operation difference between the cuboid entity and the tunnel entity. Three-dimensional solid structure generated by extrusion for gates and flip bucket The grid of the computational domain is divided by hexahedral elements.

3.3. calculation condition

According to the possible flood discharge schemes of this project, by setting different reservoir water levels and gate openings, the water flow pattern, velocity and flow in the spillway tunnel are simulated, and the influence of different gate opening schemes on the water flow characteristics in the spillway tunnel is studied.

Journal of Physics: Conference Series **2271** (2022) 012031 doi:10.1088/1742-6596/2271/1/012031

Table 1 Calculation condition

Working condition	Gate opening /m	Reservoir water level /m
1	0.5	32.00
2		36.00
3	2.5	32.00
4		36.00
5	4	32.00
6		36.00
7	5	32.00
8		36.00

4. Analysis of calculation results

4.1. Flow analysis in the cave

When the gate opening is 0.5m, under the two water level conditions, the tunnel is in a state of open current. From the confluence of the two holes to the front of the narrowing section of the tunnel diameter and at the bend, the flow state is unstable and the water flow fluctuates significantly. When the gate opening is 2.5m and the reservoir water level is low, the water flow pattern is unstable and fluctuates greatly from the confluence of the two holes to the narrowing section of the tunnel diameter, which is detrimental to the safety of the tunnel. When the gate opening is 4.0m and the reservoir water level is low, there will be an alternating state of free and pressure flow in the cave. When the sluice gate is fully open, under the two water level conditions, there is a pressure flow state in the whole tunnel of the flood discharge tunnel, the flow state is stable, the flow velocity distribution is uniform, and there is no obvious bad flow state. When the water level is 32.00m, the water pressure distribution is shown in Figures 2 to 5.

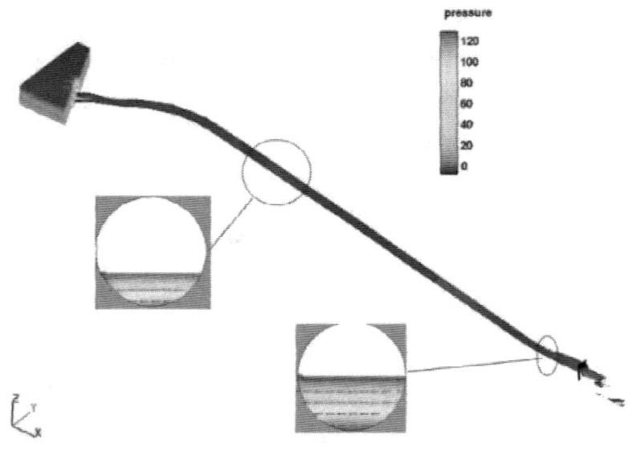

Figure 2. Condition 1 water pressure distribution diagram（kPa）

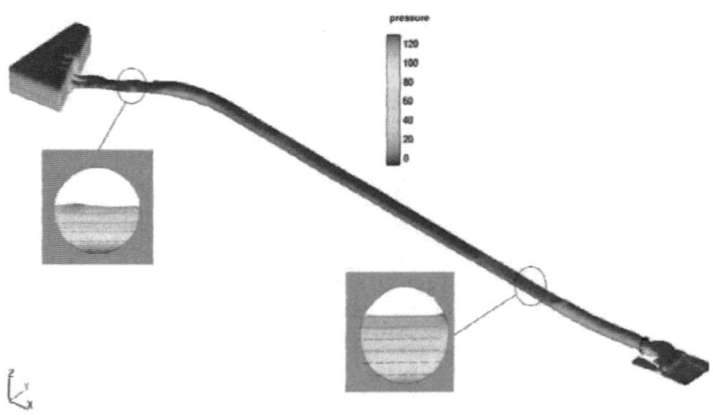

Figure 3. Condition 3 water pressure distribution diagram（kPa）

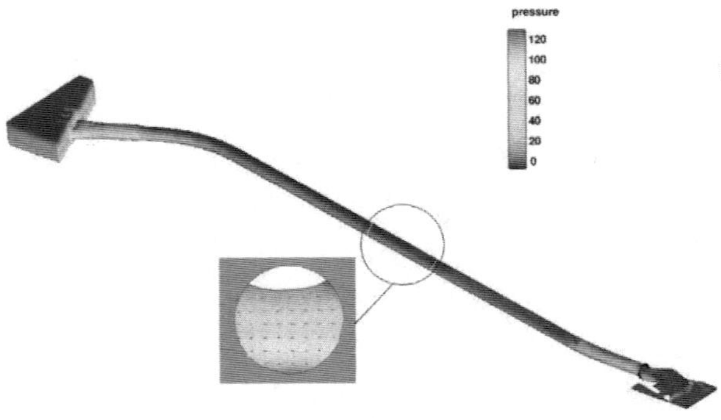

Figure 4. Condition 5 water pressure distribution diagram（kPa）

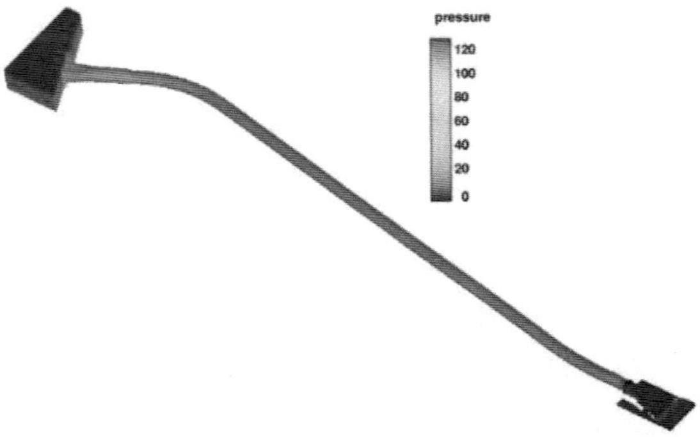

Figure 5. Condition 7 water pressure distribution diagram（kPa）

4.2. *Outlet flow analysis*

When the gate opening is 0.5m, the outlet is in a no-pressure state. Due to the obstruction of the flip bucket, the flow velocity at the outlet slows down, and the water flow backflow occurs at the initial stage of the discharge, and a hydraulic jump occurs in the hole. When the gate opening is 2.5m, due to the lower outlet at the entrance, the air supply in the cave is insufficient, and there is a vacuum above the water surface in the cave, which may lead to cavitation. When the flow velocity slows down, there will be small backflow. At the same time, the pick distance is small, which is easy to scour the slope. The gate opening is 4.0m, the exit is under pressure, the flip bucket are all rapids, and the pick distance is relatively long. When the sluice gate is fully open, the flip bucket are all rapids, and the flow state is stable. With the increase of water level, and the falling points of the water flow are far away from the flip bucket, which has little impact on the exit slope. When the water level is 32.00m, the flow velocity distribution at the outlet is shown in Figures 6 to 9.

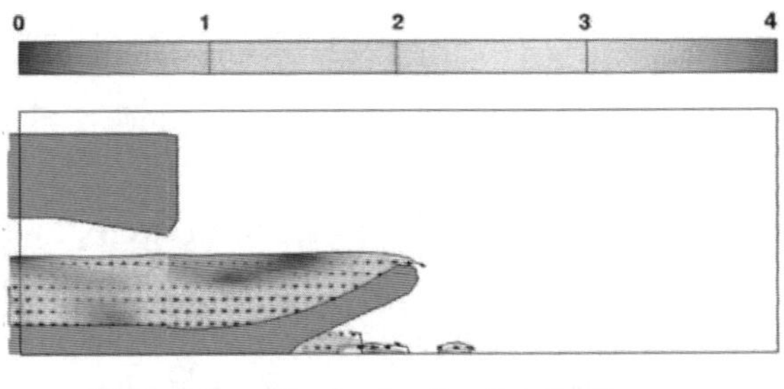

Figure 6. Flow velocity distribution in condition 1（m/s）

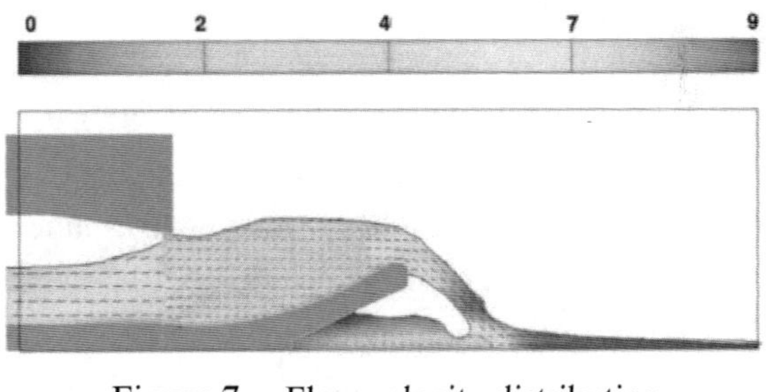

Figure 7. Flow velocity distribution in condition 3（m/s）

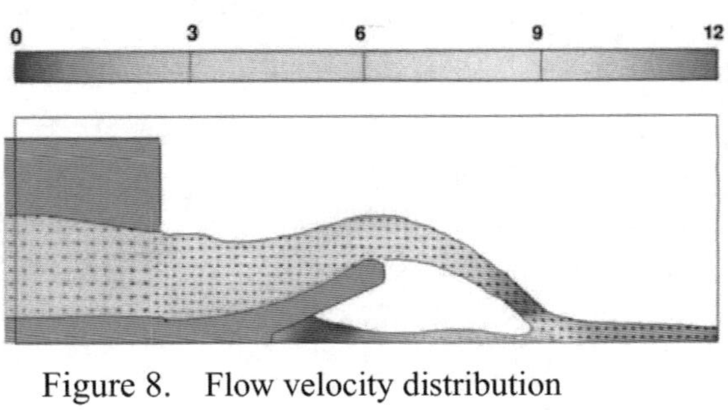

Figure 8. Flow velocity distribution in condition 5（m/s）

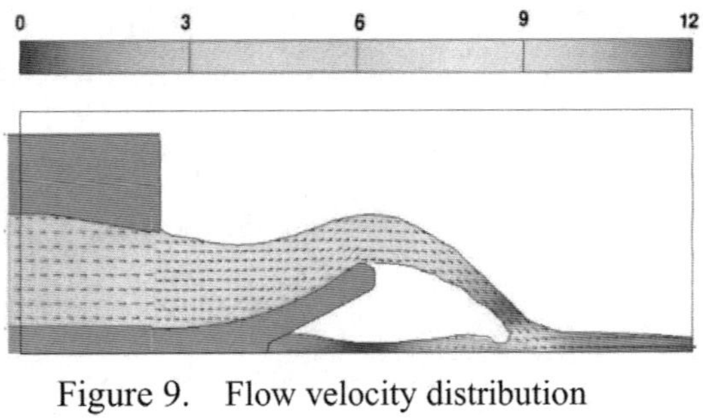

Figure 9. Flow velocity distribution in condition 7（m/s）

5. Conclusion

When the gate is fully opened, there is pressure flow in the whole tunnel, and the flow state is stable , and there is no unfavorable flow state along the tunnel. When the gate is not fully open and the water level is low, the flow state is unstable, and the free and pressure flow occur alternately, which is not conducive to the safety of the tunnel.When the gate is at a Large opening, the lower outlet height at the outlet of the spillway tunnel is beneficial to ensure that the tunnel is in a pressurized state, but at a small opening, the flip bucket raises the water level at the outlet, and the lower outlet height leads to insufficient air supply in the tunnel, which is prone to cavitation. At the same time, due to the blocking effect of the flip bucket, the flow velocity at the exit section is slowed down, and the flip distance is small, which is easy to scour the slope. The lower outlet height at the outlet of the spillway tunnel further exacerbates the complexity of the hydraulic characteristics in the reverse arc section, which should be paid attention to during design and construction, and it is necessary to carry out hydraulic model test research.

Acknowledgments

This study was supported by the Research Project of Basic Public Welfare of Zhejiang Province of China (No. GF18E090006) and Special project of Research Institute of Zhejiang Province of China (No. ZJ-A20001).

References

[1] Guo Jun, Gao Jizhang. (2013)Discussions on operation safety of discharge facilities with high head and large discharge flow. Journal of Hydroelectric Engineering, 32(5) : 169–173.

[2] Jiang Pan,Yin Jinbu, He Wuquan, et al. (2016)Numerical Simulation and Experimental Research on Pressure Characteristic of Curved Section of Pressure Spillway Tunnel. Water Power, 42(2) :49–53.

[3] Chen Ruihua, Yang Jijian, Ma Lin, et al. (2017)Numerical simulation of tunnel of Xiaowan Hydropower Station. Journal of Drainage and Irrigation Machinery Engineering, 35(6): 488–494.

[4] Liu Jiafu, Qi Xin. (2019) Study on hydraulic characteristics of ogee spillway tunnel of hydropower station. Water Resources and Hydropower Engineering,50(2) : 139–143.

[5] Wu Jingxia, Zhang Chunjin. (2021)Experimental study and numerical simulation of hydraulic characteristics of ogee spillway tunne.Water Resources and Hydropower Engineering, 52(7):123–131.

[6] Zhang Yiqi, Xie Feng, Hu Wenzhu, et al. (2021)Numerical Simulation Analysis on the Shape Design of Double-expanded Flip Bucket. Journal of Yangtze River Scientific Research Institute, 38(3):70–76.

[7] Li Guangning, Sun Shuangke,Liu Haitao, et al. (2020)Study on the distribution law of hydraulic characteristic index at the end of the anti-arch section of spillway tunnel. South-to-North Water Transfers and Water Science & Technology,18(3) :174–183.

[8] Liu Chong, Zhang Zongxiao, Chen Xiaowei. (2017)Numerical Simulation of Spillway Tunnel in Jet Bucket. Journal of Water Resources and Architectural Engineering,15(3) :40–44.

Numerical investigation for CO_2/N_2 mixture gas - liquid Taylor flow in microchannel by CFD simulation

Qiuwen Zhong[1,2], Jiancheng Zhou[3], Zhen Jiao[2,3]*

[1]School of Energy and Environment, Southeast University, Nanjing, Jiangsu 211189, PR China

[2]Joint Research Institute of Southeast University and Monash University, Suzhou, Jiangsu 215123, PR China

[3]School of Chemistry and Chemical Engineering, Southeast University, Nanjing, Jiangsu 211189, PR China

*Email: jiaozhen@seu.edu.cn

Abstract: Taking CO_2/N_2 and water as a system, the influence of microchannel inlet structure on the hydrodynamics of gas-liquid two-phase Taylor flow is numerically simulated by using VOF and species transport model. The flow properties of Taylor flow in perpendicular inlet structure are firstly explored, and the simulation results are compared with the inlet structures with different angles. The microchannel inlet structure has an obvious effect on the flow properties of Taylor flow. The simulation results provide some guidance for the design of micro reactor.

1. Introduction

The main component of the micro reaction system is the microchannel reactor, which is composed of a characteristic size of about 10μm-3mm capillary. Due to the viscous force and surface tension dominating the movement, the motion patterns in the micro reactor have numerous unique characteristics compared with the traditional macroscopical flow. The common flow patterns are bubbly flow, Taylor flow, Taylor-annular flow, and annular flow[1], which is mainly affected by the gas-liquid superficial velocity. With the popularization of micro reactor, Taylor flow has received extensive attention from researchers. Taylor flow is the most widely distributed flow pattern in microchannels where the liquid phase separates the gas phase to form bullet-shaped bubble slugs due to the mutual extrusion of multiphase [2]. The inside of the gas and liquid slug are internal circulation flows[3], which immensely enhance the mass and heat transfer capacity between phases[4]. Because of the distinctive properties, Taylor flow has tremendous application potential in chemical engineering.

Experiments and numerical methods are primarily applied to study the changes of gas-liquid slug length, bubble velocity, pressure, gas hold-up, bubble shape, liquid film thickness under different conditions [5-7]. Some scholars have shown that the microchannel inlet structure also has a crucial impact on the properties of Taylor flow, however, the conclusions are quite different. Fries and von Rohr[8] considered that the shear force at the gas phase inlet angle will affect the length of the slug, so reducing the inlet angle will shorten the slug. But Tan et al. [9] show that when the inlet structure is a right angle, the slug will be significantly shorter. Therefore, the influence of inlet structure on the properties of Taylor flow needs to be further explored.

The purpose of this research is to explore the hydrodynamic performance of Taylor flow formed by two phase flow in the T-shaped capillary by FLUENT package. The mixture of CO_2 and N_2 is adopted as the input gas rather than the pure gas in the most previous study. Meanwhile, the influence of microchannel inlet structure on the slug length of Taylor flow is further discussed.

2. Mathematical model

2.1 Model and material parameters

Compared with the common macroscopic flow, gas dispersed phase and liquid dispersed phase in capillary reactor has unique properties. In this study, the simplest T-shaped capillary reactor is choosen to reflect the transport phenomena of Taylor flow while saving calculation time. The T-junction consists of two inlet tubes and a relatively long main tube. The specific dimensions are shown in Figure 1.

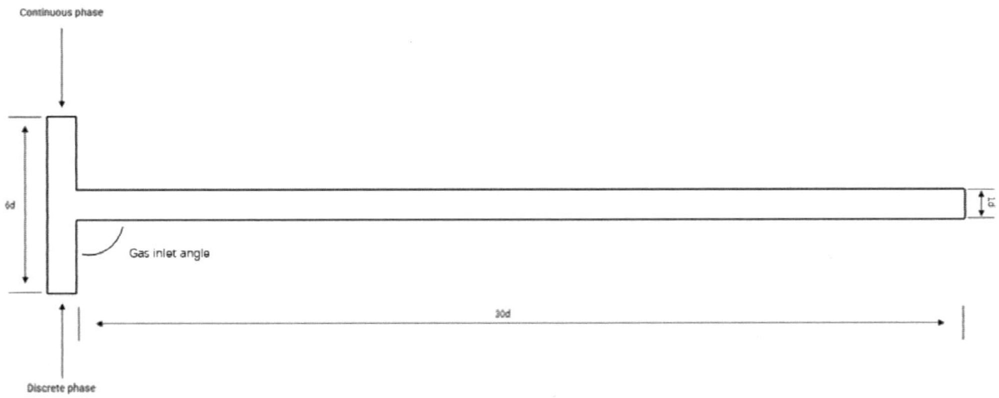

Figure 1. Microchannel model in the simulation.

The simulation process includes two phases. The gas phase is characterized by Species transport model, including CO_2 and N_2. Material properties are shown in Table 1.

Table 1. Material properties for two phase

Material	ρ (kg/m³)	μ (pa·s)	σ(n/m)
CO_2	1.7878	1.37×10^{-5}	
N_2	1.138	1.663×10^{-5}	
water	998.2	0.001003	0.072

2.2 Governing equation

Commercial CFD software package Fluent 19.2 is applied to perform the numerical simulations. The volume of fluid (VOF) model is chosen to capture clear phase boundary. For the mixture of CO_2 and N_2, the species transport model needs to be turned on. The governing equation related to hydrodynamic performance is given as follows.

The continuity equation is

$$\frac{\partial \rho}{\partial t} = \nabla \cdot (\rho \vec{u}) = 0 \quad (1)$$

The motion equation is

$$\frac{\partial(\rho \vec{u})}{\partial t} = \nabla \cdot (\rho \vec{u}\vec{u}) = -\nabla p + \nabla \cdot [\mu(\nabla \vec{u} + \nabla \vec{u}^T)] + \rho \vec{g} + \vec{F} \quad (2)$$

The volume fraction equation is

$$\frac{\partial \alpha_i}{\partial t} = \vec{u} \cdot \nabla \alpha_i = 0 \tag{3}$$

The sum of the volume fraction of the each phase in control volume is defined as 1. The physical parameters are obtained by the weighted average of volume fraction.

$$\alpha_L + \alpha_G = 1 \tag{4}$$

The continuous surface force model proposed by Brackbill [10] is adopted.

$$\vec{F} = \sigma \frac{2\rho\kappa_G \nabla \alpha_G}{(\rho_G + \rho_L)} \tag{5}$$

where, α is the phase volume fraction. κ_G is interface curvature, and σ represents surface tension coefficient.

The species transport equation is

$$\frac{\partial(\rho y_i)}{\partial t} = \nabla \cdot (\rho \vec{u} y_i) = -\nabla \cdot \vec{J}_i \tag{6}$$

The mixing between gases can be expressed by the component transport equation。

2.3 Numerical approach

The transient calculation is applied in the simulation process. The specific settings of boundary conditions are described below. The velocity-inlet is adopted, so the gas and liquid feed velocity needs to be determined. For the gas inlet, the proportion of the mixture needs to be specified. The outlet is usually treated as pressure – outlet. In the wall adhesion condition, the value of contact angle will affect the internal hydrodynamic, where is set to 0 °. For the calculation method, the PISO algorithm is applied in this research. The second-order upwind differencing scheme is selected for the solution of the momentum equation. The PRESTO! (pressure staggering option) scheme is for pressure interpolation. The geo-reconstruction method is employed to ensure the clarity and accuracy of the gas-liquid interface. In the VOF model, the Courant number is set to 0.25. The time step of the transient calculation is set to $1*10^{-5}$s to ensure its stable convergence.

3. Results & Discussion

Figure 2 shows the volume fraction distribution in a 1mm microchannel in which gas inlet angle is 90°. In the microchannel, the gas-liquid two phases squeeze each other to form Taylor flow. It proves that the length of the slug can be controlled by adjusting the superficial flow rate.

(a) (b) (c) (d)

Figure 2. Taylor flow slug distrinbution in microchannel model. (a. U_G = 0.02 m/s; b. U_L = 0.04s, U_G = 0.03m/s; c. U_L = 0. 05m/s, U_G = 0.01m/s; d. U_L = 0.02m/s, U_G = 0.05m/s)

The numerical quantitative study proves that the length of gas slug is positively correlated with the superficial velocity of gas, however, the increase of the superfical velocity of liquid phase will accelerate the gas phase break under larger pressure during extrusion, resulting in the decrease of bubble length, as observed in Figure 3 (a), (b). The increase of gas superficial velocity leads to more gas entering the main channel before the bubble break off, which is opposite to the change of liquid slug. Figure 3 (c) shows the gas holdup increases with the increase of gas velocity.

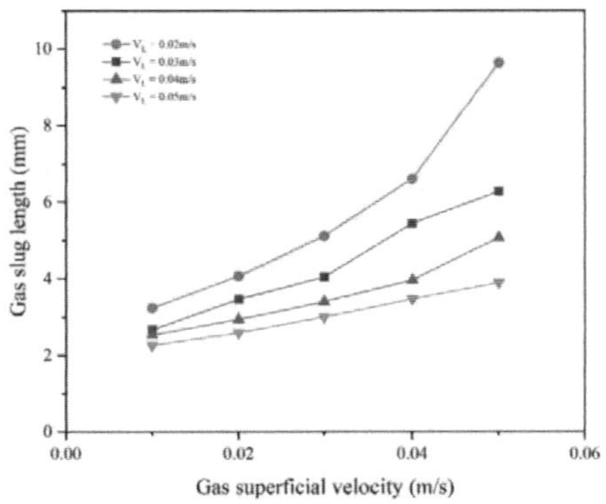

(a)

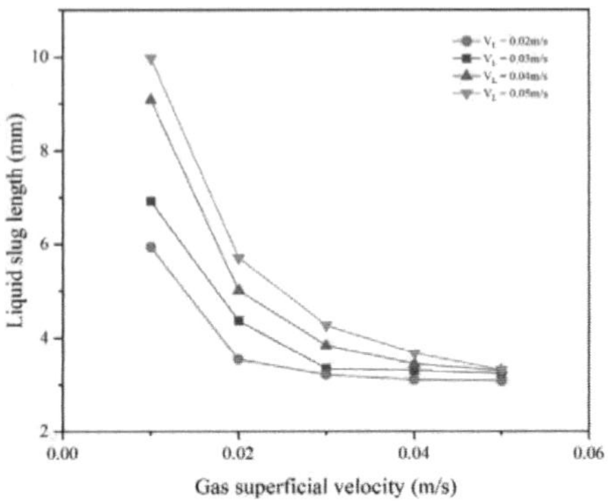

(b)

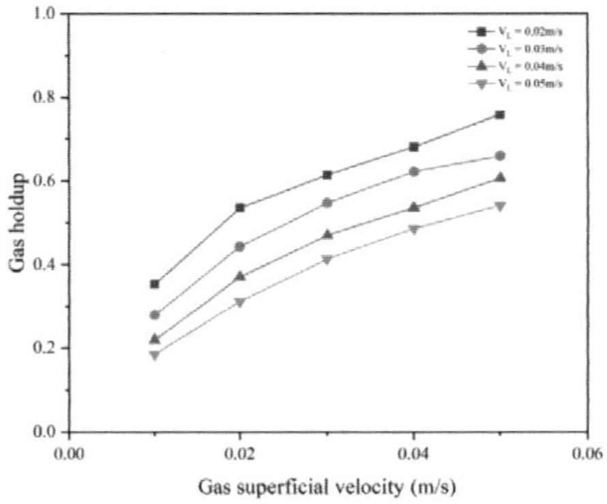

(c)

Figure 3. Effect of two-phase velocity on flow properties .(a. Taylor bubble length; b. Liquid slug length; c. Unit length)

Five different gas inlet angles are considered, which are 30°, 60°, 90°, 120°, 150 °respectively. Figure 4 displays the two-phase flow slug distribution in microreactor under different gas-phase inlet angles.

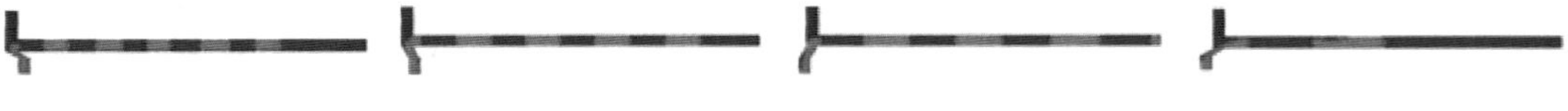

Figure 4. Slug distribution in microchannels under various gas phase inlet angles. (U_L = 0.05m/s U_G = 0.05m/s)

Obviously, in Figure 5, when α is 30 °, the Taylor bubble is the shortest in the five conditions. With the increased inlet angle, the length of gas-liquid slugs also enlarges accordingly. Therefore, the inlet structure has an important impact on the flow properties of Taylor flow.

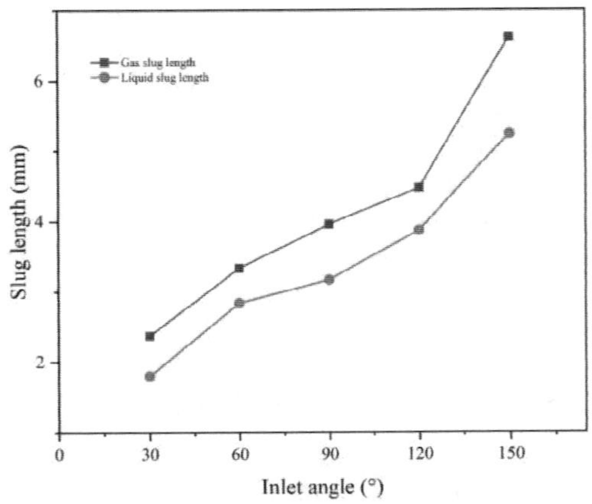

Figure 5. Variation of two phase slug length under differernt inlet angle. (U_L = 0.05m/s U_G = 0.05m/s)

4. Conclusion

The following conclusions are obtained through the numerical simulation of a T-shaped capillary reactor with the characteristic diameter of 1mm. (1) The length of the Taylor flow gas-liquid slug is affected by the superficial velocity of gas-liquid. (2) The bubble length can be controlled by two-phase flow velocity. (3) The inlet angle structure of the microchannel has an important influence on the flow properties. The length of the bubble decreases with the enhancing inlet angle.

References

[1] Lei, Z.; Guo, Y.; Dai, C.; Zi, L.; Chen, B., Simulation of hydrodynamic and mass transfer performances in monolith channel. *Catalysis Today* **2016,** *276*, 150-160.

[2] Taylor, G. I., DEPOSITION OF A VISCOUS FLUID ON THE WALL OF A TUBE. *Journal of Fluid Mechanics* **1961,** *10* (2), 161-165.

[3] Bordbar, A.; Taassob, A.; Zarnaghsh, A.; Kamali, R., Slug flow in microchannels: Numerical simulation and applications. *Journal of Industrial and Engineering Chemistry* **2018,** *62*, 26-39.

[4] Abiev, R. S., Hydrodynamics and Heat Transfer of Circulating Two-Phase Taylor Flow in Microchannel Heat Pipe: Experimental Study and Mathematical Model. *Industrial & Engineering Chemistry Research* **2020,** *59* (9), 3687-3701.

[5] Gregorc, J.; Zun, I., Inlet conditions effect on bubble to slug flow transition in mini-channels. *Chemical Engineering Science* **2013,** *102*, 106-120.

[6] Shao, N.; Gavriilidis, A.; Angeli, P., Effect of Inlet Conditions on Taylor Bubble Length in Microchannels. *Heat Transfer Engineering* **2011,** *32* (13-14), 1117-1125.

[7] Shao, N.; Salman, W.; Gavriilidis, A.; Angeli, P., CFD simulations of the effect of inlet conditions on Taylor flow formation. *International Journal of Heat and Fluid Flow* **2008,** *29* (6), 1603-1611.

[8] Fries, D. M.; von Rohr, P. R., Impact of inlet design on mass transfer in gas-liquid rectangular microchannels. *Microfluidics and Nanofluidics* **2009,** *6* (1), 27-35.

[9] Tan, J.; Du, L.; Xu, J. H.; Wang, K.; Luo, G. S., Surfactant-Free Microdispersion Process of Gas in Organic Solvents in Microfluidic Devices. *Aiche Journal* **2011,** *57* (10), 2647-2656.

[10] Brackbill, J. U.; Kothe, D. B.; Zemach, C., A CONTINUUM METHOD FOR MODELING SURFACE-TENSION. *Journal of Computational Physics* **1992,** *100* (2), 335-354.

Design of power system excitation auxiliary controller based on Multi-Agent

Qiuhua Hou, Fanmin Meng*, Xiaoyu Liu, Xiaoyan Qi, Huifang Ding and Xuju Chen

[1]State Grid Shandong Electric Power Company Laiwu Power Supply Company., Laiwu City, Shandong Province, 271100, China

Corresponding author's e-mail: 17854280839@163.com

Abstract—The continuous access of new energy to the power grid increases the complexity of the structure of the power system, which brings hidden dangers to the security of new energy power system. In order to further enhance the operation stability and security of new energy grid, a generator excitation auxiliary controller based on multi-agent is proposed in this paper. Firstly, the multi-agent model of power system is established, and then the distributed output feedback controller is designed for the generator model. The parameters of the controller are calculated and solved by H ∞ theory. Finally, the effectiveness of the controller designed is verified by simulation analysis.

1. Introduction

To solve the problems of unbalanced energy consumption[1], uncoordinated development between power supply and power grid and unreasonable power supply structure, the concept of power grid interconnection is gradually put forward[2]. However, power grid interconnection will cause low-frequency oscillation of the system. The higher the multiple of generator excitation[3], the greater the probability of low-frequency oscillation. Under the background of power grid interconnection, low-frequency oscillation can not be ignored and must be solved. There are many problems and challenges in the application of some traditional additional excitation controllers for conventional systems in large-scale complex systems, such as unable to obtain some states of the system[4], complex system calculation[5], network traffic exceeding the requirements of the controller and so on. Multi-agent system has the characteristics of autonomy and flexibility, which provides a new solution for the safety and stability control of large-scale and distributed interconnected power system.

In this paper, a distributed output feedback controller is designed on the multi-agent consistency theory. The parameters of the controller are adjusted and calculated through H∞ theory and matrix inequality, so that the controller designed in this paper meets the H ∞ performance index. Finally, the effectiveness of the controller designed is verified by simulation analysis.

2. Power System Modeling Based on Multi-Agent

By comparing that the second order swing equation of conventional synchronous generator is consistent with the form of multi-agent, the generator in each system is regarded as a multi-agent, and the principal variables in the rotor motion equation correspond to the principal variables of multi-agent as follows:

$$\begin{aligned}\frac{dx(t)}{dt}&=\frac{d\Delta\delta_i(t)}{dt}=\Delta\omega_i(t)\\ \frac{dv(t)}{dt}&=\frac{d\Delta\omega_i(t)}{dt}=-\frac{\omega_N}{T_i}\Delta P_{ei}(t)+w_i\\ \frac{dz(t)}{dt}&=\frac{d\Delta E'_{qi}(t)}{dt}=\frac{\Delta E_{qei}(t)-\Delta E_{qi}(t)}{T_{d0}}\end{aligned}\tag{1}$$

Where, δ is power angle of synchronous generator, w_i is external disturbance to generator, ω is rotor speed of generator, P_{ei} is the electromagnetic power of synchronous generator, E_{qei} is forced no-load potential of synchronous generator, T_i is the inertia time constant of synchronous generator, E'_{qi} is transient potential of synchronous generator

Formula (1) can be represented by the following formula (2):

$$\begin{cases}\dfrac{d\Delta\delta_i(t)}{dt}=\Delta\omega_i(t)\\ \dfrac{d\Delta\omega_i(t)}{dt}=-\sum\limits_{j\neq i}^{n}P_{ij}(\Delta\delta_i(t)-\Delta\delta_j(t))+w_i(t)\\ \dfrac{d\Delta E'_{qi}(t)}{dt}=-\sum\limits_{j=1}^{n}Q_{ij}(\Delta\delta_i(t)-\Delta\delta_j(t))+u_i(t)-\dfrac{\Delta E'_{qi}(t)}{T_{d0}}\end{cases}\tag{2}$$

Where, $P_{ij}=-\sum\limits_{j\neq i}^{n}E_iE_j[G_{ij}\sin\delta_{ijo}(t)-B_{ij}\cos\delta_{ijo}(t)]$, $Q_{ij}=\sum\limits_{j=1}^{n}E'_{qj}[B_{ij}\sin\delta_{ijo}(t)+G_{ij}\cos\delta_{ijo}(t)]$

Formula (2) is expressed in the form of equation of state, as shown in equation (3):

$$\begin{aligned}\dot{x}_i(t)&=(\Delta\mathrm{A}_i+\mathrm{A})x_i(t)+\mathrm{B}_2u_i(t)+\mathrm{E}_{a1}\mu_{i1}(x)+\mathrm{E}_{a2}\mu_{i2}(x)+\mathrm{B}_1w_i(t)\\ y_i(t)&=\mathrm{N}x_i(t)\end{aligned}\tag{3}$$

Where, $x_i(t)=\begin{bmatrix}\Delta\delta_i(t)\\ \Delta\omega_i(t)\\ \Delta E'_{qi}(t)\end{bmatrix}$, $\mathrm{N}=\begin{bmatrix}1&0&0\\0&1&0\\0&0&1\end{bmatrix}$, $\mathrm{B}_1=\begin{bmatrix}0\\1\\0\end{bmatrix}$, $\mathrm{A}=\begin{bmatrix}0&1&0\\0&0&0\\0&0&0\end{bmatrix}$, $\mathrm{E}_{a1}=\begin{bmatrix}0\\-1\\0\end{bmatrix}$, $\mathrm{E}_{a2}=\begin{bmatrix}0\\0\\-1\end{bmatrix}$

The state equation of multi-agent system can be represented by the following formula (4):

$$\begin{aligned}\dot{x}(t)&=(I_n\otimes\mathrm{A})x(t)+(I_n\otimes\mathrm{B}_2)u(t)+(P\otimes\mathrm{E}_{a1})x(t)\\ &\quad+(Q\otimes\mathrm{E}_{a2})x(t)+(I_n\otimes\mathrm{B}_1)w_1(t)+diag(\Delta\mathrm{A}_i)x(t)\\ y(t)&=(I_n\otimes\mathrm{N})x(t)\end{aligned}\tag{4}$$

3. Design of Multi-agent Excitation Auxiliary Controller

The power system dynamic output feedback control strategy is designed for the power system model based on Multi-Agent in this paper, as shown in formula (5):

$$\begin{aligned}\dot{v}_i(t)&=\mathrm{H}_{1c}v_i(t)+\mathrm{H}_{2c}\sum_{j\in\mathrm{N}_j}a_{ij}[y_i(t)-y_j(t)]\\ u_i(t)&=\mathrm{H}_{3c}v_i(t)+\mathrm{H}_{4c}\sum_{j\in\mathrm{N}_j}a_{ij}[y_i(t)-y_j(t)]\end{aligned}\tag{5}$$

The global controller of multi-agent system can be expressed by equation (6):

$$\begin{aligned}\dot{v}(t) &= (I_n \otimes H_{1c})v(t) + (L \otimes H_{2c})(I_n \otimes \mathrm{N})x(t) \\ u(t) &= (I_n \otimes H_{3c})v(t) + (L \otimes H_{4v})(I_n \otimes \mathrm{N})x(t)\end{aligned} \tag{6}$$

Substituting equation (6) into equation (4), we can get:

$$\begin{aligned}\begin{bmatrix}\dot{x}(t)\\ \dot{v}(t)\end{bmatrix} &= I_n \otimes \begin{bmatrix}\mathrm{A}+\Delta\mathrm{A}_i & \mathrm{B}_2 H_{3c}\\ 0 & \mathrm{A}_k\end{bmatrix}\begin{bmatrix}x(t)\\ v(t)\end{bmatrix} + P \otimes \begin{bmatrix}\mathrm{E}_{a1} & 0\\ 0 & 0\end{bmatrix}\begin{bmatrix}x(t)\\ v(t)\end{bmatrix} + I_n \otimes \begin{bmatrix}\mathrm{B}_1\\ 0\end{bmatrix} w_1(t) \\ &+ L \otimes \begin{bmatrix}H_{2c}H_{4c}\mathrm{N} & 0\\ H_{2c}\mathrm{N} & 0\end{bmatrix}\begin{bmatrix}x(t)\\ v(t)\end{bmatrix} + Q \otimes \begin{bmatrix}\mathrm{E}_{2a} & 0\\ 0 & 0\end{bmatrix}\begin{bmatrix}x(t)\\ v(t)\end{bmatrix} + I_n \otimes \begin{bmatrix}\mathrm{B}_1\\ 0\end{bmatrix} w_1(t) \\ y(t) &= (I_n \otimes \mathrm{N}, 0_n \otimes 0_{m\times m_k})\begin{bmatrix}x(t)\\ v(t)\end{bmatrix}\end{aligned} \tag{7}$$

Rewrite equation (7) as:

$$\begin{aligned}\dot{\zeta} &= (\tilde{\mathrm{A}} + \Delta\tilde{\mathrm{A}})\zeta + \tilde{\mathrm{B}} w_1 \\ \tilde{y} &= \tilde{\mathrm{N}}\zeta\end{aligned} \tag{8}$$

The following objective functions are defined:

$$W = \int_0^{\infty} (-\gamma^2 w_1^T w_1 + \tilde{y}^T \tilde{y})dt \tag{9}$$

When $\|\zeta\|_2 \to 0$ and $W < 0$ are satisfied, the system achieves consistency and meets the H∞ index.

Define Lyapunov function as:

$$V = \xi^T (I \otimes \hat{P})\xi \tag{10}$$

The derivative of equation (10) can be obtained:

$$\begin{aligned}\dot{V} &= 2\zeta^T (L \otimes \hat{P}\Gamma_2)\zeta + \zeta^T (I \otimes \Gamma_1^T \hat{P} + I \otimes \hat{P}\Gamma_1)\zeta + 2\zeta^T (Q \otimes \hat{P}\Gamma_4)\zeta \\ &+ 2\zeta^T (H \otimes \hat{P}\Gamma_3)\zeta + 2\zeta^T (I \otimes \hat{P}\Delta\Gamma)\zeta + 2\zeta^T (I \otimes \hat{P}\Gamma_5) w_1\end{aligned} \tag{11}$$

For any non-zero w_1, formula (11) can be represented by the following formula (12):

$$W = V(0) - V(\infty) + \int_0^{\infty} (\dot{V} + \tilde{y}^T \tilde{y} - \gamma^2 w_1^T w_1)dt \tag{12}$$

If $\dot{V} + \tilde{y}^T \tilde{y} - \gamma^2 w_1^T w_1 < 0$, we can get $W < 0$.

$$\begin{aligned}W_T &\le (4+\delta)\zeta^T \zeta + \left[\tau^2(L)\zeta^T (I \otimes \hat{P}\Gamma_2\Gamma_2^T \hat{P})\zeta\right] + \left[\gamma^{-2}\zeta^T (I \otimes \hat{P}\Gamma_5\Gamma_5^T \hat{P})\zeta\right] \\ &+ \left[\tau^2(H)\xi^T (I \otimes \hat{P}\Gamma_3\Gamma_3^T \hat{P})\zeta\right] + \left[\tau^2(Q)\zeta^T (I \otimes \hat{P}\Gamma_4\Gamma_4^T \hat{P})\zeta\right] \\ &+ \left[\zeta^T (I \otimes \Gamma_1^T \hat{P} + I \otimes \hat{P}\Gamma_1)\zeta\right] + \left[\zeta^T (I \otimes \hat{P}^2)\zeta\ \right] \\ &= \zeta^T [I \otimes (\Gamma_1^T \hat{P} + \hat{P}\Gamma_1 + (4+\delta)I + \hat{P}^2 + \tau^2(L)\hat{P}\Lambda_2\Lambda_2^T \hat{P}\end{aligned} \tag{13}$$

Based on LMI theory, formula (13) can be represented by the following formula (14):

$$\begin{bmatrix} \tilde{\Gamma}_0 P + P\tilde{\Gamma}_0 + \tilde{\Gamma}_2^T \tilde{Q}^T \\ +\tilde{Q}\tilde{\Gamma}_2 + (5+\delta)V^{-T}V^{-1} & \tilde{Q}F_3 & \tilde{P}\tilde{\Gamma}_3 & \tilde{P}\tilde{\Gamma}_4 & \tilde{P}\tilde{\Gamma}_5 \\ (\tilde{Q}\tilde{\Gamma}_3)^T & -\tau^{-2}(L)I & 0 & 0 & 0 \\ (\tilde{P}\tilde{\Gamma}_3)^T & 0 & -\tau^{-2}(\tilde{P})I & 0 & 0 \\ (\tilde{P}\tilde{\Gamma}_4)^T & 0 & 0 & -\tau^{-2}(\tilde{Q})I & 0 \\ (\tilde{P}\tilde{\Gamma}_5)^T & 0 & 0 & 0 & -\gamma^2 I \end{bmatrix} < 0 \tag{14}$$

Where, $\tilde{\Gamma} = V\Gamma_0 V^{-1}$, $\tilde{\Gamma}_1 = V\Gamma_1$, $\tilde{\Gamma}_2 = \tilde{\Gamma}_2 V^{-1}$, $\tilde{Q} = \tilde{P}\tilde{\Gamma}_1 H_C$

4. Case Study

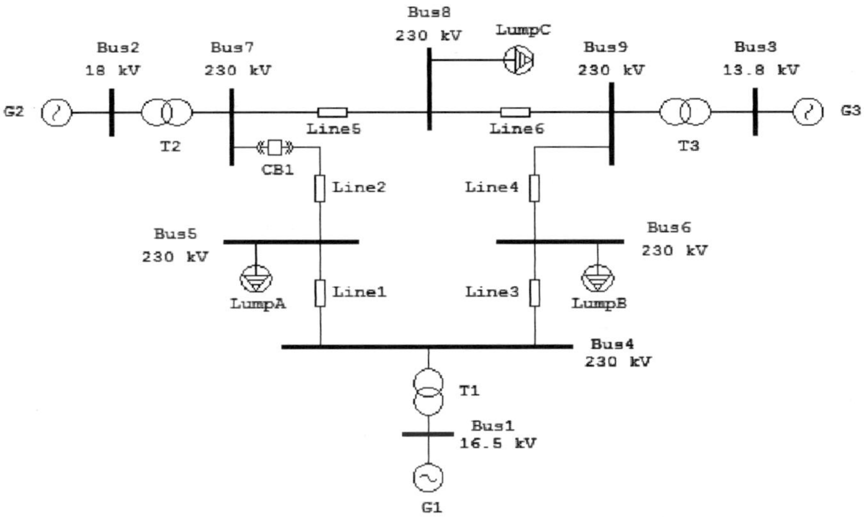

Fig.1 Power system diagram including wind power

The simulation analysis of the power system with wind power in Figure 1 to verify the effectiveness of the controller designed.

It can be obtained through the controller (6)

$$\dot{V} = H_{1c} \bullet V + LH_{2ck} \bullet \Delta X \tag{15}$$

$$u_1 = \mathrm{N} \bullet V + LH_{4c} \bullet \Delta X \tag{16}$$

Add disturbance to the system at t = 20s, and variation trend of power angle difference between traditional and conventional generators in power system is shown in Figure 1 and 2 respectively. The solid line and dotted line in the figure are shown as adding multi-agent auxiliary excitation controller, and not adding multi-agent auxiliary excitation controller. By comparing the control effects before and after adding the controller, it can be concluded that the multi-agent controller improves the operation stability of the power grid with renewable energy.

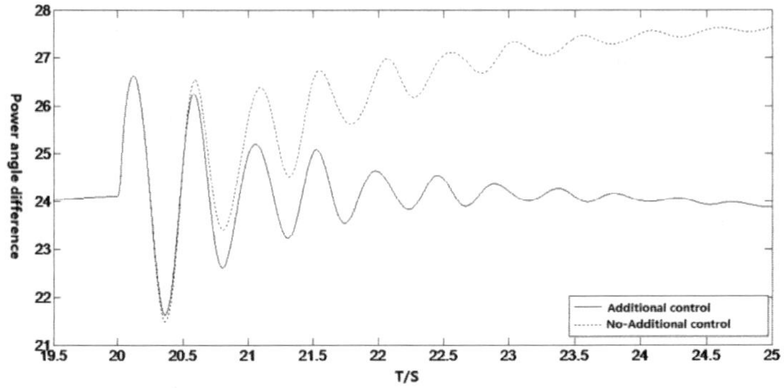

Fig.2 Under different controllers variation trend of power angle difference between G_1 and G_2

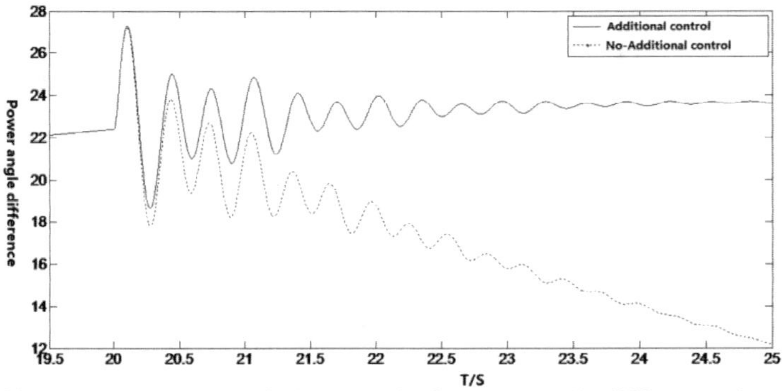

Fig.3 Under different controllers variation trend of power angle difference between G_1 and G_3

5. Conclusion

This paper designs an auxiliary excitation controller based on multi-agent. The controller parameters are adjusted by using H ∞ theory. Through simulation analysis, it is verified that the multi-agent auxiliary excitation controller can enhance the power angle stability of generator in power grid with renewable energy.

Acknowledgments

This work was supported by State Grid Shandong Electric Power Company Science and Technology Project (Project No.52061219006Q).

References

[1] C. Chen and X. Li, "Configuration Method and Multi-Functional Strategy for Embedding Energy Storage into Wind Turbine," *Energies,* vol. 14, no. 17, Sep 2021, Art. no. 5354.

[2] S. Xu, H. Sun, and Z. Zhang, "MAS-Based Decentralized Coordinated Control Strategy in a Micro-Grid with Multiple Microsources," *Energies,* vol. 13, no. 9, pp. 1-18, 2020.

[3] H. Eman, F. Abdallah, and K. Deepa, "On Effective Virtual Inertia of Storage-Based Distributed Control for Transient Stability," *IEEE Transactions on Smart Grid,* vol. 10, no. 1, pp. 327-336, 2019.

[4] T. Zhao and P. Alessandra, "Distributed Control of Battery Energy Storage Systems for Improved Frequency Regulation," *IEEE Transactions on Power Systems* vol. 35, no. 5, pp. 3729-3738, 2020.

[5] B. Ahmed and Y. Naoto, "Optimal Voltage Control Strategy for Voltage Regulators in Active Unbalanced Distribution Systems Using Multi-Agents," *IEEE Transactions on Power Systems,* vol. 35, no. 2, pp. 1023-1035, 2020.

AUTHOR INDEX